The Collingridge

DICTIONARY OF PLANT NAMES

Allen J. Coombes

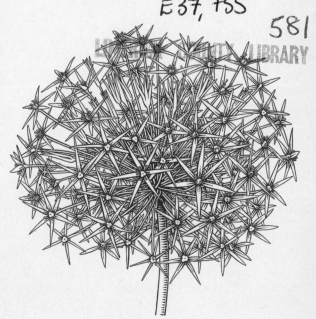

COLLINGRIDGE

Cover illustration:
Oxalis adenophylla painted by Cynthia Pow
Line drawings:
Allium christophii (title page), *Sternbergia lutea* (page 5),
Catalpa bignonioides (page 14), all by Charles Stitt

Published by Collingridge Books
an imprint of Newnes Books
The Hamlyn Publishing Group Limited
84-88 The Centre, Fetham, Middlesex, TW13 4BH
and distributed for them by
Hamlyn Distribution Services
Rushden, Northants, England

ISBN 0 600 35770 8

Printed in Finland

Contents

Introduction

The aim of this book is to provide a guide to the derivation, meaning and pronunciation of the scientific names of the more commonly grown plants. The term scientific name is preferable here to Latin name as many names derive from languages other than Latin, for example many derive from Greek or personal names. Whatever their origin, all names are treated as Latin. Generic names are treated as nouns and, as in Latin, have a gender (i.e. masculine, feminine or neuter). Names of species and varieties are adjectival and their endings follow the rules of Latin grammar, e.g. the Latin word for white can be rendered as *albus* (masculine), *alba* (feminine) or *album* (neuter) depending on the gender of the generic name. The use of Latin for plant names can certainly be confusing when first encountered, producing many words of unfamiliar form and uncertain pronunciation. It should be remembered, however, that when a scientific attitude was first taken towards the naming of plants in the 16th and 17th centuries, Latin was a common language among the intellectuals of Europe and it was second nature for many to use it. Today, although Latin has evolved from the Latin used in classical and medieval times to meet the needs of botany it forms a method of communication between botanists of all nationalities.

How to use this book

Generic names and common names are listed alphabetically. The name of each genus (plural genera) is given first e.g. *Acer*, followed by the suggested pronunciation, the family in which it is placed (related genera are placed in the same family) and the derivation of the name, whether from Latin (L.), Greek (Gk.) or other sources. Following this is a short statement of the main use in gardens of the plants listed e.g. as herbaceous perennials, shrubs, etc. and a general guide to hardiness. It should be noted that these are general statements about the genus as represented in gardens and some of the plants listed may be exceptions. Usually, unless otherwise stated, plants are at least fairly hardy in many parts of the country, semi-hardy plants require winter protection in most areas and tender plants are generally only suitable for growing under cover or for summer bedding. A common name is given if it is applied to the whole genus. If a country of origin is given here it applies to the plants listed and not necessarily to the whole genus.

Species are listed alphabetically under each genus. Each name is followed by the suggested pronunciation, the meaning of the name, the common name (if any) and the country of origin unless this is given above. If, instead of a country of origin, the abbreviation cult. (cultivated) is given this indicates that this plant is known only in gardens.

Categories below the rank of species are listed under the relevant specific name. These include those recognised botanically (sub-species, varieties and forms) which are given in small letters (e.g. *Pinus nigra maritima*) and those recognised horticulturally (cultivars and groups). A cultivar is a form of a species or hybrid selected either in the wild or in cultivation and maintained in gardens. These are given with capital initials and single inverted commas e.g. *Hedera hibernica* 'Deltoidea'. Plants which form part of a variable group are given a group name which is given with capital initials but without inverted commas e.g. *Fagus sylvatica* Purpurea.

Common names are cross referenced to the correct scientific name and are given in Roman type. They are not listed if they are the same as or very similar to the generic name.

Hybrids arise both in the wild and in cultivation when two species cross or are crossed to give rise to a distinct plant. When the parents are in different genera the plant is known as an intergeneric hybrid and the 'generic' name is preceded by a multiplication sign e.g. × *Laeliocattleya*. When the parents are in the same genus the 'specific' name is preceded by a multiplication sign e.g. *Abutilon* × *suntense*. Names of graft-hybrids are preceded by a plus sign. Sometimes when parentage is complex or unknown, or a scientific name has not been given to the plant, it is more convenient to place the cultivar or group name of a hybrid directly under the generic name e.g. *Rhododendron* Loderi. Parents of hybrids are given where known. The hybrids listed are found only in cultivation unless otherwise stated.

A synonym is a scientific name that is no longer accepted for the plant in question. It may be that the name has been mis-applied in gardens and that the plant grown under this name is really something else (e.g. *Helichrysum microphyllum*); it may be that a name has been rejected because there is an earlier name available for the same plant (e.g. *Magnolia conspicua*), or it may be that a plant is better placed in a different genus (e.g. *Alyssum maritimum*). Synonyms are cross-referenced unless they occur in the same genus and there are less than ten entries in that genus; they are indicated by an equation sign =. The abbreviation hort. (*hortorum*, of gardens) following a name means that the plant is known under that name in gardens but there may be another plant to which that name correctly applies.

Allen J. Coombes

Pronunciation

Unlike the use of scientific names, their pronunciation is not governed by rules. The majority of people who use scientific names treat them as if they are in their own language. Where pronunciation is ambiguous by this method, it is common to encounter several ways of saying a word. Ideally pronunciation should be based on the origin of the word in question. Thus names derived from Latin or Greek should be given a Latin pronunciation, names derived from personal or place names should follow the original pronunciation of that name with a Latin pronunciation for any endings, changing the position of stress where necessary. The following table illustrates the classical pronunciation of Latin letters and dipthongs which are often said differently when an English pronunciation is given to the word. It should only be applied to words of Latin or Greek origin.

Latin letter or dipthong	Classical pronunciation
a	Short *a* is variously given as either in c*a*t or in *a*bove. Long *a* as in r*a*ther, never as in g*a*te.
ae	As *i* in m*i*te.
au	As *ou* in *ou*t.
c	Always hard as in *c*at.
e	Short *e* as in l*e*t, long *e* as *a* in g*a*te.
ei	As *a* in g*a*te.
eu	As *oy* in b*oy*.
g	Always hard as in *g*ate.
i	Short *i* as in t*i*n, long *i* as *ee* in k*ee*n.
j	As *y* in *y*es.
s	As in thi*s*, not as in tho*s*e.
u	Short *u* as in f*u*ll, long *u* as *oo* in sh*oo*t, never as in t*u*b.
v	Classically pronounced as *w*.
y	Occurs in words of Greek origin and was pronounced as the French *u*.

Each scientific name is followed by a suggested pronunciation. This is often not the same as the English pronunciation of the name but does not necessarily conform to a complete classical pronunciation. Each word is divided into symbols, each symbol representing one syllable. The stressed syllable is given in italics. The pronunciation of each symbol is as in English, with the ambiguous letters being explained in the following table. For a normal English pronunciation, the scientific name can be pronounced as if it were English, for a more classical pronunciation consult the table on p. 10 for words of Latin or Greek origin.

Symbol or letter	Pronounce as in
a	apart, canal
ă	cat
e	let
ew	few
ewr	pure
g	gate
i	in
ie	kite
j	jam
o	hot
ō	note
oi	usually as oy in boy but classically as o-i
our	French pour
ow	how
s	this
th	thin
tH	this
u	full
ū	tub
zh	French je

Glossary

The use of technical terms has been kept to a minimum but some have been used for brevity and accuracy. These are defined below.

acuminate	Tapered to a long point.
anther	The portion of the stamen (q.v.) which bears the pollen.
areole	In cacti a cluster of spines.
auricle	An ear-like lobe.
awn	A bristle-like tip.
bi-pinnate	Twice pinnate, i.e. pinnate (q.v.) with the divisions pinnate.
bract	A leaf-like structure beneath a flower or group of flowers.
carpel	Part of the female portion of a flower consisting of an ovary (or part of one) and stigma.
column	In orchids, the combined style and stigma.
compound	Consisting of several parts.
corolla	The part of the flower made up of the petals.
corymb	A type of raceme (q.v.) with the flowers at more or less the same height making a rounded or flat-topped inflorescence.
dioecious	Bearing male and female flowers on separate plants.
disc	An organ at the base of the ovary.
entire	Not lobed or toothed.
epiphytic	A plant growing non-parasitically on another one.
filament	The stalk of the stamen on which the anther is borne.
glabrous	Without hairs.
glaucous	With a blue-grey bloom.
graft-hybrid	A plant which originated when one plant was grafted onto a different one. The tissues combine to create a different plant although there is no mixing of the genes.
indusium	in ferns, the covering of the sorus.
inflorescence	The arrangement of individual flowers on a stem.
involucre	A whorl of bracts beneath an inflorescence.
linear	Narrow, with parallel sides.
lip	In orchids, a modified petal.
monocarpic	Fruiting once then dying.
mucronate	With a short, abrupt point.
node	A point on the stem bearing one or more leaves.
obovate	Egg-shaped in outline, broadest above the middle.
ovate	Egg-shaped in outline, broadest below the middle.
panicle	A raceme (q.v.) with the branches branched.
pappus	A tuft of hairs or bristles on the fruit.
pedicel	The stalk of a single flower in an inflorescence.
peltate	With the stalk of a leaf inserted below the blade and not on the margin.
perianth	The part of a flower comprising the petals and sepals, generally used when they are similar.
petiole	The leaf stalk.
phyllode	A leaf-like structure consisting of a flattened petiole.

pinna (plural pinnae)	The primary division of a pinnate leaf.	spadix	Unstalked flowers densely arranged on a fleshy stem.
pinnule	The secondary division of a bi-pinnate leaf.	spathe	A large bract surrounding an inflorescence.
pistil	The female part of a flower.	sporangium (plural sporangia)	A capsule containing spores.
pollinium (plural pollinia)	A mass of pollen grains transported as a whole during pollination.	stamen	The male part of a flower consisting of an anther and a filament.
proliferous	Reproducing vegetatively.		
pseudobulb	The thickened stem of an orchid.	strobilus	A cone.
raceme	An inflorescence bearing single, stalked flowers from a central axis, the youngest at the apex.	style	The female part of a flower which bears the pollen-receiving organ (stigma) at its tip.
receptacle	The apex of a flower stem on which the floral parts are borne.	type specimen	A specimen from which a plant was originally described and which defines the application of the name given.
scape	A leafless flower stalk.		
sepal	The parts of the perianth outside of the petals.	umbel	An inflorescence in which the individual flower stalks all arise from one point at the apex of the stem, sometimes compound and then with the primary stalks bearing umbels.
sessile	Unstalked.		
simple	Not divided or lobed.		
sorus (plural sori)	A cluster of sporangia (q.v.).		

DICTIONARY OF
PLANT NAMES

A

Aaron's Beard see *Hypericum calycinum*.
Aaron's Rod see *Verbascum thapsus*.
Abele see *Populus alba*.

Abelia a-*bel*-ee-a *Caprifoliaceae*. After Dr
Clarke Abel (1780–1826) who introduced *A.
chinensis*. Deciduous and evergreen shrubs.
   chinensis chin-*en*-sis. Of China.
   engleriana eng-gla-ree-*ah*-na. After
Heinrich Engler (1844–1930). China.
   floribunda flō-ri-*bun*-da. Profusely
flowering. Mexico.
   × *grandiflora* grän-di-*flō*-ra. *A. chinensis* ×
A. uniflora. Large-flowered.
   schumannii shoo-*mahn*-ee-ee. After K. M.
Schumann (1851–1904), German botanist.
C China.
   triflora tri-*flō*-ra. Three-flowered, the
flowers are in clusters of three. Himalaya.

Abeliophyllum a-bel-ee-ō-*fil*-lum *Oleaceae*.
From *Abelia* q.v. and Gk. *phyllon* (a leaf)
referring to the similar leaves. Deciduous
shrub.
   distichum dis-tik-um. Two-ranked (the
leaves). Korea.

Abelmoschus ǎ-bel-*mos*-kus *Malvaceae*.
From Arabic *abu-l-mosk* (father of musk)
referring to the musk-scented seeds. Tender
annuals.
   esculentus es-kew-*len*-tus. Edible (the fruit).
Ladies' Fingers, Okra. Tropical Asia.
   manihot mah-nee-hot. From *manioc* the
Brazilian name for cassava (*Manihot
esculenta*). China.

Abies ǎ-bee-ayz *Pinaceae*. The classical
name. Evergreen conifers. Fir.
   alba äl-ba. White (the bark of old trees).
European Silver Fir. C and S Europe.
   amabilis a-*mah*-bi-lis. Beautiful. Red Silver
Fir. Pacific Silver Fir. NW United States.
   balsamea bäl-*săm*-ee-a. Balsam-producing
(the bark). Balsam Fir. Canada, E United
States.
   bracteata brăk-tee-*ah*-ta. With bracts (on
the cone scales) conspicuous in this
species. Santa Lucia Fir. California.
   cephalonica kef-a-*lon*-i-ka. Of Cephalonia
(now Kefallinia) a Greek Island. Greek
Fir. SE Europe.

Abies (continued)
   concolor kon-ko-lor. Of the same colour (the
leaf surfaces). White Fir. SW United
States.
   delavayi del-a-*vay*-ee. After its discoverer
the Abbé Jean Marie Delavay (1838–95),
French missionary in China who
introduced many notable plants to
cultivation. China.
     forrestii fo-*rest*-ee-ee. After its discoverer
and introducer George Forrest
(1873–1932), a Scottish plant collector
who made several expeditions to China.
China.
   grandis grän-dis. Large. Giant Fir. W N
America.
   homolepis ho-mō-*lep*-is. With similar scales
(on the cone). Nikko Fir. Japan.
   koreana ko-ree-*ah*-na. Of Korea. Korean
Fir. S Korea.
   lasiocarpa lǎ-see-ō-*kar*-pa. With rough
cones. Alpine Fir. W N America.
   magnifica mahg-*ni*-fi-ka. Magnificent. Red
Fir. W United States.
   nordmanniana nord-mahn-ee-*ah*-na. After
its discoverer Alexander von Nordmann
(1803–66), German botanist. Caucasian
Fir. Caucasus, W Asia.
   pindrow pin-drō. The native name.
W Himalaya.
   procera prō-*kay*-ra. Tall. Noble Fir.
W United States.
   spectabilis spek-*tah*-bi-lis. Spectacular.
Himalayan Fir. Himalaya.
   squamata skwah-*mah*-ta. Flaking (the
bark). Flaky Fir. China.

Abronia a-*brō*-nee-a *Nyctaginaceae*. From
Gk. *abros* (delicate) referring to the bracts.
Annual herb.
   umbellata um-bel-*lah*-ta. The flowers
appear to be in umbels. Pink Sand
Verbena. W N America.

Abutilon a-*bew*-ti-lon *Malvaceae*. From the
Arabic name for a similar plant. Tender and
semi-hardy shrubs.
   hybridum hort. *hib*-rid-um. Hybrid. Name
used for various hybrids.
   insigne in-*sig*-nee. Remarkable. Colombia,
Venezuela.
   megapotamicum meg-a-pot-*ǎm*-ik-um. From
near the Rio Grande, Brazil.

Abutilon (continued)
× *milleri* mil-la-ree. *A. megapotamicum* ×
A. pictum. After Philip Miller (1691–1771)
curator of Chelsea Physic Garden.
pictum pik-tum. (= *A. striatum*). Painted
(the flowers). Brazil.
× *suntense* sūn-*ten*-see. *A. ochsenii* × *A.*
vitifolium. From Sunte House, Sussex
where it was raised in 1967.
vitifolium vee-ti-*fo*-lee-um. *Vitis*-leaved.
Chile.

Abyssinian Feathertop see *Pennisetum*
villosum.

Acacia a-*kay*-see-a or classically a-*kah*-kee-a
Leguminosae. Gk. name for *A. arabica* from
akis (a sharp point) referring to the thorns.
Tender and semi-hardy evergreen trees and
shrubs. Wattle.
baileyana bay-lee-*ah*-na. After F. M. Bailey
(1827–1915) Australian botanist.
Cootamundra Wattle. New South Wales.
cultriformis cul-tree-*form*-is. Knife-shaped
(the phyllodes). Knife Acacia. E Australia.
dealbata dee-al-*bah*-ta. Whitened (the
young shoots and leaves). Silver Wattle.
SE Australia.
farnesiana far-neez-ee-*ah*-na. From the
garden of the Farnese Palace, Rome.
Tropical America.
longifolia long-gi-*fo*-lee-a. With long leaves
(phyllodes). Sydney Golden Wattle.
Australia.
melanoxylon mel-a-*noks*-i-lon. With black
wood. Blackwood. S Australia, Tasmania.
mucronata mew-kron-*ah*-ta. Mucronate (the
phyllodes). Victoria, Tasmania.
podalyriifolia pōd-a-li-ree-i-*fo*-lee-a. With
leaves (phyllodes) like *Podalyria* (S
African shrub). Pearl Acacia. NE
Australia.
pravissima prah-*vis*-im-a. Very crooked (the
phyllodes). Ovens Wattle. SE Australia.
riceana ries-ee-*ah*-na. After Mr T. Spring-
Rice, Chancellor of the Exchequer
1835–9. Rice's Wattle. Tasmania.
verticillata ver-tik-il-*lah*-ta. Whorled (the
phyllodes). Star Acacia, Prickly Moses.
Victoria, Tasmania.

Acaena a-*see*-na or classically a-*kie*-na
Rosaceae. From Gk. *akaina* (a thorn)
referring to the spiny fruits. Herbaceous
perennials and sub-shrubs. New Zealand
Burr.
adscendens ăd-*send*-enz. Ascending (the
branches from procumbent stems). New
Zealand.

Acaena (continued)
buchananii bew-kăn-*ăn*-ee-ee. After John
Buchanan (1819–98), botanist with the
Geological Survey in New Zealand. New
Zealand, Australia.
inermis in-*er*-mis. Not spiny. New Zealand.
microphylla mik-rō-*fi*-la. Small-leaved. New
Zealand.
novae-zeelandiae no-vie-zee-*lahn*-dee-ie. Of
New Zealand.

Acalypha ă-ka-*lee*-fa *Euphorbiaceae*. Classical
name for the nettle from the similar leaves.
Tender shrubs.
hispida *his*-pid-a. Bristly. Chenille Plant,
Red-hot Cat's Tail. Malay Archipelago.
wilkesiana wilks-ee-*ah*-na. After Admiral
Charles Wilkes (1798–1877), American
explorer of the Pacific. Pacific Islands.

Acantholimon a-kănth-ō-*lee*-mon
Plumbaginaceae. From Gk. *akanthos* (a
thorn) and *limonium* (sea lavender). Spiny
sub-shrubs. Prickly Thrift.
glumaceum gloo-mah-*kee*-um. Having
glumes, the papery bracts around the
flowers. Caucasus, W Asia.
olivieri o-liv-ee-*e*-ree. After G. A. Olivier
(1756–1814) French naturalist. Iran.
ulicinum ew-li-*kee*-num. (= *A.*
androsaceum). Like *Ulex*. SE Europe,
W Asia.
creticum *kray*-ti-kum. (= *A. creticum*). of
Crete. Crete, W Asia.
venustum ven-*us*-tum. Charming. W. Asia.

Acanthopanax a-kănth-ō-*păn*-ăks *Araliaceae*.
From Gk. *akanthos* (a thorn) and *Panax* a
related genus from *panakes* (a panacea)
referring to the medicinal properties of
Panax ginseng. Spiny shrubs.
henryi hen-ree-ee. After Augustine Henry,
see *Illicium henryi*. Japan.
sieboldianus see-bōld-ee-*ah*-nus. After
Philipp Franz von Siebold (1796–1866),
German doctor who introduced and named
many Japanese plants. China, Japan.

Acanthus a-*kănth*-us *Acanthaceae*. From Gk.
akanthos (a thorn) referring to the often spiny
leaves and flower bracts. Perennial herbs.
Bear's Breeches.
hungaricus hun-*gah*-ri-kus. (= *A.*
balcanicus. *A. longifolius*). Of Hungary. SE
Europe.
mollis *mol*-lis. Soft, the leaves are not spiny.
SW Europe.

Acanthus (continued)
'Latifolius' lah-tee-*fo*-lee-us. Broad-leaved.
spinosus speen-ō-sus. (= *A. spinosissimus*).
Spiny (the leaves). SE Europe.

Acer ay-ser or classically *ă*-ker *Aceraceae.*
The classical L. name. Deciduous trees.
Maple.
campestre kăm-*pes*-tree. Of fields. Common
Maple, Field Maple. Europe, N Africa,
W Asia.
capillipes ka-*pil*-lee-pays. Slender-stalked
(the flowers). Japan.
cappadocicum kăp-a-*do*-kik-um. Of
Cappadocia, W Asia. Caucasus, W Asia.
carpinifolium kar-peen-i-*fo*-lee-um.
Carpinus-leaved. Hornbeam Maple. W N
America.
circinatum ker-kin-*ah*-tum. Rounded (the
leaves). Vine Maple. W N America.
crataegifolium kra-tieg-i-*fo*-lee-um.
Crataegus-leaved. Japan.
davidii dă-*vid*-ee-ee. After Armand David,
see *Davidia.*
distylum di-*stil*-um. With two styles (in
some species they are united). Japan.
ginnala gin-*nah*-la. The native name. Amur
Maple. NE Asia.
griseum gris-ee-um. Grey (the lower leaf
surface). Paper-bark Maple. C China.
grosseri grō-sa-ree. After Grosser. China.
hersii herz-ee-ee. After Joseph Hers who
discovered it in 1919.
japonicum ja-*pon*-i-kum. Of Japan.
'Aconitifolium' ă-kon-ee-ti-*fo*-lee-um.
Aconitum-leaved.
'Vitifolium' vee-ti-*fo*-lee-um. *Vitis*-leaved.
lobelii lō-*bel*-ee-ee. After M. Lobel
(1538–1616) Flemish botanist. S Italy.
macrophyllum măk-rō-*fil*-um. Large-leaved.
Oregon Maple. W N America.
monspessulanum mon-spes-ew-*lah*-num. Of
Montpelier. Montpelier Maple. Europe,
N Africa, W Asia.
negundo ne-gun-do. From the native name
of *Vitex negundo* because of a supposed
similarity of leaf. Ash-leaved Maple. Box
Elder. N America.
nikoense nik-ō-*en*-see. Of Nikko, Japan.
Nikko Maple. Japan, C China.
palmatum pahl-*mah*-tum. Hand-like (the
leaves). Japanese Maple. Japan.
'Atropurpureum' aht-rō-pur-*pewr*-ree-um. Dark purple (the leaves).
'Dissectum' dis-*sek*-tum. Finely cut (the
leaves).

Acer (continued)
Heptalobum hep-ta-*lō*-bum. Seven-lobed
(the leaves).
'Linearilobum' lin-ee-ah-ri-*lō*-bum. With
linear lobes (the leaves).
pensylvanicum pen-sil-*vahn*-i-kum. Of
Pennsylvania. Moosewood. E N America.
'Erythrocladum' e-rith-rō-*klă*-dum.
With red shoots.
platanoides plă-ta-*noi*-deez. Like *Platanus.*
Norway Maple. Europe.
pseudoplatanus sood-ō-*plă*-ta-nus. False
Platanus. Sycamore. Europe.
rubrum rub-rum. Red (autumn colour). Red
Maple. E N America.
'Schlesingeri' shlez-*ing*-ga-ree. After Mr
Schlesinger in whose grounds it was
found.
rufinerve roof-i-*ner*-vee. Red-veined,
referring to the reddish hairs on the leaf
veins. Japan.
saccharinum săk-a-*ree*-num. Sugary (the
sap). Silver Maple. E N America.
saccharum sa-*kah*-rum. L. name for sugar
cane. Maple syrup and sugar are made
from the sap of this tree. Sugar Maple. E N
America.
tataricum ta-*tah*-ri-kum. Of Tatary, C Asia.
SE Europe, W Asia.
× *zoeschense* zur-*shen*-see. *A. campestre* ×
A. cappadocicum. Of the Zoeschen
nursery, Germany.

Achillea a-*kil*-lee-a or classically ă-kil-*lee*-a
Compositae. After Achilles of Gk. mythology
who is said to have used it medicinally.
Herbaceous perennials.
ageratifolium a-ge-ra-ti-*fo*-lee-um.
Ageratum-leaved. SE Europe.
ageratum a-*ge*-ra-tum. From the
resemblance to *Ageratum.* Mediterranean
Region.
chrysocoma kris-*o*-ko-ma. With golden hair,
referring to the yellow flower heads. SE
Europe.
clavennae kla-*ven*-ie. After Niccola
Chiavera, died 1617, Italian apothecary.
Alps, Balkans.
clypeolata kli-pee-ō-*lah*-ta. Shield-shaped
(the flower heads). SE Europe.
filipendulina fi-li-pen-dew-*lee*-na. Like
Filipendula. W Asia, Caucasus.
millefolium meel-lee-*fo*-lee-um. Thousand-leaved, referring to the finely divided
leaves. Yarrow. Europe, W Asia.
ptarmica tar-mi-ka. Gk. name for a plant
which caused sneezing, referring to its use
for snuff. Sneezewort. Europe, Asia.

Achillea (continued)
  taygetea tay-*gee*-tee-a. From the Taygetos
  mountains, Greece. The true plant (now
  A. aegyptiaca) is rarely grown.
  umbellata um-bel-*lah*-ta. The flowers
  appear to be in umbels.

Achimenes ă-kim-*ee*-neez *Gesneriaceae*.
Possibly from Gk., meaning not over-
wintering referring to their tender nature,
after King Hakhamash of Turkey (Gk.
Achaemenes) or from *achaemenis*, Gk. name
of a plant. Tender perennials. Hot-water
Plant.
  erecta e-*rek*-ta (= *A. coccinea*). Erect.
  C America, Mexico, Jamaica.
  grandiflora grăn-di-*flō*-ra. Large-flowered.
  Mexico, C America.
  heterophylla he-te-rō-*fi*-la. Variably leaved
  (the leaves of a pair are unequal). Mexico,
  Guatemala.
  longiflora long-gi-*flō*-ra. Long-flowered.
  Mexico to Panama.
  mexicana meks-i-*kah*-na. Of Mexico.

Achnatherum ăk-na-*the*-rum *Gramineae*.
From Gk. *achna* (chaff) and *ather* (an awn).
Perennial grass.
  calamagrostis kăl-a-ma-*gros*-tis. (= *Stipa
  calamagrostis*). Name of a related genus
  from Gk. *calama* (a reed) and *agrostis* (a
  grass). S Europe.

Acidanthera a-kid-an-*the*-ra *Iridaceae*. From
Gk. *akis* (a point) and *anthera* (an anther)
referring to the pointed anthers. Tender or
semi-hardy, cormous perennials.
  bicolor bi-ko-lor. Two-coloured (the
  flowers). Ethiopia.
  murielae mewr-ree-*el*-ie. After Muriel.
  Ethiopia.

Acinos a-*kee*-nos *Labiatae*. From *akinos*, Gk.
name for an aromatic herb. Perennial herb.
  alpinus ăl-*peen*-us. (= *Calamintha alpina*).
  Alpine. C and S Europe.

Acokanthera a-kō-kan-*the*-ra *Apocynaceae*.
From Gk. *akoke* (a point) and *anthera* (an
anther) referring to the pointed anthers.
Tender, evergreen shrub.
  oblongifolia ob-long-gi-*fo*-lee-a. (= *A.
  spectabilis*). With oblong leaves. SW Africa.

Aconitum ă-kon-*ee*-tum *Ranunculaceae*. The
L. name. Poisonous perennial herbs.
Monkshood.
  × *bicolor* bi-ko-lor. (= *A.* × *cammarum*

Aconitum (continued)
  hort.). *A. napellus* × *A. variegatum*. Two-
  coloured (the flowers).
  carmichaelii kar-mie-*keel*-ee-ee. (= *A.
  fischeri* hort. *A. wilsonii*). After J. R.
  Carmichael (1838–70). E Asia.
  napellus na-*pel*-lus. A small turnip,
  referring to the roots. Common
  Monkshood. W and C Europe.
  paniculatum pa-nik-ew-*lah*-tum. With
  flowers in panicles. C and E Europe.
  variegatum vă-ree-a-*gah*-tum. Variegated
  (the flowers). C and E Europe.
  volubile vol-*ew*-bi-lee. Twining. E Asia.
  vulparia vul-*pah*-ree-a. (= *A. lycoctonum*
  hort.). Of Wolves. Wolf's Bane, Badger's
  Bane. Europe.

Acorus a-*ko*-rus or classically ă-ko-rus
Araceae. L. name for *Iris pseudacorus*.
Aquatic or marginal herbaceous perennials.
  calamus kă-la-mus. Gk. *calamus* (a reed)
  referring to the foliage. Sweet Flag. N
  Hemisphere.
  gramineus grah-*min*-ee-us. Grass-like (the
  foliage). E Asia.

Acradenia ăk-ra-*deen*-ee-a *Rutaceae*. From
Gk. *akros* (at the tip) and *adenia* (a gland)
referring to the glands at the tips of the
carpels. Semi-hardy, evergreen shrub.
  frankliniae frank-*lin*-ee-ie. After Lady
  Franklin, wife of the Governor of
  Tasmania in 1842. W Tasmania.

Actaea ăk-*tee*-a *Ranunculaceae*. From Gk.
aktea (elder) referring to the similar leaves.
Herbaceous perennials with poisonous fruits.
Baneberry.
  rubra rub-ra. Red (the fruits). Red
  Baneberry. N America.
  spicata spee-*kah*-ta. With flowers in spikes.
  Herb Christopher. Europe, Asia.

Actinidia ăk-tin-*id*-ee-a *Actinidiaceae*. From
Gk. *aktinos* (a ray) referring to the radiating
styles. Deciduous, twining shrubs.
  arguta ar-*gew*-ta. Sharply toothed (the
  leaves). NE Asia.
  chinensis chin-*en*-sis. Of China. Chinese
  Gooseberry, Kiwi Fruit.   •
  kolomikta ko-lō-*mik*-ta. The native name.
  China, Japan.
  polygama po-*lig*-a-ma. Polygamous i.e. with
  some flowers male, some female and some
  hermaphrodite. Silver Vine. C Japan.
  purpurea pur-*pewr*-ree-a. Purple (the
  fruits). China.

Ada *ah*-da *Orchidaceae*. After Ada, sister of Artemisia of Caria. Greenhouse orchid.
aurantiaca ow-răn-tee-*ah*-ka. Orange (the flowers). Colombia.

Adam's Needle see *Yucca filamentosa*

Adenocarpus a-deen-ō-*kar*-pus *Leguminosae*. From Gk. *aden* (a gland) and *karpos* (a fruit) referring to the sticky pods. Deciduous shrub.
decorticans day-*kor*-ti-kănz. With peeling bark. Spain.

Adenophora ă-den-*o*-fo-ra *Campanulaceae*. From Gk. *aden* (a gland) and *phorea* (to bear) referring to the glandular disc at the base of the style. Perennial herbs.
liliifolia lee-lee-i-*fo*-lee-a. *Lilium*-leaved. C Europe, N Asia.
tashiroi tăsh-ee-rō-ee. After Tashiro.

Adiantum ă-dee-*ăn*-tum *Adiantaceae*. From Gk. *adiantos* (unwetted) referring to the way the fronds repel water. Tender ferns. Maidenhair Fern.
capillus-veneris ka-*pil*-lus-*ven*-e-ris. Venus's hair, referring to the delicate foliage. Common Maidenhair Fern. Worldwide.
caudatum kaw-*dah*-tum. With a tail (the pinnae). Trailing Maidenhair Fern. Old World tropics.
concinnum kon-*kin*-um. Elegant. S America.
formosum for-*mō*-sum. Beautiful. Australian Maidenhair Fern. Australia, New Zealand.
hispidulum his-*pid*-ew-lum. Finely bristly. Rough Maidenhair Fern. Old World tropics.
pedatum ped-*ah*-tum. Like a bird's foot (the fronds). N American Maidenhair Fern. N America, E Asia.
  japonicum ja-*pon*-i-kum. Of Japan. Rose-fronded Maidenhair Fern.
raddianum rah-dee-*ah*-num. (= *A. cuneatum*). After Giuseppe Raddi (1770–1829). Delta Maidenhair Fern. Brazil.
trapeziforme tra-peez-ee-*for*-mee. Unequally four-sided (the segments of the fronds). Tropical America.
venustum ven-*us*-tum. Handsome. Kashmir Maidenhair Fern. Himalaya.

Adonis a-*dō*-nis *Ranunculaceae*. The Gk. name from the god Adonis, said to have been

Adonis (continued)
changed into this plant after death. Annual and perennial herbs.
aestivalis ie-sti-*vah*-lis. Of summer (flowering). Europe.
amurensis ăm-ew-*ren*-sis. Of the Amur region, NE Asia.
  'Plena' *play*-na. With double flowers.
annua ăn-ew-a (=*A. autumnalis*). Annual. Pheasant's Eye. S Europe, W Asia.
pyrenaica pi-ray-*nah*-i-ka. Of the Pyrenees. Pyrenees, Maritime Alps.
vernalis ver-*nah*-lis. Of spring (flowering). C and S Europe.

Adromischus ă-drō-*mis*-kus *Crassulaceae*. From Gk. *hadros* (stout) *mischos* (a stalk) referring to the stout, short pedicels. Tender succulents, S Africa.
cooperi *koo*-pa-ree. After Thomas Cooper (1815–1913). Plover Eggs.
cristatus kris-*tah*-tus. Crested (the leaves). Crinkle-leaf Plant.
maculatus măk-ew-*lah*-tus. Spotted (the leaves). Calico Hearts.
rotundifolius ro-tund-i-*fo*-lee-us. With round leaves.

Aechmea iek-*mee*-a *Bromeliaceae*. From Gk. *aichme* (a point) referring to the pointed sepals. Tender, evergreen herbaceous perennials. Brazil, unless stated otherwise.
bracteata brăk-tee-*ah*-ta. With bracts (on the scape). Mexico to Colombia.
fasciata făs-kee-*ah*-ta. Banded (the leaves). Urn Plant.
fulgens *ful*-genz. Shining (the red bracts).
racineae ră-*seen*-ee-ie. After Racine, wife of M. B. Foster, see *Cryptanthus fosterianus*. Christmas Jewels.

Aeonium ie-ō-nee-um *Crassulaceae*. The L. name of one species. Tender or semi-hardy succulents.
arboreum ar-*bo*-ree-um. Tree-like. Morocco.
  'Foliis Purpureis' *fo*-lee-is pur-*pewr*-ree-is. With purple leaves.
canariense ka-nah-ree-*en*-see. Of the Canary Islands.
domesticum see *Aichryson* × *domesticum*.
haworthii hay-*werth*-ee-ee. After Haworth, see *Haworthia*. Canary Islands.
simsii simz-ee-ee. After John Sims (1749–1831). Canary Islands
tabuliforme tăb-ew-lee-*form*-ee. Table-shaped i.e. flat-topped. Canary Islands.

Aeonium (continued)
undulatum un-dew-*lah*-tum. Wavy-edged (the leaves). Saucer Plant. Canary Islands.

Aerides ah-*e*-ree-deez *Orchidaceae*. From Gk. *aer* (air), they are epiphytic, appearing to derive nourishment from the air. Tender orchids. Fox-tail Orchid.
crispa kris-pa. Finely wavy, referring to the fringed central lobe of the lip. Indonesia.
falcata fål-*kah*-ta. Sickle-shaped (the leaves). Burma.
japonica ja-*pon*-i-ka. Of Japan.
lawrenceae lo-*rens*-ee-ie. After Lady Lawrence wife of Sir Trevor Lawrence of Burford Lodge. Philippines.
multiflora mul-tee-*flō*-ra. Many-flowered. Himalaya, India, Burma.
odorata o-dō-*rah*-ta. Scented (the flowers). Himalaya, India, SE Asia.
quinquevulnera kwing-kwee-*vul*-ne-ra. With five marks (on the flower). Philippines.
vandarum vån-*dah*-rum. Like *Vanda*. Himalaya.

Aeschynanthus ie-skee-*năn*-thus *Gesneriaceae*. From Gk. *aischune* (shame) and *anthos* (a flower) referring to the red flowers. Greenhouse climbers. Basket Plant.
boschianus bosh-ee-*ah*-nus. After J. van der Bosch (1807–54), Governor-General of the Dutch East Indies. Java.
marmoratus mar-mo-*rah*-tus. Marbled (the leaves). Zebra Basket Vine. SE Asia.
pulcher pul-ker. Pretty. Royal Red Bugler. Java.
radicans rah-di-kånz. (= *A. lobbianus*). With rooting stems. Lipstick Vine. Malaysia, Java.

Aesculus ies-ku-lus *Hippocastanaceae*. L. name for an oak with edible acorns. Deciduous trees. Horse-chestnut, Buckeye.
californica kål-i-*forn*-i-ka. Of California. California Buckeye.
× *carnea* kar-nee-a. *A. hippocastanum* × *A. pavia*. Flesh-coloured (the flowers). Red Horse-chestnut.
flava flah-va. Yellow (the flowers). Yellow Buckeye. United States.
hippocastanum hip-ō-*kå*-sta-num. The L. name. Common Horse-chestnut. N Greece, N Albania.
'Baumannii' bow-*mahn*-ee-ee. After A. N. Baumann who found the original sport.
indica in-di-ka. Indian. Indian Horse-chestnut. NW Himalaya, Afghanistan.

Aesculus (continued)
× *mutabilis* mew-*tah*-bi-lis. *A. discolor* × *A. neglecta*. Changeable (the colour of the flowers).
'Induta' in-*dew*-ta. Clothed, referring to the densely hairy undersides of the leaves.
parviflora par-vi-*flō*-ra. Small-flowered. SE United States.
pavia pah-vee-a. After Peter Paaw (Petrus Pavius), Dutch botanist, died 1616. Red Buckeye. S United States.
turbinata tur-bin-*ah*-ta. Top-shaped (the fruit). Japanese Horse-chestnut. Japan.

Aethionema ieth-ee-ō-*nee*-ma *Cruciferae*. Possibly from Gk. *aethes* (unusual) and *nema* (a thread) referring to the stamens. Sub-shrubby perennials. Stone Cress.
armenum ar-*meen*-um. Of Armenia. Turkey.
coridifolium ko-ri-di-*fo*-lee-um. With leaves like *Coris*. W Asia.
grandiflorum grån-di-*flō*-rum. Large-flowered. W Asia.
iberideum i-be-*ri*-dee-um. Like *Iberis*. Greece, Turkey.
oppositifolium op-o-si-ti-*fo*-lee-um. With opposite leaves. Lebanon.

African Boxwood see *Myrsine africana*
African Daisy see *Arctotis*
African Hemp see *Sparmannia africana*
African Lily see *Agapanthus*
African Violet see *Saintpaulia*

Agapanthus åg-a-*pănth*-us *Amaryllidaceae*. From Gk. *agape* (love) and *anthos* (a flower). Hardy and semi-hardy perennial herbs. African Lily. S Africa.
africanus åf-ri-*kah*-nus. (= *A. umbellatus*). African.
campanulatus kåm-pån-ew-*lah*-tus. Bell-shaped (the flowers).
caulescens kaw-*les*-enz. With a stem.
inapertus in-a-*per*-tus. Closed, referring to the narrow-mouthed flowers.
orientalis o-ree-en-*tah*-lis. From the east (of S Africa).
praecox prie-koks. Early (flowering).

Agapetes åg-a-*peet*-eez *Ericaceae*. From Gk. *agapetos* (desirable). Evergreen, semi-hardy shrub.
serpens ser-penz. (=*Pentapterygium serpens*). Creeping. Himalaya.

Agastache a-*gah*-sta-kee *Labiatae*. From Gk. *agan* (very much) and *stachys* (a spike) referring to the numerous flower spikes. Perennial herb.
  mexicana mek-si-*kah*-na. Of Mexico.

Agathis a-*gah*-this, classically ă-ga-this *Araucariaceae*. From Gk. *agathis* (a ball of thread) referring to the cones. Semi-hardy conifer.
  australis ow-*strah*-lis. Southern. Kauri Pine. New Zealand.

Agave a-*gah*-vee *Agavaceae*. From Gk. *agave* (noble) referring to the tall inflorescence of *A. americana*. Tender and semi-hardy succulents.
  americana a-me-ri-*kah*-na. American. Century Plant. Mexico.
  'Marginata' mar-gin-*ah*-ta. Margined (the leaves).
  angustifolia an-gust-i-*fo*-lee-a. Narrow-leaved. Cult.
  attenuata a-ten-ew-*ah*-ta. Drawn out (the inflorescence). Mexico.
  filifera fee-*li*-fe-ra. Bearing threads (on the leaf margins). Thread Agave. Mexico.
  parviflora par-vi-*flō*-ra. Small-flowered. Arizona, N Mexico.
  potatorum pō-tah-*to*-rum. Of drinkers, a drink is made from the plant. Mexico.
  stricta strik-ta. Upright (the leaves). Mexico.
  victoriae-reginae vik-*tor*-ree-ie-ray-*geen*-ie. After Queen Victoria. Mexico.

Ageratum a-ge-ra-tum *Compositae*. From Gk. *a* (not) and *geras* (age) referring to the long-lasting flowers. Annual herbs. Floss Flower.
  conyzoides kon-ee-*zoi*-deez. Like *Conyza* a related genus. W Indies, Mexico, C America.
  houstonianum hew-stōn-ee-*ah*-num. After William Houston (1695–1733), a Scottish physician who collected in C America. Mexico, C America.

Aglaonema a-glah-ō-*nee*-ma *Araceae*. From Gk. *aglaos* (bright) and *nema* (a thread) referring to the stamens. Tender, evergreen, foliage perennials.
  commutatum kom-ew-*tah*-tum. Changeable. Philippines, Celebes.
  'Treubii' *troy*-bee-ee. (= *A. treubii*). After Melchior Treub, Director of the Botanic Garden, Buitenzorg (now Bogor) Java.
  costatum kos-*tah*-tum. Ribbed, referring to

Aglaonema (continued)
  the white midrib. Spotted Evergreen. Malay Peninsula.
  crispum kris-pum. Finely wavy, perhaps referring to the irregular variegation. Painted Drop-tongue. Philippines.
  modestum mod-*es*-tum. Modest, it is not variegated like many species. Chinese Evergreen. SE Asia.
  nitidum ni-tid-um. Shining (the leaves). SE Asia.
  pictum pik-tum. Painted – the variegated leaves. Sumatra.

Agrostemma ăg-rō-*stem*-a *Caryophyllaceae*. From Gk. *agros* (a field) and *stemma* (a garland). Annual herb.
  githago gi-*thah*-go. The L. name. Corn Cockle. E Mediterranean region.

Aichryson ie-*kris*-on *Crassulaceae*. The Gk. name for *Aeonium arboreum*. Tender, succulent annuals and perennials.
  × *domesticum* dom-*es*-ti-kum. *A. punctatum* × *A. tortuosum*. (= *Aeonium domesticum*). Cultivated. Youth and Old Age.
  laxum lăks-um. Open (the inflorescence). Canary Islands.
  villosum vil-*lō*-sum. Softly hairy. Azores, Madeira.

F 37,7 55

Ailanthus ie-*lan*-thus *Simaroubaceae*. From *ailanthos* (tree of heaven) the Indonesian name for *A. moluccana*. Deciduous tree.
  altissima ăl-*tis*-si-ma. Tallest. Tree of Heaven. N China.

Air Plant see *Kalanchoe pinnata*

Ajuga a-*joo*-ga *Labiatae*. Derivation obscure. Perennial herbs.
  genevensis gen-e-*ven*-sis. Of Geneva. Europe, Asia.
  pyramidalis pi-ra-mid-*ah*-lis. Pyramidal (the inflorescence). Europe.
  reptans rep-tănz. Creeping. Bugle. Europe.

Akebia a-*kee*-bee-a *Lardizabalaceae*. From the Japanese name. Woody climbers.
  quinata kwi-*nah*-ta. In fives (the leaflets). Japan, China, Korea.
  trifoliata tri-fo-lee-*ah*-ta. With three leaves (leaflets). China, Japan.

Albizia ăl-*biz*-ee-a *Leguminosae*. After F. del Albizzi who introduced the following. Semi-hardy tree.
  julibrissin yoo-lee-*bris*-sin. From the native

Albizia (continued)
name. Pink Siris, Pink Mimosa, Silk Tree.
W Asia.

Alcea ăl-kee-a *Malvaceae*. From *alkaia*, Gk.
name for a kind of mallow. Herbaceous
perennials or biennials.
ficifolia fee-ki-*fo*-lee-a. (= *Althaea ficifolia*).
Ficus-leaved. Fig-leaved Hollyhock. Cult.
rosea ro-see-a. (= *Althaea rosea*). Rose-
coloured (the flowers). Hollyhock. Cult.

Alchemilla ăl-ke-*mil*-la *Rosaceae*. From an
Arabic name. Herbaceous perennials. Lady's
Mantle.
alpina ăl-*peen*-a. Alpine. Alpine Lady's
Mantle. Europe.
conjuncta kon-*yunk*-ta. Joined, it resembles
A. alpina but the leaflets are joined at the
base. Alps.
erythropoda e-rith-rō-*pō*-da. Red-stalked
(the leaves of some plants). E Europe,
Caucasus.
mollis mol-lis. Softly hairy (the leaves).
E Carpathians.

Alder see *Alnus*
  Common see *A. glutinosa*
  Grey, see *A. incana*
  Italian, see *A. cordata*
Alder Buckthorn see *Rhamnus frangula*
Alecost see *Balsamita major*

Aletris a-*let*-ris *Liliaceae*. From Gk. *aleitris*,
a female slave who ground corn, referring to
the mealy perianth. Herbaceous perennials.
farinosa fă-ri-*nō*-sa. Mealy (the perianth).
Unicorn Root. E United States.

Alexandrian Laurel see *Danae racemosa*
Algerian Statice see *Limonium bonduellii*

Alisma a-*lis*-ma *Alismataceae*. Gk. name for
a water plant. Aquatic or marginal perennial
herbs.
gramineum grah-*min*-ee-um. Grass-like (the
leaves). N temperate regions.
lanceolatum lăn-kee-ō-*lah*-tum. Lance-
shaped (the leaves). Europe.
natans see *Luronium natans*
plantago-aquatica plăn-*tah*-gō-a-*kwah*-ti-ka.
Literally water plantain – the common
name. N temperate regions.

Allamanda ăl-a-*măn*-da *Apocynaceae*. After
Frederick Allamand, 18th century Swiss
botanist who introduced *A. cathartica*.
Tender, evergreen shrubs.

Allamanda (continued)
cathartica ka-*thar*-ti-ka. Purging. Golden
Trumpet. S America.
neriifolia ne-ree-i-*fo*-lee-a. *Nerium*-leaved.
S America.

Allegheny Spurge see *Pachysandra
procumbens*

Allium ă-lee-um *Amaryllidaceae*. The Gk.
name for garlic, *A. sativum*. Perennial herbs
aflatunense ă-fla-tun-*en*-see. Of Aflatun,
C Asia.
albo-pilosum see *A. christophii*.
ascalonicum hort. see *A. cepa* Aggregatum.
beesianum beez-ee-*ah*-num. After Bees
Nursery, the introducers. China.
cepa ke-pa. The L. name. Onion. Cult.
  Aggregatum ag-re-*gah*-tum (= *A.
  ascalonicum* hort.). Clustered (the
  bulbs). Shallot. Cult.
  Proliferum prō-*li*-fe-rum. Proliferous
  referring to the production of bulbs in
  the flower heads. Tree Onion. Cult.
christophii kris-*tof*-ee-ee (= *A. albo-
pilosum*). After? Stars of Persia. N Iran,
C Asia.
cyaneum see-*ah*-nee-um. Blue (the flowers).
NW China.
cyathophorum see-ă-tho-*fo*-rum. Cup-
bearing, referring to the cup-shaped tube
formed by the united filaments. C Asia.
  farreri fă-ra-ree. After Reginald Farrer,
  see *Viburnum farreri*. NW China.
fistulosum fis-tew-*lō*-sum. Hollow-stemmed.
Welsh Onion. Cult.
flavum flah-vum. Yellow (the flowers).
S Europe.
giganteum gi-*găn*-tee-um. Very large. Iran,
Afghanistan, C Asia.
karataviense kă-ra-tah-vee-*en*-see. Of the
Kara Tau, Kazakhstan. C Asia.
moly mo-lee. Gk. name of a herb. E Spain,
SW France.
narcissiflorum nar-kis-i-*flō*-rum. Narcissus-
flowered. SW Alps.
neapolitanum nee-ah-pol-i-*tah*-num. Of
Naples. Mediterranean region.
oreophilum o-ray-o-fi-lum. (= *A.
ostrowskianum*). Mountain-loving. C Asia,
E Turkey, Caucasus.
ostrowskianum see *A. oreophilum*
paradoxum pă-ra-*doks*-um. Unusual,
perhaps referring to the single leaf.
Caucasus, N Iran.
porrum po-rum. The L. name. Leek. Cult.
pulchellum pul-*kel*-lum. Pretty. SE Europe,
W Asia.

Allium (continued)
roseum ro-see-um. Rose-coloured (the flowers, usually). Mediterranean region.
sativum sa-*tee*-vum. Cultivated. Garlic. Cult.
schoenoprasum skoyn-ō-*prah*-sum. From Gk. *schoinos* (a rush) and *prasum* (leek) referring to the rush-like leaves. Chives. Europe.
scorzonerifolium skorz-on-e-ree-i-*fo*-lee-um. *Scorzonera*-leaved.
siculum see *Nectaroscordium siculum*.
triquetrum tri-kwee-trum. Three-angled (the scape). W Mediterranean region.
tuberosum tew-be-*rō*-sum. Tuberous, referring to the rhizome. Chinese Chives. Himalaya, Assam, China.
zebdanense zeb-dan-*en*-see. Of Zebdani, Syria. E Mediterranean region.

Alloplectus ă-lō-*plek*-tus *Gesneriaceae*. From Gk. *allos* (diverse) and *pleco* (to plait) referring to the overlapping calyx lobes. Tender perennials.
capitatus kăp-i-*tah*-tus. In a dense head (the flowers). Velvet Alloplectus. N Andes.

Allspice, Californian see *Calycanthus occidentalis*
Carolina see *C. floridus*
Almond see *Prunus dulcis*
Dwarf Russian see *P. tenella*

Alnus ăl-nus *Betulaceae*. The L. name. Deciduous trees. Alder.
cordata kor-*dah*-ta. Heart-shaped (the leaves). Italian Alder. S Italy, Corsica.
glutinosa gloo-ti-*nō*-sa. Sticky (the young shoots and leaves). Common Alder. Europe, N Africa, W Asia.
incana in-*kah*-na. White (the undersides of the leaves). Grey Alder. Europe, Caucasus.
japonica ja-*pon*-i-ka. Of Japan. NE Asia, Japan.

Alocasia ă-lō-*kah*-see-a *Araceae*. From Gk. *a* (without, not) and *Calocasia* a related genus. Tender foliage herbs.
cuprea kew-pree-a. Coppery (the leaves). Giant Caladium. Borneo.
lindenii lin-*den*-ee-ee. After Lucien Linden, Belgian nurseryman. New Guinea.
macrorrhiza măk-rō-*ree*-za. (= *A. indica*). With a large root. Giant Elephant's Ear. SE Asia.
micholitziana mich-o-litz-ee-*ah*-na. AfterW. Micholitz who collected for Sander in

Alocasia (continued)
New Guinea. Philippines.
sanderiana sahn-da-ree-*ah*-na. After the Sander nursery. Kris Plant. Philippines.

Aloe ă-lō-ee *Liliaceae*. From the native name. Tender succulents.
arborescens ar-bo-*res*-enz. Tree-like. Candelabra Plant. S Africa.
aristata ă-ris-*tah*-ta. With a long, bristle-like tip (the leaves). Lace Aloe. S Africa.
brevifolia brev-i-*fo*-lee-a. With short leaves. S Africa.
ciliaris ki-lee-*ah*-ris. Fringed with hairs. S Africa.
ferox fe-rōks. Spiny (the leaves). Cape Aloe. S Africa.
humilis hum-i-lis. Low-growing. S Africa.
striata stree-*ah*-ta. Striped (the leaves). S Africa.
variegata vă-ree-a-*gah*-ta. Variegated (the leaves). Partridge-breasted Aloe. S Africa.
vera ve-ra (= *A. barbadensis*). True. Mediterranean region (naturalised).

Alonsoa a-*lon*-zō-a *Scrophulariaceae*. After Alonzo Zanoni, Secretary of State of Colombia in the 18th century. Tender annuals. Mask Flower. Peru.
acutifolia a-kewt-i-*fo*-lee-a. With pointed leaves.
warscewiczii var-sha-*vich*-ee-ee. After Joseph Warscewicz (1812–66) who collected in S America.

Alopecurus ă-lo-pe-*kew*-rus *Gramineae*. From *alopekouros*, Gk. name for a grass with an inflorescence like a fox's tail. Perennial grass.
pratensis prah-*tayn*-sis. Growing in meadows. Foxtail Grass. Europe, Caucasus, N Asia.

Aloysia ă-lō-*is*-ee-a *Verbenaceae*. After Maria Louisa, Princess of Parma, died 1819. Semi-hardy deciduous shrub.
triphylla tri-*fil*-la. (= *Lippia citriodora*). Three-leaved, the leaves are in whorls of three. Lemon Verbena. Chile.

Alstroemeria ahl-strurm-*e*-ree-a *Alstroemeriaceae*. After Baron Claus Alstroemer (1736–94). Hardy and semi-hardy perennial herbs. Chile.
aurantiaca ow-răn-tee-*ah*-ka. Orange (the flowers).
ligtu lig-too. The Chilean name.

Alternanthera ăl-ter-nan-*the*-ra
Amaranthaceae. From L. *alternans*
(alternating) and *anthera* (an anther), alternate
anthers are sterile. Tender foliage herbs.
　dentata den-*tah*-ta. Toothed. W Indies,
　S America.
　ficoidea fee-*koi*-dee-a. Like *Ficus*. Mexico,
　C and S America.
　　'Amoena' a-*moy*-na. Pleasant.
　　'Bettzickiana' bet-zik-ee-*ah*-na. After
　　Bettzick.
　　'Versicolor' ver-*si*-ko-lor. Variously
　　coloured (the leaves).

Althaea ăl-*thie*-a *Malvaceae*. The classical
name from Gk. *althaine* (to heal) referring to
medicinal properties. Perennial herb.
　ficifolia see *Alcea ficifolia*.
　officinalis o-fik-i-*nah*-lis. Sold as a herb.
　Marsh Mallow. Europe.
　rosea see *Alcea rosea*.

Aluminium Plant see *Pilea cadierei*
Alum Root see *Heuchera*

Alyssum a-*lis*-sum *Cruciferae*. From Gk. *a*
(not) and *lyssa* (madness), it was said to cure
rabies. Annual and perennial herbs.
　alpestre ăl-*pes*-tree. Of the lower mountains.
　C Europe.
　idaeum ee-*die*-um. Of Mt. Ida of ancient
　Greece. Crete.
　maritimum see *Lobularia maritima*.
　montanum mon-*tah*-num. Of mountains.
　Europe.
　murale mew-*rah*-lee. (*A. argenteum* hort.).
　Growing on walls. E Europe.
　saxatile see *Aurinia saxatilis*
　serpyllifolium ser-pil-li-*fo*-lee-um. Thyme-
　leaved. SW Europe.
　spinosum see *Ptilotrichum spinosum*
　wulfenianum wul-fen-ee-*ah*-num. After
　Wulfen. see *Wulfenia*. S Europe.

Amaranthus ăm-a-*răn*-thus *Amaranthaceae*.
From Gk. *amarantos* (unfading) referring to
the long-lasting flowers. Annual herbs.
　caudatus kaw-*dah*-tus. With a tail, referring
　to the slender, drooping inflorescence.
　Love-lies-Bleeding. Tropics.
　hybridus hib-ri-dus. Hybrid. Widespread.
　　erythrostachys e-rith-ro-*stă*-kis. (= *A.*
　　hypochondriacus). With red spikes.
　　Prince's Feather.
　tricolor tri-ko-lor. Three-coloured (the
　leaves). Joseph's Coat. Tropics.

Amaranthus (continued)
　'Salicifolius' să-lik-i-*fo*-lee-us. *Salix*-
　leaved. Fountain Plant.

Amaryllis ăm-a-*ril*-lis *Amaryllidaceae*. After a
shepherdess in Gk. mythology. Semi-hardy,
bulbous herb. The large-flowered bulbs
commonly called *Amaryllis* are *Hippeastrum*
hybrids.
　belladonna bel-a-*don*-a. Beautiful lady, an
　extract was used to dilate the pupils.
　Belladonna Lily. S Africa.

Amelanchier ă-me-*lan*-kee-er *Rosaceae*.
From the French name for *A. ovalis*.
Deciduous trees and shrubs. Serviceberry.
　canadensis kan-a-*den*-sis. Of Canada or
　America. E N America.
　florida flō-ri-da. Flowering. WN America.
　laevis lie-vis. Smooth (the leaves). E N
　America.
　lamarckii la-*mar*-kee-ee. After Lamarck
　(1744–1829), French naturalist. ? E N
　America.
　ovalis ō-*vah*-lis. Oval (the leaves). Snowy
　Mespilus. C and S Europe.
　spicata spee-*kah*-ta. In spikes (the flowers).
　E N America.
　stolonifera sto-lō-*ni*-fe-ra. Producing
　stolons. E N America.

Ammannia a-*mah*-nee-a *Lythraceae*. After
Paul Ammann (1634–91), German botanist.
Aquarium plants.
　baccifera bah-*ki*-fe-ra. Berry-bearing. Old
　World tropics.
　senegalensis sen-e-gahl-*en*-sis. Of Senegal.

Amorpha a-*mor*-fa *Leguminosae*. From Gk.
amorphos (deformed) the flowers have only
one petal. Deciduous shrub.
　fruticosa froo-ti-*kō*-sa. Shrubby. False
　Indigo. S United States.

Amorphophallus a-mor-fo-*făl*-lus *Araceae*.
From Gk. *amorphos* (deformed) and *phallus*
referring to the shape of the tubers. Tender
perennial herbs.
　bulbifer bul-bi-fer. Bulb-bearing (the
　leaves). Himalaya, India, Burma.
　rivieri ri-vee-*e*-ree. After M. Riviere, head
　gardener at Luxembourg Palace Gardens,
　Paris. SE Asia.

Ampelopsis ăm-pel-*op*-sis *Vitaceae*. From
Gk. *ampelos* (vine) and *-opsis* indicating
resemblance. Deciduous, woody climbers.
　aconitifolia ă-kon-ee-ti-*fo*-lee-a. *Aconitum*-

Ampelopsis (continued)
leaved. China.
brevipedunculata brev-ee-pe-dunk-ew-*lah*-ta. With short peduncles. NE Asia.
'Citrulloides' kit-rul-*loi*-deez. Like *Citrullus*, the water melon, with deeply divided leaves.
'Elegans' *ay*-le-gahnz. Elegant, the variegated leaves.
megalophylla meg-a-lō-*fil*-la. Large-leaved. W China.

Amsonia ăm-*son*-ee-a *Apocynaceae*. After Dr Charles Amson, 18th century Virginia physician. Herbaceous perennial.
tabernaemontani tă-ber-nie-mon-*tah*-nee. After Jakob Theodor von Bergzabern (Tabernaemontanus), 16th century physician and herbalist. E United States.

Anacampseros ăn-a-*kămp*-se-ros *Portulacaceae*. Gk. name for a herb. Tender succulent.
rufescens roof-*e*-senz. Reddish (the leaves). S Africa.

Anacyclus ăn-a-*sik*-lus *Compositae*. From Gk. *an* (without) *anthos* (a flower) and *kuklos* (a ring) referring to the circle of ovaries surrounding the disc. Perennial herb.
depressus day-*pres*-sus. Depressed (the habit). Mount Atlas Daisy. Morocco.

Anagallis ăn-a-*gah*-lis *Primulaceae*. Gk. name of a plant. Annual and perennial herbs.
arvensis ar-*ven*-sis. Of cultivated fields. Scarlet Pimpernel. Europe.
caerulea kie-*roo*-lee-a. Blue (the flowers).
monelli mon-el-ee. After ? SW Europe.
linifolia leen-i-*fo*-lee-a. (= *A. linifolia*). *Linum*-leaved.
tenella ten-*el*-la. Dainty. Bog Pimpernel. W Europe, N Africa.

Ananas ă-na-nas *Bromeliaceae*. From the native name. Tender evergreen perennial herbs.
bracteatus brăk-tee-*ah*-tus. With bracts (on the fruit). Wild Pineapple. Brazil.
comosus kom-ō-sus. With a tuft of leafy bracts (on the fruit). Pineapple. Cult.

Anaphalis a-*nah*-fa-lis *Compositae*. From the Gk. name of a similar plant. Herbaceous perennials. Pearly Everlasting.
margaritacea mar-ga-ri-*tah*-kee-a. (= *A. cinnamomea*). Pearl-like (the flower heads). Himalaya.

Anaphalis (continued)
triplinervis tri-plee-*ner*-vis. Three-veined (the leaves). Himalaya, China.
monocephala mon-ō-*kef*-a-la. (= *A. nubigenus* hort.). One-headed.
yedoensis yed-ō-*en*-sis. Of Yeddo (now Tokyo), Japan.

Anchusa ăn-*kew*-sa *Boraginaceae*. From the L. name. Annual, biennial or perennial herbs.
azurea a-*zewr*-ree-a. (= *A. italica*). Deep blue (the flowers). S Europe.
caespitosa kie-spi-*tō*-sa. Tufted. Crete.
capensis ka-*pen*-sis. Of the Cape of Good Hope.

Andromeda ăn-*drom*-e-da *Ericaceae*. After Andromeda of Gk. mythology. Evergreen shrubs. Bog Rosemary.
glaucophylla glow-kō-*fil*-la. Glaucous-leaved. N America.
polifolia pol-i-*fo*-lee-a. Grey-leaved. N temperate regions.

Androsace ăn-*dros*-a-kee *Primulaceae*. Gk. name for another plant from *aner* (a man) and *sakos* a shield. Alpine perennial herbs. Rock Jasmine.
alpina ăl-*peen*-a. Alpine. Alps.
carnea kar-nee-a. Flesh-coloured (the flowers). Alps, Pyrenees.
chamaejasme kăm-ie-*yăs*-mee. Dwarf jasmine. S Europe to N Russia.
helvetica hel-*vay*-ti-ka. Swiss. Alps.
lanuginosa lah-noo-gi-*nō*-sa. Woolly. Himalaya.
leichtlinii liekt-*lin*-ee-ee. After Max Leichtlin (1831–1910).
primuloides preem-ew-*loi*-deez. Like *Primula*. Himalaya.
chumbyi chum-bee-ee. Of the Chumby Valley, N India.
pyrenaica pi-ray-*nah*-i-ka. Of the Pyrenees.
sarmentosa sar-men-*tō*-sa. Producing runners. Himalaya.
sempervivoides sem-per-vee-*voi*-deez. Like *Sempervivum*. Himalaya.
villosa vil-*lō*-sa. Softly hairy. Europe, W Asia.
arachnoidea ă-răk-*noi*-dee-a. With hairs like a spider's web.

Anemone a-*nem*-o-nee *Ranunculaceae*. Derivation as for Adonis who was also known as Naamen. His blood is said to have given rise to the blood-red flowers of *A. coronaria*. Perennial herbs.

Anemone (continued)
alpina see *Pulsatilla alpina*
apennina ă-pen-*nee*-na. Of the Apennines.
S Europe.
blanda blăn-da. Pleasant. SE Europe,
W Asia.
coronaria ko-rō-*nah*-ree-a. Used in
garlands. Mediterranean region, C Asia.
halleri see *Pulsatilla halleri*
hortensis hor-*ten*-sis. Of gardens.
Mediterranean region.
× *fulgens ful*-genz. *A. hortensis* × *A.
pavonina*. Shining (the flowers).
C Mediterranean region.
hupehensis hew-pee-*hen*-sis. Of Hupeh (now
Hubei) China.
× *hybrida hib*-ri-da. *A. hupehensis* × *A.
vitifolia*. Hybrid.
× *lesseri les*-sa-ree. *A. multifida* × *A.
sylvestris*. After L. Lesser.
narcissiflora nar-kis-i-*flō*-ra. Narcissus-
flowered. S and C Europe, W Asia.
nemorosa nem-o-*rō*-sa. Of woods. Wood
Anemone. N Europe, NW Asia.
pulsatilla see *Pulsatilla vulgaris*
rivularis reev-ew-*lah*-ris. Growing by
streams. Himalaya, India, China.
vernalis see *Pulsatilla vernalis*
virginiana vir-jin-ee-*ah*-na. Of Virginia.
N America.

Anemonopsis a-nem-o-*nop*-sis
Ranunculaceae. From *Anemone* q.v. and
Gk. *-opsis* indicating resemblance. Perennial
herb.
macrophylla măk-rō-*fil*-la. Large-leaved.
Japan.

Anethum a-*nay*-thum *Umbelliferae*. The Gk.
name. Annual herb.
graveolens gra-*vay*-o-lenz. Strong smelling.
Dill. SW Asia.

Angelica ăn-*gel*-i-ka *Umbelliferae*. From the
angelic medicinal properties. Biennial or
perennial herb.
archangelica ark-ăn-*gel*-i-ka. After the
Archangel Raphael. Angelica. Europe,
Asia.

Angel's Fishing Rod see *Dierama
pulcherrimum*
Angel's Tears see *Billbergia nutans, Narcissus
triandrus*
Golden see *Narcissus triandrus pallidulus*
Angel's Trumpet see *Datura arborea*
Angel's Wings see *Caladium*

Angraecum ăn-*griek*-um *Orchidaceae*. From
angurek the Malayan name for epiphytic
plants. Greenhouse orchids.
bilobum bi-*lō*-bum. Two lobed (the leaf
apex). S Africa.
distichum dis-ti-kum. Two-ranked (the
leaves). Tropical Africa.
eichlerianum iek-la-ree-*ah*-num. After
August Wilhelm Eichler (1839–1887),
German botanist. Tropical W Africa.
falcatum făl-*kah*-tum. Sickle-shaped (the
leaves). China, Japan.
kotschyi kot-shee-ee. After Kotschy.
E Africa.
sesquipedale ses-kwee-ped-*ah*-lee. 1½ feet
long (the spur of the flower). Star of
Bethlehem Orchid. Madagascar.
superbum soo-*perb*-um. (= *A. eburneum*).
Superb. Madagascar.

Anguloa ăn-gew-*lō*-a *Orchidaceae*. After Don
Francisco de Angulo, 18th century Spanish
botanist. Greenhouse orchids. Cradle Orchid.
clowesii clowz-ee-ee. After the Rev. John
Clowes (1777–1846), who built up a large
collection of orchids. Colombia.
ruckeri ruk-a-ree. After Mr Rucker, an
orchid grower. Colombia.
uniflora ew-ni-*flō*-ra. One-flowered (as are
all species). Peru.

Anigozanthus a-nee-gō-*zăn*-thus
Haemadoreaceae. From Gk. *anoigo* (to open)
and *anthos* (a flower), referring to the widely
open flowers. Kangaroo's Paw. SW
Australia.
flavidus flah-vi-dus. Yellow (the flowers).
manglesii mang-*galz*-ee-ee. After Robert
Mangles (died 1860) who introduced and
grew many W Australian plants,
presumably receiving seed from his two
brothers who both visited W Australia.
viridis vi-ri-dis. Green (the flowers).

Animated Oat see *Avena sterilis*

Anisodontea a-nee-sō-*dont*-ee-a *Malvaceae*.
From Gk. *anisos* (unequal) and *odon* (a tooth).
Tender shrub.
scabrosa skă-*brō*-sa. (= *Malvastrum
scabrosum*). Rough. Hairy Mallow.
S Africa.

Annona a-*no*-na *Annonaceae*. From the
native name. Tender evergreen shrubs.
cherimola ke-ree-*mō*-la. From *cherimoya* the
native name. Custard Apple. N Andes.

Annona (continued)
  muricata mew-ri-*kah*-ta. With sharp points (the fruit). Tropical America.
  reticulata ray-tik-ew-*lah*-ta. Netted (the carpels). Custard Apple. Tropical America.
  squamosa skwah-*mō*-sa. Scaly (the fruit). Sweet Sop. Tropical America.

Anoectochilus a-neek-tō-*keel*-us *Orchidaceae*. From Gk. *anoektos* (open) and *cheilos* (a lip), the blades of the lip spread, giving it an open appearance. Warm greenhouse orchids.
  argyroneurus arg-i-rō-*newr*-rus. Silver-veined (the leaves). Java.
  discolor dis-ko-lor. Of two colours (the leaves). China.
  elwesii el-*wez*-ee-ee. After H. J. Elwes, see *Galanthus elwesii*. Himalaya.
  regalis ray-*gah*-lis. Royal. King Plant. Sri Lanka.

Antennaria ǎn-ten-*ah*-ree-a *Compositae*. From L. *antenna*, the pappus hairs of the male flowers have swollen tips resembling a butterfly's antennae. Perennial herb.
  dioica dee-ō-*ee*-ka. Dioecious. Cat's Foot. Arctic and alpine N hemisphere.

Anthemis ǎn-them-is *Compositae*. The Gk. name for *Chamaemelum nobile*. Perennial herbs.
  arabica see *Cladanthus arabicus*
  cupaniana kew-pahn-ee-*ah*-na. After Francesco Cupani (1657–1711), Italian monk and naturalist. Italy.
  marschalliana mar-shǎl-ee-*ah*-na. (= *A. biebersteiniana*, *A. rudolphiana*). After Bieberstein, see *Cerastium biebersteinii*. Caucasus.
  nobilis see *Chamaemelum nobile*
  sancti-johannis sank-tee-yō-*hǎn*-is. For its flowering in mid-summer, referring to St John's day, June 24th. Bulgaria.
  tinctoria tink-*to*-ree-a. Used in dyeing, a yellow dye can be extracted from the flowers. Dyer's Chamomile, Yellow Chamomile. Europe, W Asia.

Anthericum ǎn-*the*-ri-kum *Liliaceae*. From *antherikon* Gk. name for asphodel. Perennial herbs.
  liliago lee-lee-*ah*-go. Like *Lilium*. St Bernard Lily. S Europe.
  ramosum rah-*mō*-sum. Branched (the inflorescence). W and S Europe.

Antholyza paniculata see *Crocosmia paniculata*

Anthriscus ǎn-*thris*-kus *Umbelliferae*. L. name for a plant. Annual herb.
  cerefolium kay-ree-*fo*-lee-um. Waxen-leaved. Chervil. SE Europe, W Asia.

Anthurium ǎn-*thewr*-ree-um *Araceae*. From Gk. *anthos* (a flower) and *oura* (a tail) referring to the slender, tail-like spadix which bears the flowers. Tender perennial herbs.
  andreanum on-dray-*ah*-num. After Edouard Francis André (1840–1911). Tail Flower, Painter's Palette. Colombia.
  crystallinum kris-tal-*lee*-num. Crystalline (the appearance of the leaves). Strap Flower. Colombia.
  scherzerianum skairts-a-ree-*ah*-num. After Herr Scherzer of Vienna who discovered it. Flamingo Flower, Flame Plant. Guatemala, Costa Rica.

Anthyllis ǎn-*thil*-lis *Leguminosae*. The Gk. name. Shrubs.
  barba-jovis bar-ba-*yo*-vis. Jupiter's beard, referring to the silky hairy leaves and shoots. Jupiter's Beard. SW Europe, Mediterranean region.
  hermanniae her-*mah*-nee-ie. From the resemblance to *Hermannia*. Mediterranean region.
  montana mon-*tah*-na. Of mountains. S Europe.

Antigonon ǎn-*ti*-go-non *Polygonaceae*. From Gk. *anti* (like) and *polygonon* (knotweed) referring to its relationship to *Polygonum*. Tender perennial climber.
  leptopus lep-to-pus. Slender-stalked. Coral Vine. Mexico.

Antirrhinum ǎn-tee-*ree*-num *Scrophulariaceae*. From Gk. *anti* (like) and *rhis* (a snout) from the appearance of the flower. Perennial herb.
  asarina see *Asarina procumbens*
  majus mah-yus. Larger. Snapdragon. Mediterranean region.

Anubias a-*new*-bee-as *Araceae*. From Anubis the ancient Egyptian god. Aquarium plants.
  afzelii ǎf-*zee*-lee-ee. After Adam Afzelius (1750–1837) who collected in W Africa. Sierra Leone.
  congensis kong-*gen*-sis. Of the Congo. W Africa.

Anubias (continued)
nana nah-na. Dwarf. Cameroons.

Aphelandra åf-el-*ån*-dra *Acanthaceae*. From Gk. *apheles* (simple) and *aner* (male) referring to the one-celled anthers. Tender evergreen shrubs.
aurantiaca ow-rån-tee-*ah*-ka. (= *A. nitens*). Orange (the flowers). Mexico to NS America.
'Roezlii' *rurz*-lee-ee. After Benedikt Roezl (1824–85), a plant collector.
chamissoniana sham-i-son-ee-*ah*-na. After Ludolf Adalbert von Chamisso (1781–1838). Brazil.
squarrosa skwah-*rō*-sa. With the parts spreading horizontally, referring to the flower spikes. Saffron Spike, Zebra Plant. Brazil.
'Louisae' loo-*eez*-ie. After Queen Louise of Belgium.

Apium å-pee-um *Umbelliferae*. The L. name for celery and parsnip. Biennial herb.
graveolens gra-*vay*-o-lenz. Strong smelling. Wild Celery. Widely distributed.
dulce dul-kee. Sweet. Celery.
rapaceum ra-*pah*-kee-um. Turnip-like, the swollen stem. Celeriac.

Aponogeton a-pon-ō-*gay*-ton *Aponogetonaceae*. From *Aquae Aponi* a Roman healing spring and *geiton* (a neighbour). Aquatic perennials.
crispus kris-pus. Crisped (the leaves). Sri Lanka.
distachyus di-*stå*-kee-us. With two spikes, referring to the forked inflorescence. Cape Asparagus, Cape Pondweed. S Africa.
madagascariensis måd-a-gås-kah-ree-*en*-sis. (= *A. fenestralis*). Of Madagascar. Madagascar Lace Plant.
ulvaceus ul-*vah*-kee-us. Like *Ulva* (Sea lettuce). Madagascar.
undulatus un-dew-*lah*-tus. Wavy-edged (the leaves). SE Asia.

Aporocactus a-po-rō-*kåk*-tus *Cactaceae*. From Gk. *aporos* (impenetrable) and *Cactus* q.v.
flagelliformis fla-gel-lee-*form*-is. Whip-like (the slender stems). Rat's-tail Cactus. Mexico.
× *mallisonii* see × *Heliaporus smithii*

Apple see *Malus*
Apple of Peru see *Nicandra physalodes*

Apricot see *Prunus armeniaca*
Japanese see *P. mume*

Aptenia åp-*teen*-ee-a *Aizoaceae*. From Gk. *apten* (unwinged), the capsules lack wings distinguishing the genus from *Mesembryanthemum* in which it was once included. Tender succulent.
cordifolia kor-di-*fo*-lee-a. (= *Mesembryanthemum cordifolium*). Heart-shaped (the leaves). S Africa.

Aquilegia å-kwi-*lee*-gee-a *Ranunculaceae*. From L. *aquila* (an eagle) referring to the shape of the petals. Perennial herbs. Columbine.
alpina ål-*peen*-a. Alpine. Alps.
bertolonii ber-to-*lō*-nee-ee. After Antonio Bertoloni (1775–1869), Italian botanist. S Europe.
caerulea kie-*ru*-lee-a. Deep blue (the flowers). Rocky mountains.
canadensis kån-a-*den*-sis. Of Canada or NE America. E N America.
chrysantha kris-*ånth*-a. With golden flowers. SW United States, N Mexico.
discolor dis-ko-lor. Two-coloured (the flowers). Spain.
flabellata flå-bel-*lah*-ta. Fan-shaped (the leaflets). Japan.
glandulosa glån-dew-*lō*-sa. Glandular. Siberia.
longissima long-*gis*-si-ma. Longest, referring to the very long spur. Texas.
scopulorum skop-ew-*lo*-rum. Growing on cliffs. W United States.
vulgaris vul-*gah*-ris. Common. Europe.

Arabis å-ra-bis *Cruciferae*. Derivation obscure. Perennial herbs.
blepharophylla ble-fa-rō-*fil*-la. With fringed leaves. California.
caucasica kaw-*kås*-i-ka. Of the Caucasus. SE Europe, W Asia.
ferdinandii-coburgii fer-di-*nahn*-dee-ee-kō-burg-ee-ee. After King Ferdinand of Bulgaria. Bulgaria.

Aralia a-*rah*-lee-a *Araliaceae*. From the French-Canadian name *aralie*. Perennial herbs and deciduous shrubs.
cachemerica kåsh-*me*-ri-ka. Of Kashmir. Himalaya, Tibet.
elata ay-*lah*-ta. Tall. Japanese Angelica Tree. NE Asia.
elegantissima see *Dizygotheca elegantissima*
sieboldii see *Fatsia japonica*

Aralia (continued)
spinosa spee-*nō*-sa. Spiny (the stem and petioles). Hercules Club. E United States.

Araucaria ǎ-row-*kah*-ree-a *Araucariaceae*. From the Araucani Indians who live where *A. araucana* grows. Hardy and tender conifers.
araucana ǎ-row-*kah*-na. As above. Monkey Puzzle, Chile Pine. Chile, Argentina.
heterophylla he-te-rō-*fil*-la. Variably-leaved. Norfolk Island Pine. Norfolk Islands.

Araujia a-*row*-hee-a *Asclepiadaceae*. From the Brazilian name. Evergreen climber.
sericofera se-ri-*ko*-fe-ra. Silk-bearing, referring to the white hairs on the young shoots and beneath the leaves. Cruel Plant. S America.

Arbutus ar-bu-tus *Ericaceae*. The L. name. Evergreen trees and shrubs.
andrachne ǎn-*drǎk*-nee. The Gk. name. SE Europe.
× *andrachnoides* ǎn-drǎk-*noi*-deez. *A. andrachne* × *A. unedo*. Like *A. andrachne*. Greece.
menziesii men-*zeez*-ee-ee. After Menzies, see *Menziesia*. Madrona. WN America.
unedo ew-nee-do. The L. name. Strawberry Tree. Mediterranean region, SW Ireland.

Arctostaphylos ark-tō-*stǎ*-fil-os *Ericaceae*. From Gk. *arctos* (a bear) and *staphyle* (a bunch of grapes). Evergreen shrubs.
manzanita mǎn-za-*neet*-a. The native name for this and similar species. Manzanita. California.
nevadensis nev-a-*den*-sis. Of the Sierra Nevada, California.
uva-ursi oo-va-*ur*-see. Bear's grape. Bearberry. N temperate regions.

Arctotis ark-*tō*-tis *Compositae*. From Gk. *arctos* (a bear) and *otus* (an ear), the pappus scales are said to resemble a bear's ears. Annual and perennial herbs. African Daisy. S Africa.
acaulis a-*kaw*-lis. Stemless.
breviscapa brev-ee-*skah*-pa. With a short scape.
stoechadifolia stoy-ka-di-*fo*-lee-a. With leaves like *Lavandula stoechas*.
grandis grǎn-dis. (= *A. grandis*). Large.

Ardisia ar-*dis*-ee-a *Myrsinaceae*. From Gk. *ardis* (a point) referring to the pointed anthers. Tender, evergreen shrub.

Ardisia (continued)
crispa kris-pa. Finely wavy (the leaf margins). Coral Berry. Japan, S China.

Areca a-*ree*-ka *Palmae*. From the native name. Tender palm.
catechu kah-tee-koo. From the native name. Betel Nut Palm. Malaysia.

Arenaria ǎ-ray-*nah*-ree-a *Caryophyllaceae*. From L. *arena* (sand) many species grow in sandy places. Low growing perennial herbs.
balearica bǎ-lee-*ah*-ri-ka. Of the Balearic Islands. W Mediterranean region.
caespitosa hort. see *Sagina glabra*
grandiflora grǎn-di-*flō*-ra. Large-flowered. Europe.
laricifolia see *Minuartia laricifolia*
montana mon-*tah*-na. Of mountains. W Europe.
purpurascens pur-pewr-*rǎs*-enz. Purplish (the flowers). SW Europe.
tetraquetra tet-ra-*kwee*-tra. Four-angled (the shoots). Pyrenees, Spain.

Argemone ar-ge-*mō*-nee *Papaveraceae*. Gk. name for a similar plant. Annual herbs.
grandiflora grǎn-di-*flō*-ra. Large-flowered. Mexico.
mexicana meks-i-*kah*-na. Mexican. Devil's Fig, Prickly Poppy. Florida, C America.
platyceras plǎ-tee-*ke*-ras. With broad horns (on the fruit). Crested Poppy. Mexico.

Argyroderma ar-gi-rō-*der*-ma *Aizoaceae*. From Gk. *argyros* (silver) and *derma* (skin) referring to the whitish leaves. Tender succulents.
octophyllum ok-tō-*fil*-lum. Eight-leaved, which it rarely is. S Africa.

Arisaema ǎ-ris-*ie*-ma *Araceae*. From Gk. *aron* (arum) and *haema* (blood) meaning related to *Arum*. Perennial herbs.
candidissimum kǎn-di-*dis*-si-mum. Most white (the spathe). W China.
dracontium dra-*kon*-tee-um. Dragon-like. S United States.
triphyllum tri-*fil*-lum. With three leaves (leaflets). Jack-in-the-Pulpit, Indian Turnip. E and S N America.

Arisarum a-ris-*a*-rum *Araceae*. From *arisaron* Gk. name for *A. vulgare*. Perennial herb.
proboscideum pro-bo-*ski*-dee-um. Like an elephant's trunk (the spadix). Mouse-tail Plant. S Italy, SW Spain.

Aristolochia a-ris-to-*lok*-ee-a
Aristolochiaceae. From Gk. *aristos* (best) and
lochia (childbirth) referring to supposed
medicinal properties. Hardy and tender
climbers. Birthwort.
  altissima ăl-*tis*-si-ma. (= *A. sempervirens*).
  Tallest. SE Europe, N Africa.
  elegans ay-le-gahnz. Elegant. Calico
  Flower. Brazil.
  macrophylla măk-rō-*fil*-la. (= *A. durior. A.
  sipho*). Large-leaved. Dutchman's Pipe.
  E United States.

Armeria ar-*me*-ree-a *Plumbaginaceae*. From
the L. name for a *Dianthus*. Evergreen
perennials. Thrift.
  alliacea ăl-lee-*ah*-kee-a. (= *A. plantaginea*).
  Allium-like (the leaves). Jersey Thrift.
  W Europe.
  juniperifolia yoo-ni-pe-ri-*fo*-lee-a. (= *A.
  caespitosa*). *Juniperus*-leaved. Spain.
  maritima ma-*ri*-ti-ma. Growing near the sea.
  Common Thrift. Widely distributed.
  mauritanica mo-ri-*tah*-ni-ka. Of N Africa.
  pseudarmeria sood-ar-*me*-ree-a. (= *A.
  latifolia*). False *Armeria*. Portugal.

Armoracia ar-mo-*rah*-kee-a *Cruciferae*. L.
name for a related plant. Perennial herb.
  rusticana rus-ti-*kah*-na. Of the country.
  Horse-radish. SE Europe.

Arnebia ar-*nee*-bee-a *Boraginaceae*. From the
Arabian name. Annual and perennial herbs.
  decumbens day-*kum*-benz. (= *A. cornuta*).
  Prostrate. N Africa, Russia, W and C
  Asia.
  pulchra pul-kra. (= *A. echioides* hort.).
  Pretty. Prophet Flower. W Asia.

Arnica *ar*-ni-ka *Compositae*. From Gk.
arnakis (lambskin) from the texture of the
leaves. Perennial herb.
  montana mon-*tah*-na. Of mountains.
  Mountain Tobacco, Leopard's Bane.
  Europe.

Aronia a-*rō*-nee-a *Rosaceae*. From *aria* the
Gk. name for *Sorbus aria*. Deciduous shrubs.
Chokeberry.
  arbutifolia ar-bewt-i-*fo*-lee-a. *Arbutus*-
  leaved. Red chokeberry. E N America.
  melanocarpa me-la-nō-*kar*-pa. With black
  fruits. Black Chokeberry. E N America.

Arrhenatherum ă-ren-ă-the-rum *Gramineae*.
From Gk. *arren* (male) and *ather* (a bristle)

Arrhenatherum (continued)
  referring to the bristles on the male flowers.
Perennial grass.
  elatius ay-*lah*-tee-us. Tall. Oat Grass.
  Europe, N Africa, W Asia.

Arrowhead see *Sagittaria*
Arrowhead Vine see *Syngonium angustatum*,
S. podophyllum
Arrowroot see *Maranta arundinacea*

Artemisia ar-tay-*mis*-ee-a *Compositae*. After
the Gk. goddess *Artemis*. Perennial herbs,
shrubs and sub-shrubs.
  abrotanum a-*brot*-a-num. The L. name.
  Southernwood. S Europe.
  absinthium ăb-*sin*-thee-um. The L. name.
  Wormwood. Europe.
  arborescens ar-bo-*res*-enz. Becoming tree-
  like. Mediterranean region.
  dracunculus dra-*kun*-kew-lus. A small
  dragon; the words dragon and tarragon
  have the same derivation. French
  Tarragon. S Europe, Asia.
    inodorus in-o-*dō*-rus. Unscented. Russian
    Tarragon.
  frigida fri-gi-da. Of cold regions. W N
  America, Siberia.
  lactiflora lăk-ti-*flō*-ra. Milk-flowered,
  referring to the creamy-white flowers.
  China.
  ludoviciana loo-do-vik-ee-*ah*-na. Of
  Louisiana. W United States, Mexico.
  schmidtiana shmit-ee-*ah*-na. After Schmidt.
  Japan.
  stelleriana stel-la-ree-*ah*-na. After its
  discoverer Georg Wilhelm Steller
  (1709–46). Dusty Miller, Old Woman. NE
  Asia.
  tridentata tri-den-*tah*-ta. Three-toothed (the
  leaves). Sage Brush. W United States.

Arthropodium arth-rō-*pō*-dee-um *Liliaceae*.
From Gk. *arthron* (a joint) and *podion* (a
stalk) referring to the jointed pedicels.
Perennial herb.
  cirrhatum ki-*rah*-tum. With tendrils,
  referring to the tendril-like appendages on
  the filaments. Rock Lily. New Zealand.

Artichoke, Chinese see *Stachys affinis*
  Globe see *Cynara cardunculus*
  Jerusalem see *Helianthus tuberosus*
Artillery Plant see *Pilea microphylla*

Arum ă-rum *Araceae*. From *aron* the Gk.
name. Perennial herbs. The name *Arum* is
also commonly applied to *Zantedischia*.

Arum (continued)
cornutum see *Sauromatum guttatum*
creticum kray-ti-kum. Of Crete.
italicum ee-*tǎ*-li-kum. Of Italy. Europe,
N Africa, W Asia.
maculatum mǎk-ew-*lah*-tum. Spotted (the
spathe). Cuckoo Pint, Lords and Ladies.
Europe.
orientale o-ree-en-*tah*-lee. Eastern. SE
Europe, W Asia.
pictum pik-tum. Painted, referring to the
white-veined leaves. W Mediterranean
region.

Aruncus a-*run*-kus *Rosaceae*. The Gk. name.
Herbaceous perennial often called spiraea.
dioicus dee-ō-*ee*-kus. (= *A. sylvester*).
Dioecious. Goat's Beard. Europe, Asia,
N America.

Arundinaria a-run-di-*nah*-ree-a *Gramineae*.
From L. *arundo* (a reed). Bamboos.
anceps ǎn-keps. Doubtful (the country of
origin). NW Himalaya.
japonica ja-*pon*-i-ka. Of Japan.
murielae mewr-ree-*el*-ie. After Ernest
Wilson's daughter Muriel. China.
nitida ni-ti-da. Shining (the leaves). China.
pumila pew-mi-la. Low-growing. Japan.
vagans vǎ-gǎnz. Wandering, referring to its
invasive habit. Japan.
variegata vǎ-ree-a-*gah*-ta. Variegated (the
leaves). Japan.
viridistriata vi-ri-dee-stree-*ah*-ta. Green-
striped (the leaves). Japan.

Arundo a-*run*-dō *Gramineae*. From L. *arundo*
(a reed). Semi-hardy grass.
donax do-nǎks. Gk. name for a kind of reed.
Giant Reed. Mediterranean region.

Asarina a-*sah*-ri-na *Scrophulariaceae*. Spanish
name for an *Antirrhinum*. Perennial herb.
barclayana see *Maurandya barclayana*
erubescens see *Maurandya erubescens*
procumbens prō-*kum*-benz. (= *Antirrhinum
asarina*). Prostrate. Spain, Portugal.
scandens see *Maurandya scandens*

Asarum a-*sah*-rum. *Aristolochiaceae*. The L.
and Gk. name. Perennial herbs. Wild
Ginger.
canadense kǎn-a-*den*-see. Of Canada or NE
America. E N America.
caudatum kaw-*dah*-tum. With a tail (the
calyx lobes are prolonged into slender
tails). W N America.
europaeum oy-rō-*pie*-um. Of Europe.

Asclepias a-*sklay*-pee-as *Asclepiadaceae*.
From Gk. *Asklepios*, god of medicine,
referring to medicinal properties. Herbaceous
perennials and tender shrubs. Milkweed.
curassavica ku-ra-*sah*-vi-ka. Of Curacao.
S America.
incarnata in-kar-*nah*-ta. Flesh pink (the
flowers). N America.
speciosa spe-kee-ō-sa. Showy. Canada,
W United States.
tuberosa tew-be-rō-sa. Tuberous (the root).
N America, N Mexico.

Ascocentrum ǎs-kō-*ken*-trum *Orchidaceae*.
From Gk. *ascos* (a bag) and *kentron* (a spur)
referring to the large spur hanging from the
lip. Greenhouse orchids.
ampullaceum ǎm-pul-*ah*-kee-um. Flask-like
(the lip). Himalaya, Burma.
miniatum min-ee-*ah*-tum. Cinnabar red (the
flowers). Java, Philippines.

Ash see *Fraxinus*
Arizona see *F. velutina*
Common see *F. excelsior*
Manna see *F. ornus*
Narrow-leaved see *F. angustifolia*
White see *F. americana*

Asimina a-*si*-mi-na *Annonaceae*. From the N
American Indian name *assimin*. Deciduous
shrub or small tree.
triloba tri-*lō*-ba. Three-lobed (the calyx).
Pawpaw. E N America. Not to be
confused with the tropical American
pawpaw, *Carica papaya*.

Asparagus a-*spǎ*-ra-gus *Liliaceae*. The
classical name. Hardy and tender herbs.
asparagoides a-spǎ-ra-*goi*-deez. Asparagus-
like, it was originally placed in another
genus. Smilax (of florists). S Africa.
densiflorus dens-i-*flō*-rus. Densely-flowered.
S Africa.
    'Sprengeri' *spreng*-a-ree. After Carl L.
    Sprenger (1846–1917), German
    nurseryman who introduced it.
falcatus fǎl-*kah*-tus. Sickle-shaped. Africa,
Sri Lanka.
officinalis o-fi-ki-*nah*-lis. Sold as a herb.
Garden Asparagus.
scandens skǎn-denz. Climbing. S Africa.
setaceus say-*tah*-kee-us. (= *A. plumosus*)
Bristled. Asparagus Fern. S Africa.

Aspasia a-*spah*-see-a *Orchidaceae*. After
Aspasia wife of Pericles, from Gk. *aspasies*

Aspasia(continued)
(delightful). Greenhouse orchids.
   lunata loon-*ah*-ta. Crescent-shaped, (? the
   lip). Brazil.
   variegata vă-ree-a-*gah*-ta. Variegated (the
   flowers). S America.

Aspen see *Populus tremula*

Asperula a-*spe*-ru-la *Rubiaceae*. From L.
asper (rough) referring to the roughly hairy
stems. Annual and perennial herbs.
   hexaphylla heks-a-*fil*-la. Six-leaved, the
   leaves are in whorls of six. SW Alps.
   hirta hir-ta. Hairy. Pyrenees.
   lilaciflora li-lah-ki-*flŏ*-ra. (= *A. caespitosa*).
   With lilac flowers. W Asia.
   odorata see *Galium odoratum*
   orientalis o-ree-en-*tah*-lis. (= *A. azurea*).
   Eastern. Europe, Asia.
   suberosa soo-be-*rŏ*-sa. Corky-stemmed.
   Greece, Bulgaria.

Asphodel see *Asphodelus*
   White see *A. albus*
   Yellow see *Asphodeline lutea*

Asphodeline a-sfod-e-*lee*-nee *Liliaceae*. Like
Asphodelus q.v. Perennial herbs.
   liburnica li-*burn*-i-ka. (= *Asphodelus
   liburnicus*). From Croatia (Liburnia) on
   the Yugoslav coast. SE Europe.
   lutea loo-tee-a. Yellow (the flowers). Yellow
   Asphodel, King's Spear. Mediterranean
   region.

Asphodelus a-*sfod*-e-lus *Liliaceae*. Gk. name
for *A. ramosus*. Perennial herbs. Asphodel.
   albus ăl-bus. White (the flowers). White
   Asphodel. S Europe.
   liburnicus see *Asphodeline liburnica*
   ramosus rah-*mŏ*-sus. (= *A. cerasiferus*).
   Branched (the inflorescence). S Europe.

Aspidistra ă-spi-*di*-stra *Liliaceae*. From Gk.
aspideon (a small, round shield) referring to
the shape of the stigma. Tender, evergreen
herb.
   elatior ay-*lah*-tee-or. Taller. Cast Iron
   Plant. Japan.

Asplenium a-*splay*-nee-um *Aspleniaceae*.
From Gk. *a* (not) and *splen* (spleen) referring
to supposed medicinal properties. Hardy and
tender ferns. Spleenwort.
   adiantum-nigrum ă-dee-ăn-tum-*nig*-rum.
   Black Adiantum. Black Spleenwort.
   N hemisphere.

Asplenium (continued)
   bulbiferum bul-*bi*-fe-rum. Producing bulbs
   (plantlets are borne on the leaves). Mother
   Spleenwort, Hen and Chicken Fern. New
   Zealand, Australia, India.
   caudatum kaw-*dah*-tum. With a tail (the
   pinnae). Tropics.
   daucifolium dow-ki-*fo*-lee-um. *Daucus*-
   leaved. Mauritius, Reunion Islands.
   falcatum făl-*kah*-tum. Sickle-shaped (the
   pinnae).
   flabellifolium fla-bel-lee-*fo*-lee-um. With
   fan-shaped leaves. Necklace Fern. New
   Zealand, Australia.
   flaccidum flăk-ki-dum. Flaccid (the fronds).
   Hanging Spleenwort. Australia,
   Tasmania, Pacific Islands.
   marinum ma-*reen*-num. Growing near the
   sea. Sea Spleenwort. W Europe.
   nidus nee-dus. A nest. Bird's Nest Fern.
   Tropical Asia, Polynesia.
   platyneuron plă-tee-*newr*-ron. With broad
   veins. Ebony Spleenwort. E N America.
   ruta-muraria roo-ta-mew-*rah*-ree-a. Rue of
   walls. Wall Rue. E United States, Europe,
   Asia.
   trichomanes tri-kŏ-*mah*-neez. Gk. name for
   a fern. Maidenhair Spleenwort.
   N America, Europe, Asia.
   viride vi-ri-dee. Green. Green Spleenwort.
   N America, Europe, Asia.

Aster ă-ster *Compositae*. From L. *aster* (a
star) referring to the flower heads.
Herbaceous perennials.
   acris see *A. sedifolius*
   albescens ăl-*bes*-enz. (= *Microglossa
   albescens*). Whitish (the undersides of the
   leaves). Himalaya, W China.
   alpinus ăl-*pee*-nus. Alpine. Europe, Asia.
   amellus a-*mel*-lus. The L. name. Europe,
   W Asia.
   ericoides e-ri-*koi*-deez. Like *Erica* (the
   foliage). N America, Mexico.
   farreri fă-ra-ree. After Farrer, see *Viburnum
   farreri*. W China, Tibet.
   × *frikartii* fri-*kart*-ee-ee. *A. amellus* × *A.
   thomsonii*. After Carl Ludwig Frikart
   (1879–1964).
   natalensis see *Felicia rosulata*
   nemoralis ne-mo-*rah*-lis. Of woods.
   N America.
   novae-angliae no-vie-*ang*-glee-ie. Of New
   England. N America.
   novi-belgii nŏ-vee-*bel*-gee-ee. Of New York
   (previously called New Belgium).
   Michaelmas Daisy. E N America.
   pappei see *Felicia pappei*

Aster (continued)
  puniceus pew-*ni*-kee-us. Reddish-purple (the stems). EN America.
  sedifolius say-di-*fo*-lee-us. (= *A. acris*). *Sedum*-leaved. S Europe.
  spectabilis spek-*tah*-bi-lis. Spectacular. N America.
  thomsonii tom-*son*-ee-ee. After Thomas Thomson (1817–78), Scottish physician and superintendant of Calcutta Botanic Garden. W Himalaya.

Asteranthera a-ste-ran-*the*-ra *Gesneriaceae*. From L. *aster* (a star) and *anthera* (an anther) referring to the star-like arrangement of the anthers. Semi-hardy, evergreen climber.
  ovata ō-*vah*-ta. Ovate (the leaves). Chile, Argentina.

Astilbe a-*stil*-bee *Saxifragaceae*. From Gk. *a* (without) and *stilbe* (brilliance), the individual flowers are very small. Perennial herbs.
  × *arendsii* ah-*rendz*-ee-ee. After Georg Arends of Ronsdorf (1862–1952), who hybridised astilbes.
  chinensis chin-*en*-sis. Of China. China, Japan.
      davidii dă-*vid*-ee-ee. After Armand David, see *Davidia*.
  × *crispa* kris-pa. Finely wavy (the leaves).
  grandis grăn-dis. Large. China.
  japonica ja-*pon*-i-ka. Of Japan.
  rivularis reev-ew-*lah*-ris. Growing near streams. Nepal.
  × *rosea* ros-ee-a. *A. chinensis* × *A. japonica*. Rose-coloured (the flowers).
  simplicifolia sim-pli-ki-*fo*-lee-a. With simple leaves. Japan.
  thunbergii thun-*berg*-ee-ee. After Thunberg, see *Thunbergia*.

Astrantia a-*străn*-tee-a *Umbelliferae*. Derivation obscure, possibly from L. *aster* (a star) referring to the star-like flowers. Herbaceous perennials. Masterwort.
  carniolica kar-nee-*o*-li-ka. Of Carniola, N Yugoslavia. Europe.
  major mah-yor. Larger. Europe.
  maxima mahk-si-ma. Largest. S Europe, Caucasus.
  minor min-or. Smaller. Europe.

Astrophytum a-*stro*-fi-tum *Cactaceae*. From Gk. *astron* (a star) and *phyton* (a plant) referring to the often star-shaped body.
  asterias a-*ste*-ree-as. Star-like. Sea Urchin Cactus, Silver Dollar Cactus. S Texas, N Mexico.

Astrophytum (continued)
  capricorne kă-pri-*kor*-nee. Like a goat's-horn. Goat's Horn Cactus. Texas, Mexico.
  myriostigma mi-ree-ō-*stig*-ma. With many stigmas. Bishop's Cap Cactus. Mexico.
  ornatum or-*nah*-tum. Ornamental. Bishop's Cap Cactus. Mexico.

Athrotaxis ăth-rō-*tăks*-is *Taxodiaceae*. From Gk. *athroos* (crowded) and *taxis* (arrangement) referring to the densely arranged leaves. Evergreen conifers. Tasmania.
  cupressoides kew-pres-*oi*-deez. *Cupressus*-like.
  laxifolia lăks-i-*fo*-lee-a. Loose-leaved, the leaves are more spreading than in *A. cupressoides*.
  selaginoides se-lah-gi-*noi*-deez. Like *Lycopodium selago* (the foliage). King William Pine.

Athyrium a-*thi*-ree-um *Athyriaceae*. Derivation obscure possibly from Gk. *athyros* (doorless) referring to the late opening indusium of *A. filix-femina*. Lady Fern.
  distentifolium dis-ten-ti-*fo*-lee-um. (= *A. alpestre*). With distended leaves. Alpine Lady Fern. Europe, Asia.
  filix-femina fi-liks-*fay*-mi-na. Lady fern, referring to the delicate fronds as compared with the male fern (*Dryopteris filix-mas*). Common Lady Fern. N hemisphere.
  goeringianum gur-ring-gee-*ah*-num. After Goering. Japan.
  'Pictum' *pik*-tum. Painted (the leaves). Japanese Painted Fern.

Atraphaxis ăt-ra-*făks*-is *Polygonaceae*. Gk. name for a species of *Atriplex*. Deciduous shrub.
  frutescens froo-*tes*-enz. Shrubby. SE Europe, Caucasus, C Asia.

Atriplex ah-tri-plex *Chenopodiaceae*. Gk. name for *A. hortensis*. Herbs and shrubs.
  canescens kah-*nes*-enz. Grey-hairy (the leaves and shoots). Grey Sage Brush. W N America.
  halimus hă-li-mus. The Gk. name from *halimos* (maritime) it grows near the sea. Tree Purslane. S Europe.
  hortensis hor-*ten*-sis. Of gardens. Orache. Asia.

Aubergine see *Solanum melongena*

Aubrieta ŏ-bree-*aý*-ta *Cruciferae*. After Claude Aubriet (1668–1743), French botanical artist. Perennial herb.
  deltoidea del-*toi*-dee-a. Deltoid i.e. shaped like the Gk. letter delta (Δ) referring to the petals. SE Europe, W Asia.

Aucuba ow-*kew*-ba *Cornaceae*. From the Japanese name. Evergreen shrub.
  japonica ja-*pon*-i-ka. Of Japan.
    'Crotonifolia' krŏ-ton-i-*fo*-lee-a. With leaves like croton i.e. *Codiaeum*.
    'Longifolia' long-gi-*fo*-lee-a. Long-leaved.
    'Nana Rotundifolia' *nah*-na ro-tun-di-*fo*-lee-a. Dwarf, round-leaved.
    'Picturata' pik-tew-*rah*-ta. Variegated.
    'Salicifolia' să-li-ki-*fo*-lee-a. *Salix*-leaved.

Auricula see *Primula auricula*

Aurinia ow-*rin*-ee-a *Cruciferae*. From L. *aureus* (golden) referring to the flowers. Sub-shrubby perennial.
  saxatilis săks-*ah*-ti-lis. (= *Alyssum saxatile*). Growing among rocks. S and C Europe, Turkey.

Australian Bluebell Creeper see *Sollya heterophylla*
Australian Honeysuckle see *Banksia*
Austrian Copper Briar see *Rosa foetida* 'Bicolor'

Austrocedrus ow-stro-*ked*-rus *Cupressaceae*. From L. *australis* (southern) and *Cedrus* q.v. The only species is found in the Southern Hemisphere. Evergreen conifer.
  chilensis chil-*en*-sis. (= *Libocedrus chilensis*) of Chile. Chile, Argentina.

Austrocylindropuntia cylindrica see *Opuntia cylindrica*

Avena a-*vay*-na *Gramineae*. The classical name. Annual grass.
  candida hort. see *Helictotrichon sempervirens*
  sempervirens see *Helictotrichon sempervirens*
  sterilis ste-ri-lis. Barren i.e. not producing the oats of cultivation. Animated Oat. S Europe, W Asia.

Avocado Pear see *Persea americana*
Aylostera deminuta see *Rebutia deminuta*
Azalea see *Rhododendron*
  pontica see *R. luteum*

Azara a-*zah*-ra *Flacourtiaceae*. After J. N. Azara (1731–1804) Spanish scientist. Evergreen, semi-hardy shrubs.
  integrifolia in-teg-ri-*fo*-lee-a. With entire leaves. Chile, Argentina.
  microphylla mik-rŏ-*fil*-la. Small-leaved. Chile, Argentina.
  petiolaris pee-tee-ŏ-*lah*-ris. With a petiole, relatively long in this species. Chile.

Azolla a-*zol*-la *Azollaceae*. From Gk. *azo* (to dry) and *ollua* (to kill), they are killed by drying. Floating, aquatic ferns.
  caroliniana kă-ro-lin-ee-*ah*-na. Of Carolina. N America to W Indies.
  filiculoides fi-lik-ew-*loi*-deez. Like *Filicula*. N and S America

B

Babiana băb-ee-*ah*-na *Iridaceae*. From *babiaans*, Afrikaans for baboon which is said to eat the plants. Semi-hardy, cormous herbs. Baboon Root.
  plicata pli-*kah*-ta. Pleated (the leaves).
  sambucina săm-bew-*kee*-na. Like *Sambucus* (the scent of the flowers).
  stricta strik-ta. Erect (the stems).

Baboon Root see *Babiana*
Baby Blue Eyes see *Nemophila*
Baby Rubber Plant see *Peperomia obtusifolia*
Baby's Tears see *Hypoestes phyllostachya*, *Soleirolia soleirolii*

Baccharis bă-ka-ris *Compositae*. After Bacchus, god of wine. Deciduous and evergreen shrubs.
  halimifolia hă-li-mi-*fo*-lee-a. With leaves like *Atriplex halimus*. Bush Groundsel, Groundsel Tree. E N America.
  patagonica păt-a-*gon*-i-ka. Of Patagonia. S Chile, S Argentina.

Bachelor's Buttons see *Tanacetum parthenium*
  White see *Ranunculus aconitifolius*
  Yellow see *Ranunculus acris* 'Flore Pleno'
Badger's Bane see *Aconitum vulparia*
Balloon Flower see *Platycodon grandiflorus*
Balloon Vine see *Cardiospermum halicacabum*

Ballota ba-*lŏ*-ta *Labiatae*. Gk. name for *B. nigra*. Evergreen sub-shrub.
  pseudodictamnus soo-dŏ-dik-*tăm*-nus. False *Dictamnus*. Greece, Crete.

Balm of Gilead see *Populus candicans*
Balsam see *Impatiens balsamina*
 Himalayan see *I. glandulifera*

Balsamita băl-sa-*mee*-ta *Compositae*. Bearing balsam. Herbaceous perennial.
 major mah-yor. Larger. Alecost, Costmary. W Asia, Caucasus, Iran.

Bamboo see *Arundinaria, Phyllostachys, Sasa*
 Black see *Phyllostachys nigra*
Baneberry see *Actaea*
 Red see *A. rubra*

Banksia bank-see-a *Proteaceae*. After Sir Joseph Banks (1743–1820), botanist and patron of science. Australian Honeysuckle.
 collina ko-*lee*-na. Growing on hills. SE Australia.
 integrifolia in-teg-ri-*fo*-lee-a. With entire leaves. E Australia.
 serrata se-*rah*-ta. Saw-toothed (the leaves). E Australia.

Banyan Tree see *Ficus benghalensis*

Baptisia băp-*tis*-ee-a *Leguminosae*. From Gk. *bapto* (to dye), the following have been used as substitutes for indigo (*Indigofera tinctoria*). Perennial herbs.
 australis ow-*strah*-lis. Southern. False Indigo. E United States.
 tinctoria tink-*to*-ree-a. Used in dyeing. Wild Indigo. E United States.

Barbados Gooseberry see *Pereskia aculeata*
Barbados Pride see *Caesalpinia pulcherrima*
Barberry see *Berberis*
 Common see *B. vulgaris*
Barberton Daisy see *Gerbera jamesonii*

Barleria bar-*le*-ree-a *Acanthaceae*. After Jacques Barrelier (1606–73), French monk and botanist. Tender, evergreen shrub.
 lupulina lup-ew-*leen*-a. Hop-like (the flower spikes). Mauritius.

Barrenwort see *Epimedium*
Bartonia aurea see *Mentzelia lindleyi*
Basil see *Ocimum basilicum*
 Bush see *O. minimum*
Basket Grass see *Oplismenus hirtellus*
Basket Plant see *Aeschynanthus*
Bat Plant see *Tacca integrifolia*

Bauera bow-a-ra *Saxifragaceae*. After the brothers Franz (1758–1840) and Ferdinand (1760–1826) Bauer, Austrian botanical artists.

Bauera (continued)
Tender, evergreen shrub.
 rubioides roo-bee-*oi*-deez. Like *Rubia*, the trifoliolate, opposite leaves appear to be a whorl of six leaves as in *Rubia*. New South Wales.

Bauhinia bow-*hin*-ee-a *Leguminosae*. After John and Caspar Bauhin, 16th century Swiss botanists. The two brothers are represented by the usually two-lobed leaves. Tender shrubs.
 acuminata a-kew-min-*ah*-ta. Acuminate. Orchid Bush. SE Asia.
 blakeana blayk-ee-*ah*-na. After Sir Henry and Lady Blake. Sir Henry Blake was Governor of the Hong Kong Botanic Garden from where it was described. China.
 punctata punk-*tah*-ta. (= *B. galpinii*). Spotted. Tropical Africa.
 purpurea pur-*pewr*-ree-a. Purple (the flowers). Butterfly Tree. SE Asia.
 variegata vă-ree-a-*gah*-ta. Variegated (the flowers). Purple Orchid Tree. India, China.

Bay Laurel see *Laurus nobilis*
Bayberry see *Myrica pensylvanica*
 California see *M. californica*
Bead Plant see *Nertera depressa*
Bead Tree see *Melia azederach*
Bearberry see *Arctostaphylos uva-ursi*

Beaumontia bō-*mont*-ee-a *Apocynaceae*. After Lady Diana Beaumont (died 1831). Tender, woody climber.
 grandiflora grăn-di-*flō*-ra. Large-flowered. Herald's Trumpet. Himalaya.

Bear's Breeches see *Acanthus*
Beauty Berry see *Callicarpa*
Beauty Bush see *Kolkwitzia amabilis*
Bee Balm see *Monarda didyma*
Beech see *Fagus*
 Common see *F. sylvatica*
 Copper see *F. sylvatica* Purpurea
Beefsteak Plant see *Iresine herbstii*
Beetroot see *Beta vulgaris*

Begonia bay-*gon*-ee-a *Begoniaceae*. After Michael Bégon (1638–1710), Governor of French Canada. Tender perennial herbs and shrubs.
 albo-picta ăl-bō-*pik*-ta. Painted with white (the leaves). Guinea-wing Begonia. Brazil.
 × *argenteo-guttata* ar-*gen*-tee-ō-gu-*tah*-ta. *B. albo-picta* × *B. olbia*. Dotted with

Begonia (continued)
silver (the leaves). Trout Begonia.
boliviensis bo-liv-ee-*en*-sis. Of Bolivia.
boweri bow-a-ree. After Bower. Miniature
Eyelash Begonia. Mexico.
× *cheimantha* kie-*măn*-tha. *B. dregei* × *B.*
socotrana. Winter-flowering. Christmas
Begonia,
clarkei clark-ee-ee. After Major Trevor
Clarke who grew it. Bolivia.
coccinea kok-*kin*-ee-a. Scarlet (the flowers).
Angelwing Begonia. Brazil.
corallina ko-ra-*leen*-a. Coral-red (the
flowers). Brazil.
daedalea see *B. strigillosa*
davisii day-*vis*-ee-ee. After Walter Davis
(1847–1930) who collected for Veitch in
S America. Peru.
diadema dee-a-*day*-ma. A crown,
presumably referring to the red flower
sheath. Borneo.
dregei dree-gee-ee. After Johann Franz
Drege (1794–1881) a German botanist who
collected in S Africa. Maple-leaf Begonia.
S Africa.
× *erythrophylla* e-rith-rõ-*fil*-la. (= *B.* ×
feastii). *B. hydrocotylifolia* × *B. manicata*.
Red-leaved Beefsteak Begonia.
evansiana see *B. grandis*
× *feastii* see *B.* × *erythrophylla*
fuchsioides fuk-see-*oi*-deez. *Fuchsia*-like.
Fuchsia Begonia. Venezuela.
grandis grăn-dis. (= *B. evansiana*). Large.
Hardy Begonia. SE Asia.
haageana see *B. scharffii*
luxurians luk-*sewr*-ree-ănz. Luxuriant.
Palm-leaf Begonia. Brazil.
maculata măk-ew-*lah*-ta. Spotted (the
leaves). Brazil.
manicata măn-i-*kah*-ta. Long-sleeved,
referring to the long petioles with a ruff
of hairs at the apex. Mexico, Guatemala.
masoniana may-son-ee-*ah*-na. After Mr L.
Maurice Mason who introduced it from
Singapore in 1957. Iron Cross Begonia.
China.
metallica me-*tă*-li-ka. Metallic (the
appearance of the leaves). Metallic-leaf
Begonia. Brazil.
rex reks. King. Assam.
Rex-cultorum reks-kul-*to*-rum. The
cultivated *B. rex* which is grown mainly
as hybrids.
'Richmondensis' rich-mond-*en*-sis. Of
Richmond.
scharffii sharf-ee-ee. (= *B. Laageana*). After
Carl Scharff who collected in Brazil about
1888. Elephant's-ear Begonia. Brazil.

Begonia (continued)
semperflorens sem-per-*flõ*-renz. Ever-
flowering. Brazil, Argentina.
Semperflorens-cultorum sem-per-*flõ*-renz-
kul-*to*-rum. The cultivated *B. semperflorens*
which is largely grown as hybrids.
serratipetala se-rah-tee-*pe*-ta-la. With
toothed petals. New Guinea.
socotrana so-ko-*trah*-na. Of Socotra.
strigillosa stri-gi-*lõ*-sa. (= *B. daedalea*).
With short, appressed bristles. C America.
Tuber-hybrida *tew*-ber-*hib*-ri-da. Tuberous
hybrid. The commonly grown tuberous
Begonias.

Belamcanda bel-am-*kăn*-da *Iridaceae*. From
the native name. Perennial herb.
chinensis chin-en-sis. Of China. Leopard
Flower. Himalaya, China, SE Asia, Japan.

Belladonna Lily see *Amaryllis belladonna*

Bellevalia bel-*vah*-lee-a *Liliaceae*. After
Pierre Riche de Belleval (1564–1632).
Bulbous perennial.
paradoxa pă-ra-*doks*-a. (= *Muscari*
paradoxum). Unusual. E Turkey,
Caucasus.

Bellflower see *Campanula*
  Adriatic see *C. garganica*
  Chimney see *C. pyramidalis*
  Clustered see *C. glomerata*
  Giant see *C. latifolia, Ostrowskia magnifica*
  Italian see *C. isophylla*
  Milky see *C. lactiflora*
  Spurred see *C. alliariifolia*

Bellis *bel*-is *Compositae*. From L. *bellus*
(pretty). Perennial herb.
perennis pe-*re*-nis. Perennial. Daisy.
Europe, W Asia.

Bellium *bel*-ee-um *Compositae*. From *Bellis*
q.v. which they resemble. Annual and
perennial herbs.
bellidioides bel-i-dee-*oi*-deez. Like *Bellis*.
W Mediterranean region.
minutum mi-*new*-tum. Small. Greece,
W Asia.

Bells of Ireland see *Moluccella laevis*
Bellwort see *Uvularia*
Beloperone guttata see *Justicea brandegeana*

Berberidopsis ber-be-ri-*dop*-sis
Flacourtiaceae. From *Berberis* q.v. and Gk.

Berberidopsis (continued)
-*opsis* indicating resemblance. The two genera
are not related. Evergreen, woody climber.
  corallina ko-ra-*lee*-na. Coral-red (the
  flowers). Coral Plant. Chile.

Berberis ber-be-ris *Berberidaceae*. From the
Arabic name. Deciduous and evergreen
shrubs.
  aggregata ăg-re-*gah*-ta. Clustered (the
  flowers). W China.
  buxifolia buks-i-*fo*-lee-a. *Buxus*-leaved.
  Chile, Argentina.
  candidula kăn-*did*-ew-la. Diminutive of
  candida (white), the leaves are white
  beneath. China.
  darwinii dar-*win*-ee-ee. After Charles
  Darwin who discovered it in 1835. Chile.
  gagnepainii găn-ya-*păn*-ee-ee. After
  Francois Gagnepain (1866–1952). China.
  hookeri huk-a-ree. After W. J. Hooker.
  Himalaya.
  ilicifolia ee-li-ki-*fo*-lee-a. *Ilex*-leaved. Chile.
  julianae yoo-lee-*ah*-nie. Name given by
  Schneider, after his wife Juliana. C China.
  koreana ko-ree-*ah*-na. Of Korea.
  linearifolia lin-ee-ah-ri-*fo*-lee-a. With linear
  leaves. Argentina, Chile.
  × *lologensis* lo-lo-*gen*-sis. *B. darwinii* × *B.
  linearifolia*. Of Lolog, Argentina.
  × *ottawensis* o-ta-*wen*-sis. *B. thunbergii* ×
  B. vulgaris. Of Ottawa, where it was
  raised.
  panlanensis păn-lăn-*en*-sis. Of Pan-lan,
  China.
  × *rubrostilla* rub-rō-*stil*-la. From L. *rubra*
  (red) and *stilla* (a drop of liquid) referring
  to the red fruits.
  sargentiana sar-jen-tee-*ah*-na. After
  Sargent, see *Prunus sargentii*. China.
  × *stenophylla* sten-ō-*fil*-la. *B. darwinii* × *B.
  empetrifolia*. Narrow-leaved.
  thunbergii thun-*berg*-ee-ee. After Thunberg,
  see *Thunbergia*. Japan, China.
  valdiviana văl-di-vee-*ah*-na. Of Valdivia,
  Chile.
  verruculosa ve-roo-kew-*lō*-sa. With small
  warts (on the shoots). China.
  vulgaris vul-*gah*-ris. Common. Common
  Barberry. Europe, N Africa, temperate
  Asia.
  wilsoniae wil-*so*-nee-ie. After Ernest
  Wilson's wife, Ellen. W China.

Bergenia ber-*gen*-ee-a *Saxifragaceae*. After
Karl August von Bergen (1704–68).
Herbaceous perennials.
  ciliata ki-lee-*ah*-ta. Fringed with hairs (the

Bergenia (continued)
  leaves). Himalaya.
  ligulata lig-ew-*lah*-ta. (= *B. ligulata*).
  Strap-like. Himalaya, Assam, Tibet.
  cordifolia kor-di-*fo*-lee-a. With heart-
  shaped leaves. NE Asia.
  crassifolia kră-si-*fo*-lee-a. Thick-leaved. NE
  Asia.
  purpurascens pur-pew-*răs*-enz. (= *B.
  delavayi*). Purplish (the young leaves).
  Himalaya, China.
  × *schmidtii* shmit-ee-ee. *B. ciliata* × *B.
  crassifolia*. After Ernst Schmidt.
  stracheyi stray-kee-ee. After Lieutenant-
  General Sir Richard Strachey (1817–1908)
  who collected in the Himalaya. Himalaya.

Bergeranthus ber-ga-*răn*-thus *Aizoaceae*.
After Alwyn Berger (1871–1931), German
botanist and Gk. *anthos* (a flower). Tender
succulents. S Africa.
  multiceps mul-tee-keps. With many heads.
  scapiger skah-pi-ger. With a scape.
  vespertinus ves-per-*teen*-us. With the flower
  opening in the evening.

Bertolonia ber-to-lō-nee-a *Melastomataceae*.
After Antonio Bertoloni (1775–1869), Italian
botanist. Tender herbs. Brazil.
  maculata măk-ew-*lah*-ta. Spotted (the
  leaves).
  marmorata mar-mo-*rah*-ta. Marbled (the
  leaves).
    aenea ie-nee-a. Bronze (the leaves).

Bessera bes-a-ra *Amaryllidaceae*. After
Wilibald von Besser (1784–1842), Austrian
botanist. Tender cormous herb.
  elegans ay-le-gahnz. Elegant. Coral Drops.
  Mexico.

Beta bay-ta *Chenopodiaceae*. The L. name.
Annual or biennial herb.
  vulgaris vul-*gah*-ris. Common. Beetroot,
  Sugarbeet. Cult.

Betula bet-ew-la *Betulaceae*. The L. name.
Deciduous trees. Birch.
  albo-sinensis ăl-bō-si-*nen*-sis. The Chinese
  B. alba (= *B. pendula*). W China.
    septentrionalis sep-ten-tree-ō-*nah*-lis.
    Northern.
  ermanii er-*mahn*-ee-ee. After Adolph
  Erman. NE Asia.
  grossa grō-sa. Very large. Japanese Cherry
  Birch. Japan.
  jacquemontii zhahk-a-*mont*-ee-ee. After
  Victor Jacquemont (1801–32), French

Betula (continued)
   naturalist. W Himalaya.
   lenta len-ta. Tough but flexible. Sweet
   Birch. E N America.
   lutea loo-tee-a. Yellow referring to the
   yellowish-brown bark. Yellow Birch. EN
   America.
   nana nah-na. Dwarf. Dwarf Birch.
   N Europe, N America.
   nigra ni-gra. Black (the bark). River Birch.
   E United States.
   papyrifera pă-pi-*ri*-fe-ra. Paper-bearing,
   referring to the papery bark. N America.
   pendula pen-dew-la. Pendulous (the
   branchlets). Silver Birch. Europe, N Asia.
      'Dalecarlica' dă-lee-*kar*-li-ka. Of Dalarno
      (Dalecarlia) Sweden.
      'Tristis' *tris*-tis. Sad, referring to the
      weeping habit.
   utilis ew-ti-lis. Useful, the wood is used for
   timber. Himalayan Birch. Himalaya,
   W China.

Bifrenaria bi-fray-*nah*-ree-a *Orchidaceae*.
From L. *bis* (twice) *frenum* (a strap) referring
to the two strap-like structures which connect
the pollinia and the glands.
   harrisoniae hă-ri-*so*-nee-ie. After Mrs
   Arnold Harrison of Aigburth who grew it
   and painted it for Hooker's Exotic Flora.
   Brazil.

Big Tree see *Sequoiadendron giganteum*

Bignonia big-*nō*-nee-a *Bignoniaceae*. After
Abbé Jean Paul Bignon (1662–1743).
Evergreen woody climber.
   capreolata kă-pree-ō-*lah*-ta. Bearing
   tendrils (from the ends of the petioles).
   Cross Vine. SE United States.

Bilberry see *Vaccinium myrtillus*

Billardiera bi-lar-dee-*e*-ra *Pittosporaceae*.
After J. J. H. de Labillardière, French
botanist who worked on the Australian flora.
Semi-hardy, evergreen climber.
   longiflora long-gi-*flō*-ra. Long-flowered.
   Tasmania.

Billbergia bil-*berg*-ee-a *Bromeliaceae*. After J.
G. Billberg (1772–1844), Swedish botanist.
Tender, epiphytic herbs.
   amoena a-*moy*-na. Pleasant. Brazil.
      rubra rub-ra. Red (the leaves).
   horrida ho-ri-da. Spiny (the leaves).
   iridifolia ee-ri-di-*fo*-lee-a. *Iris*-leaved.
   Brazil.

Billbergia (continued)
   nutans new-tănz. Nodding (the flowers).
   Angel's Tears, Friendship Plant. Brazil.
   pyramidalis pi-ra-mi-*dah*-lis. Pyramidal (the
   inflorescence). Summer Torch. Brazil.
   venezuelana ven-ez-way-*lah*-na. Of
   Venezuela.
   vittata vi-*tah*-ta. Banded (the leaves).
   Brazil.
   × *windii* vin-dee-ee. *B. decora* × *B. nutans*.
   After Wind, a gardener.
   zebrina ze-*bree*-na. Banded (the leaves).
   Brazil.

Birch see *Betula*
   Dwarf see *B. nana*
   Himalayan see *B. utilis*
   Japanese Cherry see *B. grossa*
   Paper see *B. papyrifera*
   River see *B. nigra*
   Silver see *B. pendula*
   Sweet see *B. lenta*
   Yellow see *B. lutea*
Bird of Paradise Flower see *Strelitzia reginae*
Bird's Eyes see *Gilia tricolor*
Bird's Foot Trefoil see *Lotus corniculatus*
Birthwort see *Aristolochia*
Bishop's Cap see *Mitella*
Bishop's Wort see *Stachys macrantha*
Bistort see *Polygonum bistorta*
Black-eyed Susan see *Rudbeckia hirta*,
Thunbergia alata
Black Gum see *Nyssa sylvatica*
Black Snakeroot see *Cimicifuga racemosa*
Blackthorn see *Prunus spinosa*
Blackwood see *Acacia melanoxylon*
Bladder-nut see *Staphylea*
Bladder Senna see *Colutea arborescens*
Bladderwort see *Utricularia*
Blanket Flower see *Gaillardia*
Blazing Star see *Mentzelia lindleyi*

Blechnum blek-num *Blechnaceae*. From
blechnon Gk. name for a fern. Hardy and
tender ferns.
   brasiliense bra-zil-ee-*en*-see. Of Brazil. Rib
   Fern, Brazil Tree Fern. Brazil, Peru.
   capense ka-*pen*-see. Of the Cape of Good
   Hope. Palm-leaf Fern.
   discolor dis-ko-lor. Two-coloured, the
   leaves are dark green above and cinnamon-
   hairy beneath. Crown Fern. Australia,
   New Zealand.
   gibbum gi-bum. (= *B. moorei* hort.).
   Swollen on one side. New Caledonia,
   S Pacific Islands.
   occidentale ok-ki-den-*tah*-lee. Western.
   Hammock Fern. Tropical America.

Blechnum (continued)
orientale o-ree-en-*tah*-lee. Eastern.
Himalaya to Australia.
penna-marina pen-a-ma-*reen*-a. Sea pen,
presumably from the resemblance of the
fronds to this marine animal. Dwarf Hard
Fern. New Zealand, Australia, S America.
spicant *spee*-kant. Tufted. Hard Fern.
Europe, W Asia, Japan, W N America.

Bleeding Heart see *Dicentra spectabilis*
Bleeding-heart Vine see *Clerodendrum
thomsoniae*

Bletilla ble-*til*-la *Orchidaceae*. A diminutive
of *Bletia*, a related genus, named after Louis
Blet, 18th century Spanish apothecary. Semi-
hardy orchid.
striata stree-*ah*-ta. Striped, the ribbed
leaves. China, Japan.

Blood Flower see *Haemanthus katharinae*
Blood Lily see *Haemanthus*
Bloodleaf see *Iresine herbstii*
Bloodroot see *Sanguinaria*
Blue-eyed Mary see *Collinsia verna*,
Omphalodes verna
Blue Flowered Torch see *Tillandsia lindenii*
Blue Lace Flower see *Trachymene caerulea*
Blue Lips see *Collinsia grandiflora*
Blue Thimble Flower see *Gilia capitata*
Blue Trumpet Vine see *Thunbergia
grandiflora*
Bluebell see *Hyacinthoides non-scripta*
  (Scotland) see *Campanula rotundifolia*
  Spanish see *Hyacinthoides hispanica*
Blueberry, Box see *Vaccinium ovatum*
  Highbush see *V. corymbosum*
Bluets see *Hedyotis caerulea*
Blushing Bromeliad see *Neoregelia carolinae*
Boat Lily see *Rhoeo spathacea*

Boenninghausenia bur-ning-how-*zen*-ee-a
Rutaceae. After von Boenninghausen
(1785–1864), German botanist. Deciduous
sub-shrub.
albiflora äl-bi-*flō*-ra. White-flowered.
E Asia.

Bog Arum see *Calla palustris*
Bog Bean see *Menyanthes trifoliata*
Bog Myrtle see *Myrica gale*
Bog Rosemary see *Andromeda*

Bomarea bō-*mah*-ree-a *Alstroemeriaceae*.
After Jacques Christophe Valmont de
Bomare, French patron of science. Tender,

Bomarea (continued)
twining herb.
caldasii käl-*dă*-see-ee. (= *B. kalbreyeri*).
After Francisco José de Caldas
(1771–1816), botanical explorer.
S America.

Bongardia bon-*gard*-ee-a *Berberidaceae*.
After August Bongard (1786–1839). German
botanist. Herbaceous perennial.
chrysogonum kris-*o*-go-num. A golden star,
referring to the flowers. W and C Asia.

Borago bo-*rah*-gō *Boraginaceae*. Possibly
from L. *burra* (a hairy garment) referring to
the leaves. Annual and perennial herbs.
laxiflora läks-i-*flō*-ra. With drooping
flowers. Corsica.
officinalis o-fi-ki-*nah*-lis. Sold as a herb.
Borage. Europe, N Africa.

Borecole see *Brassica oleracea* Acephala

Boronia bo-*rō*-nee-a *Rutaceae*. After
Francesca Borone (1769–94). Tender,
evergreen shrubs.
elatior ay-*lah*-tee-or. Taller. W Australia.
heterophylla he-te-rō-*fil*-la. With variable
leaves sometimes simple, sometimes
pinnate. W Australia.
megastigma meg-a-*stig*-ma. With a large
stigma. W Australia.
serrulata se-ru-*lah*-ta. With small teeth (the
leaves). SE Australia.

Borzicactus bor-zee-*kăk*-tus *Cactaceae*. After
Antonio Borzi (1832–1921), Italian botanist,
and *Cactus* q.v.
aurantiacus ow-răn-tee-*ah*-kus. Orange (the
flowers). Peru.
haynei hayn-ee-ee. After Frederich Hayne
(1763–1832). Peru.

Boston Ivy see *Parthenocissus tricuspidata*
Bottle-brush see *Callistemon*
  Crimson see *C. citrinus*
Bottle Gourd see *Lagenaria siceraria*

Bougainvillea boo-gan-*vil*-lee-a
Nyctaginaceae. After Louis Antoine de
Bougainville (1729–1811), explorer and
scientist. Tender climbers.
× *buttiana* būt-ee-*ah*-na. *B. glabra* × *B.
peruviana*. After Mrs R. V. Butt who
found it in 1910.
glabra glă-bra. Glabrous. Brazil.
spectabilis spek-*tah*-bi-lis. Spectacular.
Brazil.

Bouncing Bet see *Saponaria officinalis*

Bouvardia boo-*var*-dee-a *Rubiaceae*. After
Dr Charles Bouvard (1572–1658). Tender,
evergreen shrubs.
   longiflora long-gi-*flō*-ra. (= *B. corymbiflora*).
   Long-flowered. Sweet Bouvardia. Mexico.
   ternifolia tern-i-*fo*-lee-a. (= *B. triphylla*).
   With leaves in threes. Scarlet Trompetilla.
   Texas, Mexico.

Bower Plant see *Pandorea jasminoides*
Bowles' Golden Grass see *Milium effusum*
'Aureum'
Box see *Buxus*
   Common see *B. sempervirens*
Box Elder see *Acer negundo*
Box Thorn see *Lycium barbarum*

Brachycome brǎ-kee-*kō*-mee *Compositae*.
From Gk. *brachys* (short) and *kome* (hair)
referring to the short pappus bristles. Annual
and perennial herbs.
   iberidifolia i-be-ri-di-*fo*-lee-a. With leaves
   like *Iberis*. Swan River Daisy. W and S
   Australia.
   rigidula ri-*gi*-dew-la. Rigid. SE Australia,
   Tasmania.

Brachyglottis brǎ-kee-*glo*-tis *Compositae*.
From Gk. *brachys* (short) and *glotta* (a tongue)
referring to the short ray florets. Semi-hardy,
evergreen shrub.
   repanda re-*pǎn*-da. Wavy-margined (the
   leaves). New Zealand.

Brake, Australian see *Pteris tremula*
   Cretan see *P. cretica*
   Sword see *P. ensiformis*
Brandy Bottle see *Nuphar lutea*
Brasiliopuntia brasiliensis see *Opuntia
brasiliensis*
Brass Buttons see *Cotula coronopifolia*
Brassaia actinophylla see *Schefflera
actinophylla*

Brassavola bra-*sah*-vo-la *Orchidaceae*. After
Antonio Musa Brassavola (1500–55), Italian
botanist. Greenhouse orchids.
   cordata kor-*dah*-ta. Heart-shaped (the lip).
   W Indies.
   digbyana dig-bee-*ah*-na. After Edward St
   Vincent Digby, the first to flower it, in
   1846. C America.
   fragrans frah-granz. Fragrant (the flowers).
   Brazil.

Brassavola (continued)
   glauca glow-ka. Glaucous (the leaves).
   C America.
   nodosa nō-*dō*-sa. With conspicuous nodes.
   Lady of the Night. C and S America.
   perrinii pe-*rin*-ee-ee. After Mr Perrin,
   gardener to Mr Harrison, whose wife
   introduced it. Brazil, Paraguay.

Brassia brahs-ee-a *Orchidaceae*. After
William Brass, plant collector and botanical
artist, (died 1783). Greenhouse orchids.
   caudata kaw-*dah*-ta. Prolonged into a tail
   (the sepals). Tropical America.
   elegantula ay-le-*gǎnt*-ew-la. Elegant.
   Mexico.
   lawrenceana lo-rens-ee-*ah*-na. After Mrs
   Lawrence. N S America.
   longissima long-*gis*-si-ma. (= *B.
   lawrenceana longissima*). Longest (the
   sepals). Costa Rica.
   maculata mǎk-ew-*lah*-ta. Spotted (the
   petals). W Indies, C America.
   verrucosa ve-roo-*kō*-sa. Warty (the lip).
   Mexico to S America.

Brassica brǎ-si-ka *Cruciferae*. L. name for
cabbage. Annual and biennial herbs.
   napus nah-pus. L. name for turnip. Rape.
   Cult.
      Napobrassica nah-pō-*brǎ*-si-ka. From
      napus and *brassica*. Swede, Rutabaga.
   oleracea o-le-*rah*-kee-a. Vegetable-like.
   Wild Cabbage. W Europe.
      Acephala a-*kef*-a-la. Without a head.
      Ornamental Kale, Borecole.
      Botrytis *bot*-ri-tis. Like a bunch of
      grapes. Cauliflower, Broccoli.
      Capitata kǎ-pi-*tah*-ta. In a dense head
      (the leaves). Cabbage.
      Costata see *B. oleracea* Tronchuda.
      Gemmifera gem-*i*-fe-ra. Bearing buds.
      Brussels Sprouts.
      Gongylodes gon-gi-*lō*-deez. Swollen (the
      stem). Kohl Rabi.
      Italica ee-*tǎ*-li-ka. Of Italy. Sprouting
      Broccoli.
      Tronchuda tron-*koo*-da. The Portuguese
      name. Couve Tronchuda, Portuguese
      Cabbage.
   rapa rah-pa. The L. name. Turnip. Cult.
      Pekinensis pee-kin-*en*-sis. Of Peking.
      Chinese Cabbage.
      Perviridis per-*vi*-ri-dis. Very green.
      Tendergreen.

× *Brassocattleya* brah-sō-*kǎt*-lee-a
Orchidaceae. Intergeneric hybrids, from the

Brassocattleya (continued)
ames of the parents. *Brassavola* × *Cattleya*.
Greenhouse orchids.

× **Brassolaelia** brah-sō-*lie*-lee-a *Orchidaceae*.
ntergeneric hybrids, from the names of the
parents. *Brassavola* × *Laelia*. Greenhouse
Orchids.

× **Brassolaeliocattleya** brah-sō-lie-lee-ō-*kăt*-
ee-a *Orchidaceae*. Intergeneric hybrids, from
ne names of the parents. *Brassavola* ×
Cattleya × *Laelia*. Greenhouse orchids.

Brazilian Edelweiss see *Sinningia leucotricha*
Brazilian Plume see *Justicea carnea*
Bridal Wreath see *Francoa sonchifolia*,
piraea 'Arguta'
Bridewort see *Spiraea salicifolia*

Brimeura bri-*mewr*-ra *Liliaceae*. After Maria
e Brimeur, a lover and grower of flowers in
he time of Clusius. Bulbous perennial.
amethystinus ă-me-*this*-ti-nus. (= *Hyacinthus
amethystinus*). Violet (the flowers). SW
Europe, NW Yugoslavia.

Briza *bree*-za *Gramineae*. L. name of a grass
rown for food. Annual and perennial
rasses.
maxima mahk-si-ma. Larger. Pearl Grass.
Mediterranean region.
media me-dee-a. Intermediate. Quaking
Grass. Europe, W Asia.
minor mi-nor. (= *B. minima*). Smaller.
Europe, W. Asia.

Broad Bean see *Vicia faba*
Broad-leaved Kindling Bark see *Eucalyptus
dalrympleana*
Broccoli see *Brassica oleracea* Botrytis
Sprouting see *B. oleracea* Italica
Brodiaea brō-dee-*ie*-a *Amaryllidaceae*. After
ames Brodie (1744–1824), Scottish botanist.
emi-hardy, cormous perennials.
californica kăl-i-*for*-ni-ka. Of California.
coronaria ko-rō-*nah*-ree-a. Of garlands.
Triplet Lily. W N America.
elegans ay-le-gahnz. Elegant. W N
America.
ida-maia see *Dichelostemma ida-maia*
laxa see *Triteleia laxa*
stellaris ste-*lah*-ris. Star-like (the flowers).
California.
uniflora see *Ipheion uniflorum*

Broom see *Cytisus*, *Genista*
Common see *Cytisus scoparius*

Broom (continued)
Dalmatian see *Genista sylvestris*
Genoa see *Genista januensis*
Montpelier see *Cytisus monspessulanus*
Mt Etna see *Genista aetnensis*
Pineapple-scented see *Cytisus battandieri*
White Spanish see *Cytisus multiflorus*

Broussonetia broo-son-*ay*-tee-a *Moraceae*.
After T. N. V. Broussonet (1761–1807),
French naturalist. Deciduous tree.
papyrifera pă-pi-*ri*-fe-ra. Paper bearing, the
bark is used to make paper in Japan. Paper
Mulberry. China, Japan.

Browallia brō-*ah*-lee-a *Solanaceae*. After
John Browall (1707–55), Swedish botanist.
Tender annuals.
demissa day-*mis*-a. Weak. Renamed by
Linnaeus from *B. elata* (tall) after falling
out with Browall. C America.
speciosa spek-ee-ō-sa. Showy. Sapphire
Flower. Colombia.
viscosa vis-*kō*-sa. Sticky (the young growths
and calyx). Peru.

Bruckenthalia bruk-an-*thahl*-ee-a *Ericaceae*.
After Samuel and Michael von Bruckenthal,
18th century Austrian noblemen. Evergreen,
heath-like shrub.
spiculifolia speek-ew-lee-*fo*-lee-a. With
spiky leaves. Spike Heath. E Europe,
W Asia.

Brunfelsia brun-*fel*-see-a *Solanaceae*. After
Otto Brunfels (1489–1534), German monk
and botanist. Tender, evergreen shrubs.
americana a-me-ri-*kah*-na. American. Lady
of the Night. W Indies.
latifolia lah-tee-*fo*-lee-a. Broad-leaved.
Brazil.
pauciflora paw-si-*flō*-ra. Few-flowered.
Yesterday, Today and Tomorrow. Brazil.
calycina kă-li-*kee*-na. (= *B. calycina*).
With a well-developed calyx.
undulata un-dew-*lah*-ta. Undulate (the
corolla). White Raintree. Jamaica.

Brunnera *brun*-a-ra *Boraginaceae*. After
Samuel Brunner (1790–1844), Swiss
botanist. Perennial herb.
macrophylla măk-rō-*fil*-la. Large-leaved.
Caucasus, Siberia.

Brussels Sprouts see *Brassica oleracea*
Gemmifera
Bryophyllum see *Kalanchoe*
Buck Bean see *Menyanthes trifoliata*

Buckeye see *Aesculus*
  California see *A. californica*
  Red see *A. pavia*
  Yellow see *A. flava*
Buckthorn, Common see *Rhamnus cathartica*

Buddleja bŭd-lee-a *Loganiaceae*. After the
Rev. Adam Buddle (1660–1715). Deciduous
shrubs. Butterfly Bush.
  alternifolia ăl-tern-i-*fo*-lee-a. With alternate
leaves, they are opposite in most species.
China.
  colvilei kol-*vil*-ee-eye. After Sir James
Colvile (died 1890). Himalaya.
  crispa kris-pa. Finely wavy (the leaves).
Himalaya.
  davidii dă-*vid*-ee-eye. After Armand David
who discovered it, see *Davidia*. China.
  fallowiana fă-lō-ee-*ah*-na. After George
Fallow (1890–1915), gardener at the Royal
Botanic Garden, Edinburgh. China.
  globosa glo-*bō*-sa. Spherical (the
inflorescence). Chile, Peru.
  × *weyeriana* vay-a-ree-*ah*-na. *B. davidii
magnifica* × *B. globosa*. After Van de
Weyer who raised it in 1914.

Bugbane see *Cimicifuga*
Bugle see *Ajuga reptans*

Buglossoides bew-glos-*oi*-deez *Boraginaceae*.
Like *Buglossum*. Perennial herb.
  purpurocaerulea pur-pew-rō-kie-*ru*-lee-a. (=
Lithospermum purpurocaeruleum). Purple-
blue (the flowers). Europe, W Asia.

Bulbine bul-*bee*-nay *Liliaceae*. From *bolbine*
Gk. name for a bulbous plant. Perennial
herb.
  semibarbata se-mee-bar-*bah*-ta. Half
bearded, only the inner three filaments,
out of six, are hairy. Australia.

Bulbinella bul-bi-*nel*-a *Liliaceae*. Diminutive
of *Bulbine* q.v. Perennial herb.
  hookeri hook-a-ree. After J. D. Hooker.
New Zealand.

Bulbocodium bul-bō-*kō*-dee-um *Liliaceae*.
From Gk. *bolbos* (a bulb) and *kodion* (wool).
Cormous, perennial herb.
  vernum ver-num. Of spring (flowering).
Spring Meadow Saffron. Pyrenees, Alps
to Russia.

Bulbophyllum bul-bō-*fil*-lum *Orchidaceae*.
From Gk. *bolbos* (a bulb) and *phyllon* (a leaf)

Bulbophyllum (continued)
referring to the leaves growing from the top
of the pseudobulb. Greenhouse orchids.
  barbigerum bar-*bi*-ge-rum. Bearded (the
lip). Tropical Africa.
  careyanum kair-ree-*ah*-num. After Dr Carey
of Serampore who sent it to Liverpool
Botanic Garden before 1824. Himalaya.
  collettii ko-*let*-ee-ee. After Gen. Sir Henry
Collett of the Indian Army. Burma.
  dayanum day-*ah*-num. After John Day
(1824–88), orchid grower. Assam.
  dearei dear-ree-ee. After Col. Deare of
Engelfield Green. Borneo, Philippines.
  ericssonii e-rik-*son*-ee-ee. After Ericsson, a
Swedish plant collector who discovered it.
Borneo.
  medusae may-*dew*-sie. Of Medusa, likening
the many-flowered umbel to Medusa's
head. SE Asia.
  odoratissimum o-dō-ra-*ti*-si-mum. Most
fragrant. Himalaya, China.
  ornatissimum or-na-*ti*-si-num. Most
ornamental. Himalaya, Philippines.
  umbellatum um-bel-*ah*-tum. The flowers
appear to be in umbels. Himalaya.

Bull Bay see *Magnolia grandiflora*
Bunny Rabbits see *Linaria maroccana*

Buphthalmum buf-*thahl*-mum *Compositae*.
From Gk. *bous* (an ox) and *ophthalmos* (an
eye) referring to the flowers.
  salicifolium să-li-ki-*fo*-lee-um. *Salix*-leaved.
C Europe.
  speciosissimum see *Telekia speciosissima*
  speciosum see *Telekia speciosa*

Bupleurum boo-*plur*-rum *Umbelliferae*.
From Gk. *boupleuros* (ox rib) referring to
another plant. Evergreen shrub.
  fruticosum froo-ti-*kō*-sum. Shrubby (most
species are herbaceous). S Europe,
Mediterranean region.

Burhead see *Echinodorus*
Burning Bush see *Dictamnus albus*, *Kochia
scoparia*
Burnet see *Sanguisorba*
Bush Clover see *Lespedeza*
Bush Groundsel see *Baccharis halimifolia*
Busy Lizzie see *Impatiens walleriana*
Butcher's Broom see *Ruscus aculeatus*

Butia bew-tee-a *Palmae*. The native name.
Tender palms.
  capitata kă-pi-*tah*-ta. In a dense head (the
leaves). Jelly Palm. Brazil, Argentina.

Butia (continued
  yatay yah-tay. The native name. Yatay
  Palm. Argentina.

Butomus boo-to-mus *Butomaceae*. From Gk.
bous (an ox) and *temmo* (to cut), the sharp-
edged leaves prevent it being used for fodder.
Aquatic perennial herb.
  umbellatus um-bel-*ah*-tus. With flowers in
  umbels. Flowering Rush. Europe, Asia.

Buttercup see *Ranunculus*
  Persian see *R. asiaticus*
Butterfly Bush see *Buddleja*
Butterfly Flower see *Schizanthus*
Butterfly Lily see *Hedychium coronarium*
Butterfly Tree see *Bauhinia purpurea*
Butter-nut see *Juglans cinerea*
Butterwort see *Pinguicula*
Button Snake Root see *Liatris pycnostachya*
Buttons-on-a-String see *Crassula rupestris*

Buxus buks-us *Buxaceae*. The L. name for
B. sempervirens. Evergreen shrubs and trees.
Box.
  balearica bă-lee-*ah*-ri-ka. Of the Balearic
  Islands. Balearic Islands, SW Spain.
  microphylla mik-rō-*fil*-la. Small-leaved.
  Cult.
    japonica ja-*pon*-i-ka. Of Japan.
    koreana ko-ree-*ah*-na. Of Korea. Korea,
    China.
    sinica sin-i-ka. Of China.
  sempervirens sem-per-*vi*-renz. Evergreen.
  Common Box. Europe, N Africa, W Asia.

C

Cabbage see *Brassica oleracea* Capitata
  Chinese see *B. rapa* Pekinensis
  Portuguese see *B. oleracea* Tronchuda
  Wild see *B. oleracea*
Cabbage Gum see *Eucalyptus pauciflora*
Cabbage Tree see *Cordyline australis*

Cabomba ka-*bom*-ba *Nymphaeaceae*. From
the native name. Aquatic herbs. Fanwort.
  aquatica a-*kwah*-ti-ka. Growing in water.
  NE South America.
  caroliniana kă-ro-lin-ee-*ah*-na. Of Carolina.
  SE United States.

Cactus Gk. name for another spiny plant.
Now used only as a common name, see right.

Cactus, Baseball see *Euphorbia obesa*
  Bird's-nest see *Mammillaria camptotricha*
  Bishop's Cap see *Astrophytum myriostigma*,
    A. ornatum
  Cane see *Opuntia cylindrica*
  Chain see *Rhipsalis paradoxa*
  Chain-link see *Opuntia imbricata*
  Christmas see *Schlumbergera bridgesii*,
    S. × buckleyi
  Cinnamon see *Opuntia microdasys rufida*
  Claw see *Schlumbergera truncata*
  Cob see *Lobivia*
  Coral see *Mammillaria heyderi*
  Corncob see *Euphorbia mammillaris*
  Cotton-pole see *Opuntia vestita*
  Crab see *Schlumbergera truncata*
  Crown see *Rebutia*
  Dumpling see *Lophophora williamsii*
  Easter see *Rhipsalidopsis gaertneri*
  Eve's Pin see *Opuntia subulata*
  Feather see *Mammillaria plumosa*
  Fire Crown see *Rebutia senilis*
  Fire Cracker see *Cleistocactus
    smaragdiflorus*
  Goat's Horn see *Astrophytum capricorne*
  Gold Lace see *Mammillaria elongata*
  Golden Ball see *Notocactus leninghausii*
  Golden Barrel see *Echinocactus grusonii*
  Indian Fig see *Opuntia ficus-indica*
  Lamb's Tail see *Wilcoxia schmollii*
  Mistletoe see *Rhipsalis baccifer*
  Old Man see *Cephalocereus senilis*
  Old Woman see *Mammillaria hahniana*
  Orchid see *Nopalxochia ackermanii*
  Peanut see *Chamaecereus sylvestrii*
  Peruvian Apple see *Cereus peruvianus*
  Plain see *Gymnocalycium mihanovichii*
  Rat's-tail see *Aporocactus flagelliformis*
  Red Crown see *Rebutia minuscula*
  Ribbon see *Pedilanthes tithymaloides*
  Scarlet Ball see *Notocactus haselbergii*
  Sea Urchin see *Astrophytum asterias*,
    Echinopsis
  Silver Ball see *Notocactus scopa*
  Silver Cluster see *Mammillaria prolifera*
  Silver Dollar see *Astrophytum asterias*
  Snowball see *Mammillaria bocasana*
  Snowdrop see *Rhipsalis houlletiana*
  Spider see *Gymnocalycium denudatum*
  Star see *Astrophytum*
  Sun see *Heliocereus speciosus*
  Teddy Bear see *Opuntia bigelovii*
  Toothpick see *Stetsonia coryne*
  White Torch see *Trichocereus spachianus*

Caesalpinia kie-sal-*pee*-nee-a *Leguminosae*.
After Andreas Caesalpini (1519–1603),
Italian botanist. Tender trees.

Caesalpinia (continued)
gilliesii gi-*leez*-ee-ee. After John Gillies
(1747–1836). Argentina, Uruguay.
japonica ja-*pon*-i-ka. Of Japan.
pulcherrima pul-*ke*-ri-ma. Very Pretty.
Barbados Pride. W Indies.

Caladium ka-*lah*-dee-um *Araceae*. From
kaladi the native name. Tender herbs.
Angel's Wings, Elephant's Ears, Heart of
Jesus.
    bicolor bi-ko-lor. Two-coloured (the
    leaves). Brazil.
    × *hortulanum* hort-ew-*lah*-num. Of
    gardens.
    'Candidum' *kăn*-di-dum. White (the
    leaves).

Calamintha kăl-a-*min*-tha *Labiatae*. The Gk.
name from *kallos* (beautiful) and *minthe*
(mint). Perennial herbs. Calamint.
    alpina see *Acinos alpinus*.
    grandiflora grăn-di-*flō*-ra. Large-flowered.
    S Europe.
    nepeta ne-pe-ta. From *Nepeta* a related
    genus. Mediterranean region.

Calamondin see × *Citrofortunella mitis*

Calandrinia kă-lan-*dreen*-ee-a *Portulacaceae*.
After Jean Louis Calandrini (1703–58).
Annuals. Rock Purslane.
    ciliata ki-lee-*ah*-ta. Fringed with hairs.
    Peru, Ecuador.
    umbellata um-bel-*ah*-ta. With flowers in
    umbels. Peru, Chile.

Calanthe ka-*lăn*-thee *Orchidaceae*. From Gk.
kalos (beautiful) and *anthos* (a flower).
Greenhouse orchids.
    furcata fur-*kah*-ta. (= *C. veratrifolia*). Cleft
    (the lip). SE Asia to Australia.
    masuca ma-*soo*-ka. ? A native name.
    Himalaya.
    vestita ves-*tee*-ta. Clothed, referring to the
    hairy stem. SE Asia.

Calathea ka-*lah*-thee-a *Marantaceae*. From
Gk. *kalathos* (a basket), the flowers are
clustered as if in a basket. Tender, evergreen
herbs.
    bachemiana bah-kem-ee-*ah*-na. After Herr
    Bachem, Burgermeister of Cologne.
    Brazil.
    lancifolia lăn-ki-*fo*-lee-a. With lance-shaped
    leaves. Rattlesnake Plant. Tropical
    America.
    lindeniana lin-den-ee-*ah*-na. After J. J.

Calathea (continued)
Linden, Belgian horticulturist. Brazil.
    louisae loo-*eez*-ie. After Queen Louisa of
    Belgium.
    makoyana măk-oy-*ah*-na. After Jacob
    Makoy (1790–1873), nurseryman of
    Liège. Cathedral Windows, Peacock Plant.
    ornata or-*nah*-ta. Showy. N South America.
    picturata pik-tew-*rah*-ta. Variegated. Brazil.
    zebrina zeb-*ree*-na. Striped. Zebra Plant.
    Brazil.

Calceolaria kăl-kee-ō-*lah*-ree-a
Scrophulariaceae. After F. Calceolari, 16th
century Italian botanist. The name in L. also
means slipper-like, referring to the flowers.
Herbs and shrubs. Slipperwort.
    biflora bi-*flō*-ra. Two-flowered. Chile,
    Argentina.
    darwinii dar-*win*-ee-ee. After Charles
    Darwin who discovered it. Patagonia.
    × *herbeohybrida* herb-ee-ō-*hib*-ri-da.
    Herbaceous hybrid.
    integrifolia in-teg-ri-*fo*-lee-a. With entire
    leaves. Chile.
    mexicana meks-i-*kah*-na. Of Mexico.
    Mexico, C America.
    tenella ten-*el*-la. Dainty. Chile.
    tripartita tri-*part*-ee-ta. Three-parted (the
    leaf segments). W South America.

Calendula ka-*len*-dew-la *Compositae*. From
L. *calendae* (the first day of the month)
referring to the long flowering period. Annual
herb.
    officinalis o-fi-ki-*nah*-lis. Sold as a herb. Pot
    Marigold. S Europe.

Calico Bush see *Kalmia latifolia*
Calico Flower see *Aristolochia elegans*
Calico Hearts see *Adromischus maculatus*
California Bluebell see *Phacelia campanulata*
California Geranium see *Senecio petasites*
California Laurel see *Myrica californica*
California Nutmeg see *Torreya californica*

Calla kă-la *Araceae*. From Gk. *kallos*
(beautiful). Herbaceous perennial.
    palustris pa-*lus*-tris. Growing in bogs. Bog
    Arum. N America, Europe, Asia.

Callicarpa kă-lee-*kar*-pa *Verbenaceae*. From
Gk. *kallos* (beautiful) and *karpos* (a fruit).
They are grown for their attractive fruits.
Deciduous shrubs. Beauty Berry.
    bodinieri bo-din-ee-*e*-ree. After Emile Marie
    Bodinieri (1842–1901), French missionary
    and plant collector in China. China.

Callicarpa (continued)
  giraldii ji-*rahl*-dee-ee. After Giuseppe
    Giraldi, Italian missionary and plant
    collector in China.
  dichotoma di-*ko*-to-ma. With forking
    shoots. China, Japan.
  japonica ja-*pon*-i-ka. Of Japan.
    'Leucocarpa' loo-kō-*kar*-pa. With white
    fruits.
  rubella roo-*bel*-a. Reddish (the fruits). SE
    Asia.

Callirhoe kă-lee-rō-ee *Malvaceae*. After the
  daughter of Achelous in Gk. mythology.
  Annual and perennial herbs. Poppy Mallow.
  digitata di-gi-*tah*-ta. Fingered, the leaves
    have finger-like lobes. United States.
  involucrata in-vol-oo-*krah*-ta. With an
    involucre (around the flowers). United
    States.
  pedata ped-*ah*-ta. Like a bird's-foot (the
    leaves). S United States.

Callisia ka-*lis*-ee-a *Commelinaceae*. From Gk.
  kallis (beauty). Tender, evergreen herbs.
  Inch Plant. Mexico.
  elegans ay-le-gahnz. (= *Setcreasia striata*
    hort.). Elegant. Striped Inch Plant.
  fragrans frah-granz. Fragrant (the flowers).

Callistemon ka-*lee*-stay-mon *Myrtaceae*.
  From Gk. *kallos* (beautiful) and *stemon* (a
  stamen). The stamens are the showy part of
  the flower. Evergreen, semi-hardy shrubs.
  Bottle-brush.
  citrinus ki-*tree*-nus. Lemon-scented (the
    leaves). Crimson Bottle-brush. SE
    Australia.
  linearis lin-ee-*ah*-ris. Linear (the leaves).
    New South Wales.
  rigidus ri-gi-dus. Rigid (the leaves).
    E Australia.
  salignus sa-*lig*-nus. Willow-like (the leaves).
    SE Australia.
  speciosus spe-kee-ō-sus. Showy.
    W Australia.
  subulatus sub-ew-*lah*-tus. Awl-shaped (the
    leaves). SE Australia.

Callistephus ka-*lee*-ste-fus *Compositae*. From
  Gk. *kallos* (beautiful) and *stephanus* (a crown)
  referring to the showy, terminal flower heads.
  chinensis chin-*en*-sis. Of China. China
    Aster.

Callitriche ka-*lee*-tri-kee *Callitrichaceae*.
  From Gk. *kallos* (beautiful) and *trichos* (hair)

Callitriche (continued)
  referring to the delicate foliage. Aquatic
  herbs. Water Starwort.
  hamulata hahm-ew-*lah*-ta. Hooked (the
    ends of the leaves). Europe.
  hermaphroditica her-mă-frō-*dee*-ti-ka. (= *C.
    autumnalis*). Hermaphrodite. Europe,
    N Africa.
  palustris pa-*lus*-tris. Growing in bogs.
    N America, Europe, Asia.
  stagnalis stăg-*nah*-lis. Growing in still
    water. S Europe, N Africa.

Calluna ka-*loo*-na *Ericaceae*. From L. *kalluno*
  (to cleanse), it was used to make brooms.
  Evergreen shrub.
  vulgaris vul-*gah*-ris. Common. Heather,
    Ling. Europe, W Asia.

Calocedrus kă-lo-*ked*-rus *Cupressaceae*.
  From Gk. *kallos* (beautiful) and *Cedrus* q.v.
  Evergreen conifer.
  decurrens day-ku-renz. (= *Libocedrus
    decurrens*). With the leaf margin running
    gradually into the stem. Incense Cedar.
    WN America.

Calochortus kă-lo-*kor*-tus *Liliaceae*. From
  Gk. *kallos* (beautiful) and *chortos* (grass)
  referring to the slender leaves. Semi-hardy
  bulbous herbs. Mariposa Lily.
  albus ăl-bus. White (the flowers).
    California.
  amabilis a-*mah*-bi-lis. Beautiful. NW
    California.
  barbatus bar-*bah*-tus. Bearded (the petals).
    Mexico.
  caeruleus kie-*ru*-lee-us. Blue (the flowers).
    California.
  luteus loo-tee-us. Yellow (the flowers).
    California.
  uniflorus ew-ni-*flō*-rus. One-flowered.
    California, Oregon.

Calomeria kă-lo-*me*-ree-a *Compositae*. From
  Gk. *kallos* (beautiful) and *meris* (a part).
  Tender biennial.
  amaranthoides ăm-a-rănth-*oi*-deez.
    (= *Humea elegans*). Like *Amaranthus*.
    Plume Bush. SE Australia.

Caltha kăl-tha *Ranunculaceae*. L. name for a
  yellow-flowered plant. Aquatic and marginal
  perennials.
  asarifolia a-sah-ri-*fo*-lee-a. With leaves like
    Asarum. W N America.
  chelidonii kel-i-*dō*-nee-ee. Like
    Chelidonium. W N America.
  leptosepala lep-tō-*se*-pa-la. With slender

Caltha (continued)
sepals. W N America.
palustris pa-*lus*-tris. Growing in bogs. King
Cup, Marsh Marigold. N America,
Europe, Asia.
'Plena' *play*-na. With double flowers.
polypetala po-lee-*pe*-ta-la. With many petals
(sepals). Bulgaria.

Calycanthus kă-lee-*kanth*-us *Calycanthaceae*.
From Gk. *kalyx* (calyx) and *anthos* (a flower),
the sepals and petals are similar. Deciduous
shrubs.
fertilis fer-ti-lis. Fertile. SE United States.
floridus flō-ri-dus. Flowering. Carolina
Allspice. SE United States.
occidentalis ok-ki-den-*tah*-lis. Western.
Californian Allspice. California.

Camassia ka-*mă*-see-a *Liliaceae*. From the
N American Indian name *quamash* or *camass*.
Bulbous herbs.
cusickii kew-*sik*-ee-ee. After W. C. Cusick.
Oregon.
leichtlinii liekt-*lin*-ee-ee. After Max
Leichtlin (1831–1910). Oregon.
quamash kwah-măsh. The native name, see
above. W N America.
scilloides skil-*oi*-deez. Like *Scilla*. E N
America.

Camellia ka-*mel*-lee-a *Theaceae*. After
George Joseph Kamel (Camellus)
(1661–1706), pharmacist who studied the
Philippines flora. Evergreen shrubs.
cuspidata kus-pi-*dah*-ta. With a stiff point
(the leaves). W China.
japonica ja-*pon*-i-ka. Common Camellia.
Japan, Korea.
reticulata ray-tik-ew-*lah*-ta. Net-veined (the
leaves). China.
sasanqua sa-*sang*-kwa. The Japanese name.
Japan.
sinensis si-*nen*-sis. Of China. Tea Plant.
× *williamsii* wil-*yămz*-ee-ee. *C. japonica* ×
C. saluenensis. After J. C. Williams of
Caerhays who raised many fine forms of
this hybrid.

Campanula kăm-*pahn*-ew-la *Campanulaceae*.
From L. *campana* (a bell) referring to the
shape of the flowers. Perennial herbs.
Bellflower.
alliariifolia ă-lee-ah-ree-i-*fo*-lee-a. With
leaves like *Alliaria petiolata* (Garlic
Mustard). Spurred Bellflower. Caucasus,
Turkey.
allionii see *C. alpestris*

Campanula (continued)
alpestris ăl-*pes*-tris. (= *C. allionii*). Of the
lower mountains. SW Alps.
arvatica ar-*vă*-ti-ca. Of Arvas, N Spain.
aucheri ow-ka-ree. After P. M. R. Aucher-
Eloy (1792–1836). W Asia.
barbata bar-*bah*-ta. Bearded (the corolla).
Alps, C Europe, Norway.
bononiensis bo-nō-nee-*en*-sis. Of Bologne,
N Italy. C and E Europe, W Asia.
carpatica kar-*pă*-ti-ka. Of the Carpathians.
cochleariifolia kok-lee-ah-ree-i-*fo*-lee-a.
With leaves like *Cochlearia*. S and E
Europe.
collina ko-*leen*-a. Growing on hills.
Caucasus.
excisa eks-*kee*-sa. Cut away (the deep
corolla sinuses). Alps.
garganica gar-*gah*-ni-ka. Of Monte
Gargano, Italy. Adriatic Bellflower. SE
Italy, W Greece.
glomerata glo-me-*rah*-ta. Clustered (the
flowers). Clustered Bellflower. Europe,
W Asia.
isophylla ee-sō-*fil*-la. With equal-sized
leaves. Italian Bellflower. NW Italy.
lactiflora lăk-ti-*flō*-ra. Milky-flowered.
Milky Bellflower. Caucasus, W Asia.
latifolia lah-tee-*fo*-lee-a. Broad-leaved.
Giant Bellflower. Europe, W Asia.
latiloba see *C. persicifolia*
medium may-dee-um. Classical name for a
similar plant from Media, W Asia.
Canterbury Bells. SE France, Italy.
'Calycanthema' kă-lee-kăn-*thee*-ma.
Flowering in the calyx, the calyx
resembles the corolla. Cup and Saucer
Canterbury Bells.
morettiana mo-ret-ee-*ah*-na. After G.
Moretti (1782–1853). Dolomites (Italy).
persicifolia per-si-ki-*fo*-lee-a. (= *C.
latiloba*). With leaves like *Prunus persica*.
Europe to N Asia.
portenschlagiana por-tan-shlahg-ee-*ah*-na.
After Franz von Portenschlag-Ledermeyer
(1777–1827), Austrian botanist.
W Yugoslavia.
poscharskyana po-shar-skee-*ah*-na. After
Gustav Adolf Poscharsky (1832–1914).
W Yugoslavia.
pulla pul-la. Dark (the flowers). NE Alps
(Austria).
× *pulloides* pul-*loi*-deez. *C. carpatica* × *C.
pulla*. Like *C. pulla*.
punctata punk-*tah*-ta. Spotted (the corolla).
E Asia.
pyramidalis pi-ra-mi-*dah*-lis. Pyramidal (the
inflorescence). Chimney Bellflower.

Campanula (continued)
  N Italy, Albania, Yugoslavia.
  raineri ray-na-ree. After Rainer. SE Alps.
  rapunculoides ra-punk-ew-*loi*-deez. Like *C.*
  rapunculus. Europe, Asia.
  rotundifolia ro-tund-i-*fo*-lee-a. With
  rounded (basal) leaves. Harebell, Bluebell
  (Scotland). Europe, Asia, N America.
  sarmatica sar-*mă*-ti-ka. Of Sarmatia (E
  Europe). Caucasus.
  thyrsoides thur-*soi*-deez. Staff-like (the
  inflorescence). Alps, Balkans.
  trachelium tra-*kay*-lee-um. From Gk.
  trachelos (neck) referring to supposed
  medicinal properties. Throatwort. Europe,
  Asia.
  zoysii zoys-ee-ee. After Karl von Zoys,
  Austrian botanist. SE Alps.

Campernelle see *Narcissus* × *odorus*
Campion see *Silene*
  Alpine see *Lychnis alpina*
  Moss see *Silene acaulis*
  Rose see *Lychnis coronaria*
  Sea see *Silene vulgaris maritima*

Campsis kămp-sis Bignoniaceae. From Gk.
kampe (something bent) referring to the
curved stamens. Deciduous woody climbers.
  grandiflora grăn-di-*flō*-ra. Large-flowered.
  China.
  radicans rah-di-kănz. With rooting stems.
  Trumpet Vine. SE United States.
  × *tagliabuana* tăg-lee-ă-bew-*ah*-na. *C.*
  grandifolia × *C. radicans*. After the
  Tagliabue brothers, Italian nurserymen.

Camptosorus kămp-tō-*so*-rus Aspleniaceae.
From Gk. *kamptos* (curved) and *sorus* referring
to the curved sori. Evergreen fern.
  rhizophyllus ree-zō-*fil*-lus. Root-leaved, the
  leaves root at the tip and hence the plant
  'walks'. Walking Fern. N America.

Canada Lily see *Lilium canadense*
Canary Creeper see *Tropaeolum speciosum*
Canary Grass see *Phalaris canariensis*
Candelabra Plant see *Aloe arborescens*
Candle Plant see *Plectranthus oertendahlii*,
Senecio articulatus

Canna kă-na Cannaceae. From Gk. *kanna* (a
reed). Tender herbaceous perennials.
  edulis ed-*ew*-lis. Edible, the rhizome is
  eaten in S America. W Indies, S America.
  × *generalis* gen-e-*rah*-lis. Normal i.e. the
  commonly grown form.

Canna (continued)
  indica in-di-ka. Indian. Indian Shot.
  Tropical America.

Cannabis kă-na-bis Cannabaceae. The L.
and Gk. name. Annual herb.
  sativa sa-*tee*-va. Cultivated. Hemp. C Asia.

Canterbury Bells see *Campanula medium*
  Cup and Saucer see *C. medium*
  'Calycanthema'

Cantua kăn-tew-a Polemoniaceae. From the
Peruvian name. Tender evergreen shrub.
  buxifolia buks-i-*fo*-lee-a. *Buxus*-leaved.
  Peru.

Cape Asparagus see *Aponogeton distachyus*
Cape Cowslip see *Lachenalia*
Cape Gooseberry see *Physalis peruviana*
Cape Honeysuckle see *Tecomaria capensis*
Cape Ivy see *Senecio macroglossus*
Cape Jasmine see *Gardenia jasminoides*
Cape Leadwort see *Plumbago auriculata*
Cape Pondweed see *Aponogeton distachyus*
Caper Spurge see *Euphorbia lathyris*

Capsicum kăp-si-kum Solonaceae. From Gk.
kapto (to bite) referring to the hot taste.
Tender annuals.
  annuum ăn-ew-um. Annual. Sweet Pepper,
  Christmas Pepper. Cult.
  frutescens froo-*tes*-enz. Shrubby. Chilli
  Pepper. Tropical America.

Caragana kă-ra-*gah*-na Leguminosae. From
Caragan, the Mongolian name for *C.*
arborescens. Deciduous tree.
  arborescens ar-bo-*res*-enz. Tree-like. Pea
  Tree. Siberia, Mongolia.
  'Lorbergii' lor-*berg*-ee-ee. After
  Lorberg's nursery, Germany, where it
  was raised.

Caraway see *Carum carvi*

Cardamine kar-*dah*-mi-nee Cruciferae. From
the Gk. name of a related plant. Perennial
herbs.
  californica kăl-i-*for*-ni-ka. (= *Dentaria
  californica*). Of California. W United States.
  enneaphyllos en-ee-a-*fil*-los. (= *Dentaria
  enneaphyllos*). With nine leaves (leaflets).
  E Europe.
  heptaphylla hep-ta-*fil*-la. (= *Dentaria
  heptaphylla*). With seven leaves (leaflets).
  W and C Europe.
  laciniata la-kin-ee-*ah*-ta. Deeply cut (the

Cardamine (continued)
leaves). E United States.
lyrata li-*rah*-ta. Lyre-shaped (the leaves).
E Asia.
pentaphyllos pen-ta-*fil*-los. (= *Dentaria pentaphyllos*). With five leaves (leaflets).
W and C Europe.
pratensis prah-*tayn*-sis. Of meadows.
Ladies' Smock, Cuckoo Flower. Europe,
Asia, N America.

Cardamon see *Elettaria cardamomum*
Cardinal Flower see *Lobelia cardinalis,
Sinningia cardinalis*

Cardiocrinum kar-dee-ō-*kree*-num *Liliaceae*.
From Gk. *kardio* (heart) and *krinon* (a lily)
referring to the heart-shaped leaves. Bulbous
herbs.
cordatum kor-*dah*-tum. Heart-shaped (the
leaves). Japan.
giganteum gi-*gǎn*-tee-um. Very large.
Himalaya, Burma.

Cardiospermum kar-dee-ō-*sperm*-um
Sapindaceae. From Gk. *kardia* (heart) and
spermum (a seed) referring to the white, heart-
shaped spot on the seed. Annual climber.
halicacabum hǎ-li-*kah*-ka-bum. The L.
name for bladderwort (*Utricularia*). Love-
in-a-Puff, Balloon Vine. S United States,
Tropical America.

Cardoon see *Cynara cardunculus*

Carex kah-reks *Cyperaceae*. The L. name.
Perennial, grass-like herbs. Sedge.
morrowii mo-rō-ee-ee. After Morrow.
Japan.
pendula pen-dew-la. Pendulous (the flower
spikes). Europe, Asia, N Africa.
pilulifera pil-ew-*li*-fe-ra. Bearing little balls,
referring to the globular female
inflorescences. Europe.
pseudocyperus sood-ō-ki-*pe*-rus. False
Cyperus. Widely distributed.
riparia ree-*pah*-ree-a. Of river banks. Great
Pond Sedge. Europe, N Africa, W Asia.
scaposa ska-*pō*-sa. With a conspicuous
scape. S China.
sylvatica sil-*vǎ*-ti-ka. Of woods. Europe.

Carlina kar-*lee*-na. *Compositae*. A medieval
name. Biennial and perennial herbs. Carline
Thistle.
acanthifolia a-kǎnth-i-*fo*-lee-a. With leaves
like *Acanthus*. S Europe, W Asia.
acaulis a-*kaw*-lis. Without a stem. Europe.

Carmichaelia kar-mie-*keel*-ee-a *Leguminosae*.
After Capt. Dugald Carmichael (1772–1827),
Scottish army officer and plant collector. New
Zealand shrubs.
enysii e-*nis*-ee-ee. After John Davies Enys
(1837–1912), New Zealand magistrate and
naturalist who, with Kirk, collected the
type specimen.
petriei *pet*-ree-ee. After Petrie who collected
the type specimen in 1890, see *Coprosma
petriei*.

Carnation see *Dianthus caryophyllus*
Carolina Lupin see *Thermopsis caroliniana*

Carnegiea kar-*nee*-gee-a *Cactaceae*. After
Andrew Carnegie (1835–1919).
euphorbioides ew-for-bee-*oi*-deez. (=
Rooksbya euphorbioides). Like *Euphorbia*.
N Mexico.
gigantea gi-*gǎn*-tee-a. Very large. Saguaro.
SW United States.

Carpenteria kar-pen-*te*-ree-a
Philadelphaceae. After Prof. William M.
Carpenter, Louisiana physician. Evergreen,
semi-hardy shrub.
californica kǎl-i-*for*-ni-ka. Of California.

Carpinus kar-*peen*-us *Carpinaceae*. The L.
name. Deciduous trees. Hornbeam.
betulus bet-ew-lus. Like *Betula*. Common
Hornbeam. Europe, W Asia.
caroliniana kǎ-ro-lin-ee-*ah*-na. Of Carolina.
American Hornbeam. E N America.
japonica ja-*pon*-i-ka. Of Japan.
turczaninowii tur-chah-ni-*nov*-ee-ee. After
Nicolai Stepanovich Turczaninow
(1796–1864) who discovered it in 1831.
N China, Japan.

Carpobrotus kar-po-*brō*-tus *Aizoaceae*. From
Gk. *karpos* (a fruit) and *brotus* (edible)
referring to the edible fruit. Succulent sub-
shrub.
edulis e-*dew*-lis. Edible (the fruit).
Hottentot Fig. S Africa.

Carrot see *Daucus carota sativus*
    Wild see *D. carota*
Cartwheel Flower see *Heracleum
mantegazzianum*

Carum kǎ-rum *Umbelliferae*. From *karon* the
Gk. name. Annual or biennial herb.
carvi kar-vee. The L. name. Caraway.
W Asia.

Carya kă-ree-a *Juglandaceae*. From *karya* Gk. name of the walnut tree (*Juglans regia*). Deciduous trees. Hickory. EN America.
cordiformis kor-di-*form*-is. Heart-shaped (the nut). Bitternut Hickory.
glabra glă-bra. Glabrous (the shoots). Pignut Hickory.
ovata ō-*vah*-ta. Ovate (the leaflets). Shagbark Hickory.
tomentosa tō-men-*tō*-sa. Hairy (the young shoots). Mockernut Hickory.

Caryopteris kă-ree-*op*-te-ris *Verbenaceae*. From Gk. *karyon* (a nut) and *pteron* (a wing) referring to the winged fruits. Deciduous shrubs.
× *clandonensis* klăn-don-*en*-sis. *C. incana* × *C. mongolica*. From Clandon, Surrey where it was raised.
incana in-*kah*-na. Grey (the leaves and shoots). China, Japan.

Caryota kă-ree-*ō*-ta *Palmae*. From Gk. *karyota* (a date-shaped nut). Tender Palms. Fishtail Palm.
mitis mee-tis. Soft, not spiny. Burmese Fishtail Palm. SE Asia.
urens ew-renz. Stinging, referring to the stinging, needle-like crystals in the outer covering of the fruit. Wine Palm, Toddy Palm. Himalaya, SE Asia.

Cassia kă-see-a *Leguminosae*. Gk. name for a species of this or a related genus. Tender shrubs and herbs. Senna.
alata ah-*lah*-ta. Winged (the fruit). Tropics.
australis ow-*strah*-lis. Southern. Australia.
corymbosa ko-rim-*bō*-sa. With flowers in corymbs. S America.
marilandica mă-ri-*lănd*-i-ka. Of Maryland. American Senna. SE United States.

Cassinia ka-*seen*-ee-a *Compositae*. After Count Henri de Cassini (1781–1832), Italian botanist. Evergreen shrubs.
fulvida *ful*-vi-da. Slightly tawny (the undersides of the leaves). New Zealand.

Cassiope ka-*see*-o-pay *Ericaceae*. After Cassiope of Gk. mythology, the mother of Andromeda. Evergreen, heath-like shrubs.
fastigiata fă-stig-ee-*ah*-ta. Erect (the shoots). Himalaya.
lycopodioides lik-ō-pō-dee-*oi*-deez. Like *Lycopodium*. Japan, Alaska.
tetragona tet-ra-*gō*-na. Four-angled (the

Cassiope (continued)
shoots). Arctic N hemisphere.
wardii ward-ee-ee. After its discoverer Francis Kingdon Ward (1885–1958), plant collector, explorer and author who introduced numerous plants from China, Tibet and Burma. Himalaya.

Cast Iron Plant see *Aspidistra elatior*
Castor-oil Plant see *Ricinus communis*
  False see *Fatsia japonica*

Castanea ka-*stăn*-ee-a *Fagaceae*. The L. name from Castania, N Greece which was known for its trees. Deciduous tree.
sativa sa-*tee*-va. Cultivated. Sweet or Spanish Chestnut. S Europe, N Africa, W Asia.

Cat Thyme see *Teucrium marum*

Catalpa ka-*tăl*-pa *Bignoniaceae*. The N American Indian name. Deciduous trees.
bignonioides big-nō-nee-*oi*-deez. Like *Bignonia*. Indian Bean Tree. E United States.
× *erubescens* e-roo-*bes*-enz. *C. bignonioides* × *C. ovata*. Blushing.
  'Purpurea' pur-*pewr*-ree-a. Purple (the young shoots and leaves).
fargesii far-*geez*-ee-ee. After Farges who discovered it, see *Decaisnea fargesii*. China.
  duclouxii dew-*cloo*-ee-ee. After Monsignor Fr Ducloux who collected in China.
speciosa spe-kee-*ō*-sa. Showy. Western Catalpa. United States.

Catananche kă-ta-*năn*-kee. *Compositae*. The Gk. name. Perennial herb.
caerulea kie-*ru*-lee-a. Deep blue (the flowers). Cupid's Dart. S Europe.

Catchfly, German see *Lychnis viscaria*
  Nodding see *Silene pendula*

Catharanthus kă-tha-*răn*-thus *Apocynaceae*. From Gk. *katharos* (pure) and *anthus* (a flower). Tender perennial.
roseus ro-see-us. (= *Vinca rosea*). Rose-coloured (the flowers). Old Maid, Madagascar Periwinkle. Madagascar to India.

Cathedral Bells see *Cobaea scandens*
Cathedral Windows see *Calathea makoyana*
Catmint see *Nepeta cataria*

Catnip see *Nepeta cataria*
Cat's Foot see *Antennaria dioica*
Cat's Whiskers see *Tacca chantrieri*

Cattleya kăt-lee-a *Orchidaceae*. After William
Cattley (died 1832), horticultural patron.
  bicolor bi-ko-lor. Two-coloured (the
  flowers). Brazil.
  bowringiana bow-ring-gee-*ah*-na. After
  John Charles Bowring (1821–93), orchid
  grower. C America.
  citrina ki-*tree*-na. Lemon-yellow (the
  flowers). Mexico.
  dowiana dow-ee-*ah*-na. After Captain J. M.
  Dow of the American Packet Service.
  C America.
  intermedia in-ter-*me*-dee-a. Intermediate.
  Brazil.
  labiata lă-bee-*ah*-ta. With a lip. Brazil.
  loddigesii lod-ee-*jes*-ee-ee. After the
  Loddiges nursery of Hackney. Brazil.
  mossiae mos-ee-ie. After Mrs Moss of
  Otterspool who grew the type specimen.
  Venezuela.
  skinneri *skin*-a-ree. After Mr George Ure
  Skinner (1804–67) who collected orchids
  in South America. C America.
  trianae tree-*ah*-nie. After J. J. Triana
  (1834–90). Colombia.
  warscewiczii var-sha-*vich*-ee-ee. After
  Warscewicz, see *Alonsoa warscewiczii*.
  Colombia.

Cauliflower see *Brassica oleracea* Botrytis

Cautleya kawt-lee-a *Zingiberaceae*. After Sir
Proby Thomas Cautley (1802–71), a military
engineer who worked in India. Tender
rhizomatous perennials.
  gracilis gră-ki-lis. Graceful. Himalaya,
  W China.

Ceanothus kee-a-*nō*-thus *Rhamnaceae*. Gk.
name of a spiny shrub. Evergreen and semi-
evergreen shrubs.
  arboreus ar-*bo*-ree-us. Tree-like. Catalina
  Ceanothus. California (islands).
  'Burkwoodii' burk-*wud*-ee-ee. After
  Burkwood and Skipwith who raised it.
  dentatus den-*tah*-tus. Toothed (the leaves).
  California.
  gloriosus glō-ree-*ō*-sus. Glorious. California.
  papillosus pă-pil-*lō*-sus. With papillae,
  referring to the wart-like glands on the
  leaves. California.
    roweanus rō-ee-*ah*-nus. After E. Denys
    Rowe, Santa Barbara horticulturist

Ceanothus (continued)
  who discovered it.
  prostratus pros-*trah*-tus. Prostrate. W
  N America.
  thyrsiflorus thurs-i-*flō*-rus. With flowers in a
  thyrse (a type of inflorescence). California.
    repens ree-penz. Creeping.
  × *veitchianus* veech-ee-*ah*-nus. After
  Messrs Veitch, prominent 19th-century
  nurserymen. ? California.

Cedar see *Cedrus*
  Atlas see *C. atlantica*
Cedar of Goa see *Cupressus lusitanica*
Cedar of Lebanon see *Cedrus libani*

Cedrela ked-rel-a *Meliaceae*. Diminutive of
Cedrus q.v. from the similar wood.
Deciduous tree.
  sinensis si-*nen*-sis. Of China. Chinese Cedar
  N and W China.

Cedrus ke-drus *Pinaceae*. The L. name.
Evergreen conifers.
  atlantica ăt-*lăn*-ti-ka. Of the Atlas
  Mountains.
    Glauca *glow*-ka. Glaucous (the leaves).
  deodara dee-ō-*dah*-ra. From the N Indian
  name. Deodar. Himalaya.
  libani li-ba-nee. Of Mount Lebanon. Cedar
  of Lebanon. Lebanon, Turkey.

Celandine, Greater see *Chelidonium majus*
  Lesser see *Ranunculus ficaria*

Celastrus kel-ă-strus *Celastraceae*. From
kelastros the Gk. name for an evergreen tree.
Deciduous climbers.
  hypoleucus hi-po-*loo*-kus. White beneath
  (the leaves). China.
  orbiculatus or-bik-ew-*lah*-tus. Orbicular
  (the leaves). NE Asia.

Celeriac see *Apium graveolens rapaceum*
Celery see *Apium graveolens dulce*
  Wild see *A. graveolens*
Celery Pine see *Phyllocladus alpinus*

Celmisia kel-*mis*-ee-a *Compositae*. After
Celmisios of Gk. mythology. Evergreen
herbaceous perennials. New Zealand Daisy.
New Zealand.
  coriacea ko-ree-*ah*-kee-a. Leathery (the
  leaves).
  hieraciifolia hee-e-rah-kee-i-*fo*-lee-a.
  Hieracium-leaved.
  spectabilis spek-*tah*-bi-lis. Spectacular.

Celosia ke-*lō*-see-a *Amaranthaceae*. From Gk. *keleos* (burning) referring to the brilliantly coloured flowers. Tender annual.
 cristata kris-*tah*-ta. (= *C. argentea cristata*). Crested (the inflorescence). Cockscomb. Cult.
  'Plumosa' ploo-*mō*-sa. Feathery (the inflorescence).

Celsia arcturus see *Verbascum arcturus*
 cretica see *Verbascum creticum*

Celtis *kel*-tis *Ulmaceae*. Gk. name of a tree. Deciduous trees. Nettle Tree.
 australis ow-*strah*-lis. Southern. S Europe, W Asia.
 laevigata lie-vi-*gah*-ta. Smooth (the leaves). Sugarberry. S United States.
 occidentalis ok-ki-den-*tah*-lis. Western. Hackberry. N America.

Centaurea kent-*ow*-ree-a *Compositae*. From Gk. *kentaur* (a Centaur) which is said to have used it medicinally. Annual and perennial herbs.
 cineraria kin-e-*rah*-ree-a. (= *C. gymnocarpa*). From the resemblance to *Cineraria maritima* (= *Senecio bicolor cineraria*). W Italy, Sicily.
 cyanus see-*ah*-nus. Dark blue (the flowers). Cornflower. SE Europe, W Asia.
 dealbata dee-al-*bah*-ta. Whitened (the underside of the leaves). Caucasus.
 gymnocarpa see *C. cineraria*
 hypoleuca hi-pō-*loo*-ka. White beneath (the leaves). W Asia.
 macrocephala măk-rō-*kef*-a-la. With a large head. Caucasus.
 montana mon-*tah*-na. Of mountains. C Europe.
 moschata mos-*kah*-ta. Musk scented. Sweet Sultan. W Asia.
 pulcherrima pul-*ke*-ri-ma. Very pretty. Caucasus, W Asia.
 ragusina rah-goo-*see*-na. Of Dubrovnik (Ragusa). W Yugoslavia.
 ruthenica roo-*then*-i-ka. Of Ruthenia (SW Russia). Caucasus, S Russia, Romania.
 rutifolia roo-ti-*fo*-lee-a. *Ruta*-leaved. SE Europe.
 simplicicaulis sim-plik-ee-*kaw*-lis. With an unbranched stem. Armenia.

Centranthus ken-*trăn*-thus *Valerianaceae*. From Gk. *kentron* (a spur) and *anthos* (a flower) referring to the spurred flowers. Perennial herb. Red Valerian.

Centranthus (continued)
 ruber ru-ber. Red (the flowers). S Europe, N Africa.

Century Plant see *Agave americana*

Cephalaria kef-a-*lah*-ree-a *Dipsacaceae*. From Gk. *kephale* (a head), the flowers are borne in heads. Perennial herbs.
 alpina ăl-*peen*-a. (= *Scabiosa alpina*). Alpine. Alps.
 gigantea gi-*găn*-tee-a. (= *C. tatarica* hort.). Very large. Caucasus.

Cephalocereus kef-a-lō-*kay*-ree-us *Cactaceae*. From Gk. *kephale* (a head) and *Cereus* q.v., the flowers are borne from woolly heads.
 palmeri pahl-ma-ree. (= *Pilosocereus palmeri*). After E. Palmer (1831–1911), who collected in the S United States and Mexico. E Mexico.
 polylophus po-lee-lō-phus. (= *Neobuxbaumia polylopha*). Many crested, referring to the numerous ribs. E Mexico.
 senilis sen-*ee*-lis. Old, referring to the dense covering of long, white hairs. Old Man Cactus. C Mexico.

Cephalotaxus kef-a-lō-*tăks*-us *Cephalotaxaceae*. From Gk. *kephale* (a head) and *Taxus* q.v. from the resemblance to *Taxus*. Evergreen trees and shrubs. Plum Yew.
 fortunei for-*tewn*-ee-ee. After its introducer, Robert Fortune, see *Fortunella*. N China.
 harringtonii hă-ring-*ton*-ee-ee. After the Earl of Harrington. Cult.
  drupacea droo-*pah*-kee-a. With a fleshy fruit. Japan, Korea.

Cerastium ke-*ră*-stee-um *Caryophyllaceae*. From Gk. *keras* (a horn), referring to the shape of the seed capsule. Perennial herbs.
 alpinum ăl-*peen*-um. Alpine. N and alpine Europe.
 biebersteinii bee-ber-*stien*-ee-ee. After Friedrich August Marschall von Bieberstein (1768–1826), German botanist. Crimea.
 tomentosum tō-men-*tō*-sum. Hairy. Snow in Summer. Italy, Sicily.

Ceratophyllum ke-ra-tō-*fil*-lum *Ceratophyllaceae*. From Gk. *keras* (a horn) and *phyllon* (a leaf) from the resemblance of the leaves to antlers. Aquatic herbs. Hornwort.
 demersum day-*mer*-sum. Growing under

Ceratophyllum (continued)
water. Europe, N America.
submersum sub-*mer*-sum. Growing under
water. Europe, N Africa, Asia.

Ceratopteris ke-ra-*top*-te-ris *Parkeriaceae*.
From Gk. *keras* (a horn) and *pteris* (a fern)
from the horn-like appearance. Aquarium
fern.
thalictroides tha-lik-*troi*-deez. Like
Thalictrum (the foliage). Water Sprite. SE
Asia.

Ceratostigma ke-ra-tō-*stig*-ma
Plumbaginaceae. From Gk. *keras* (a horn)
and *stigma* referring to horn-like growths on
the stigma. Shrubs and herbs.
griffithii gri-*fith*-ee-ee. After William
Griffith (1810–45), surgeon and botanist in
India and SE Asia. E Himalaya, W. China.
plumbaginoides plum-bah-gi-*noi*-deez. Like
Plumbago. China.
willmottianum wil-mot-ee-*ah*-num. After
Miss Ellen Ann Willmott (1858–1934), a
noted gardener who introduced many
plants. She raised plants from the original
introduction by Ernest Wilson, in her
garden at Warley Place, Essex.

Cercidiphyllum ker-ki-di-*fil*-lum
Cercidiphyllaceae. From *Cercis* q.v. and Gk.
phyllon (a leaf), the leaves resemble those of
Cercis. Deciduous tree.
japonicum ja-*pon*-i-kum. Of Japan.

Cercis *ker*-kis *Leguminosae*. From *kerkis* Gk.
name of another tree. Deciduous trees.
canadensis kǎn-a-*den*-sis. Of Canada or NE
North America. Redbud. E and C United
States.
siliquastrum si-li-*kwǎ*-strum. Like a siliqua,
from the resemblance of the pod to a fruit
found in *Cruciferae*. Judas Tree. E
Mediterranean region, W Asia.

Cereus *kay*-ree-us *Cactaceae*. From L. *cereus*
(a wax taper) referring to the shape.
aethiops ie-thee-ops. (= *C. caerulescens*). Of
unusual appearance. N Argentina.
jamacaru hǎ-ma-*kah*-roo. From the native
name. Brazil.
peruvianus pe-roo-vee-*ah*-nus. Of Peru.
Peruvian Apple Cactus. SE South
America.

Ceropegia kay-rō-*pee*-gee-a *Asclepiadaceae*.
From Gk. *keros* (wax) and *pege* (a fountain)

Ceropegia (continued)
referring to the waxy flowers. Tender
succulents. S Africa.
barklyi bark-lee-ee. After Sir Henry Barkly
(1815–98), Governor of the Cape Province
and patron of botanists and plant collectors
in the colonies.
woodii wud-ee-ee. After John Medley Wood
(1827–1915), farmer and botanist in South
Africa. Rosary Vine, Hearts on a String.

Cestrum *kes*-trum *Solanaceae*. Gk. name of a
plant. Tender and semi-hardy shrubs.
aurantiacum ow-rǎn-tee-*ah*-kum. Orange
(the flowers). Guatemala.
elegans ay-le-gahnz. Elegant. Mexico.
fasciculatum fǎs-kik-ew-*lah*-tum. Clustered
(the flowers). Mexico.
'Newellii' new-*el*-ee-ee. After Mr Newell,
gardener at Ryston Hall, who raised it.
parqui par-kee. The Chilean name. Chile.

Ceterach *ke*-te-rahk *Aspleniaceae*. The
Arabic name. Fern.
officinarum o-fi-ki-*nah*-rum. Sold as a herb.
Rusty Back Fern. Europe and N Africa to
the Himalaya.

Chaenomeles kie-nō-*may*-leez *Rosaceae*.
From Gk. *chaina* (to gape) and *melon* (an
apple) referring to the belief that the fruit was
split. Deciduous, spiny shrubs. Ornamental
Quince.
cathayensis kǎ-thay-*en*-sis. Of China.
japonica ja-*pon*-i-ka. (= *Cydonia maulei*,
Cydonia japonica). Of Japan.
speciosa spe-kee-ō-sa. (= *Cydonia lagenaria*,
Cydonia speciosa). Showy. China.
× *superba* soo-*perb*-a. *C. japonica* × *C.
speciosa*. Superb.

Chaenorhinum kie-nō-*reen*-um
Scrophulariaceae. From Gk. *chaino* (to gape)
and *rhis* (a snout) referring to the open mouth
of the corolla tube. Perennial herb.
origanifolium o-ree-gahn-i-*fo*-lee-um. (=
Linaria origanifolia). *Origanum*-leaved. SW
Europe.

Chain Plant see *Tradescantia navicularis*

Chamaecereus kǎ-mie-*kay*-ree-us *Cactaceae*.
From Gk. *chamai* (on the ground) and *Cereus*
q.v. referring to the procumbent habit.
sylvestrii sil-*ves*-tree-ee. After Dr Philipo
Sylvestri who discovered it. Peanut
Cactus. Argentina.

Chamaecyparis kă-mie-*kew*-pa-ris
Cupressaceae. From Gk. *chamai* (low
growing) and *kuparissos* (cypress). Evergreen
conifers.
    lawsoniana law-son-ee-*ah*-na. After Charles
Lawson (1794–1873), Edinburgh
nurseryman who raised it from the original
introduction in 1854. Lawson Cypress.
W N America.
      'Ellwoodii' el-*wud*-ee-ee. After Ellwood,
a gardener at Swanmore Park where it
was raised.
      'Fletcheri' *flech*-a-ree. After the Fletcher
Bros. nursery who distributed it.
      'Gimbornii' gim-*born*-ee-ee. After Von
Gimborn, on whose estate in Holland it
originated.
      'Pottenii' po-*ten*-ee-ee. After the Potten
nursery, Cranbrook, where it
originated.
      'Stewartii' stew-*art*-ee-ee. After the
raisers, D. Stewart and son,
Bournemouth.
      'Wisselii' vis-*el*-ee-ee. After the raiser,
F. van der Wissel, Dutch nurseryman.
    nootkatensis nut-ka-*ten*-sis. Of Nootka
Sound, British Columbia. Nootka
Cypress. W N America.
    obtusa ob-*tew*-sa. Blunt (the leaves). Hinoki
Cypress. Japan.
      'Crippsii' *krips*-ee-ee. After the raisers,
the Cripps nursery.
    pisifera pee-*si*-fe-ra. Pea-bearing, referring
to the small cones. Sawara Cypress. Japan.
      'Filifera' fee-*li*-fe-ra. Thread-bearing,
referring to the slender shoots.
      'Plumosa' ploo-*mō*-sa. Feathery (the
foliage).
      'Squarrosa' skwah-*rō*-sa. With the leaves
spreading at right angles.
    thyoides thoo-*oi*-deez. Like *Thuja.* E N
America.
      'Andelyensis' ăn-da-lee-*en*-sis. From
Andelys, France.
      'Ericoides' e-ri-*koi*-deez. *Erica*-like (the
foliage).

Chamaedorea kă-mie-*do*-ree-a *Palmae.*
From Gk. *chamai* (on the ground) *dorea* (a
gift). Unlike many palms the fruits are borne
within reach. Tender palms.
    elegans ay-le-gahnz. (= *Collinia elegans,
Neanthe bella*). Elegant. Parlour Palm.
Mexico, Guatemala.
    erumpens ay-*rum*-penz. Breaking through,
the inflorescence ruptures the sheath as it
emerges. Bamboo Palm. C America.

Chamaemelum kă-mie-*may*-lum *Compositae.*
From Gk. *chamai* (on the ground) and *melon*
(an apple) referring to the apple-like scent
and the low habit. Perennial herb.
    nobile nō-bi-lee. (= *Anthemis nobilis*).
Notable. Chamomile. W Europe,
N Africa.

Chamaerops ka-*mie*-rops *Palmae.* From Gk.
chamai (low growing) and *rhops* (a bush)
referring to the dwarf habit compared to most
other palms. Semi-hardy palm.
    excelsa hort. see *Trachycarpus fortunei*
    humilis hum-i-lis. Low growing. Dwarf Fan
Palm. Mediterranean region.

Chamomile see *Chamaemelum nobile*
    Dyer's see *Anthemis tinctoria*
    Yellow see *Anthemis tinctoria*
Chandelier Plant see *Kalanchoe tubiflora*
Chaste Tree see *Vitex agnus-castus*
Chatham Island Forget-me-not see
Myostidium hortensia

Cheilanthes kay-*lăn*-theez *Sinopteridaceae.*
From Gk. *cheilos* (a lip) and *anthos* (a flower),
the edge of the pinnules forms a lip which
covers the sporangia. Lip Fern.
    distans dis-tănz. Widely spaced (the
pinnae). Woolly Rock Fern. New Zealand,
Australia.
    fragrans frah-granz. Fragrant (the fronds).
S Europe.
    lanosa lah-*nō*-sa. Woolly (the undersides of
the fronds). Hairy Lip Fern. E United
States.

Cheiranthus kay-*rănth*-us *Cruciferae.* From
Gk. *cheir* (hand) and *anthos* (a flower)
referring to their use in bouquets. Biennial
and perennial herbs.
    allionii ah-lee-*ōn*-ee-ee. After Carlo Allioni
(1705–1804), Italian botanist. Siberian
Wallflower. The identity and origin of
plants grown under this name is uncertain.
    cheiri kay-ree. See above. Wallflower.
S Europe.
    mutabilis mew-*tah*-bi-lis. Changeable (the
flower colour). Madeira, Canary Islands.

Chelidonium ke-li-*dō*-nee-um *Papaveraceae.*
From Gk. *chelidon* (a swallow), it is said to
start flowering as the swallow arrives.
    majus mah-yus. Larger. Greater Celandine.
Europe, W Asia.

Chelone ke-lō-nay *Scrophulariaceae*. From Gk. *chelone* (a turtle), the corolla is shaped like a turtle's head. Perennial herbs. Turtle Head.
  barbata see *Penstemon barbatus*
  glabra glǎ-bra. Glabrous. E United States.
  lyonii lie-*on*-ee-ee. After its discoverer John Lyon, see *Lyonia*. SE United States.
  obliqua o-*blee*-kwa. Oblique. E United States.

Chenille Plant see *Acalypha hispida*

Chenopodium kay-nō-*pō*-dee-um *Chenopodiaceae*. From Gk. *chen* (a goose) and *podion* (a foot) referring to the shape of the leaves. Perennial herb.
  bonus-henricus bo-nus-hen-*ree*-kus. Good Henry, from 16th century German Güter Heinrich, a goblin with knowledge of healing plants. Good King Henry. Europe.

Cherry, Bird see *Prunus padus*
  Fuji see *P. incisa*
  Sour see *P. cerasus*
  Yoshino see *P.* × *yedoensis*
Cherry Laurel see *Prunus laurocerasus*
Cherry Pie see *Heliotropium arborescens*
Cherry Plum see *Prunus cerasifera*
Chervil see *Anthriscus cerefolium*
Chestnut, Sweet or Spanish see *Castanea sativa*
Chestnut Vine see *Tetrastigma voinierianum*

Chiastophyllum kee-ǎ-stō-*fil*-lum *Crassulaceae*. From Gk. *chiastos* (arranged cross-wise) and *phyllon* (a leaf) referring to the opposite leaves. Succulent perennial.
  oppositifolium o-po-si-ti-*fo*-lee-um. (= *Cotyledon simplicifolia*). With opposite leaves. Caucasus.

Chicory see *Cichorium intybus*
Chile Pine see *Araucaria araucana*
Chilean Bellflower see *Lapageria rosea*, *Nolana*
Chilean Crocus see *Tecophilaea cyanocrocus*
Chilean Hazel see *Gevuina avellana*
Chilean Jasmine see *Mandevilla suaveolens*

Chimonanthus kee-mon-*ǎnth*-us *Calycanthaceae*. From Gk. *cheima* (winter) and *anthos* (a flower) referring to its winter-flowering habit. Deciduous shrub.
  praecox prie-koks. Early (flowering). Winter Sweet. China.

China Aster see *Callistephus chinensis*
Chincherinchee see *Ornithogalum thyrsoides*
Chinese Cedar see *Cedrela sinensis*
Chinese Evergreen see *Aglaonema modesta*
Chinese Foxglove see *Rehmannia elata*
Chinese Gooseberry see *Actinidia chinensis*
Chinese-hat Plant see *Holmskioldia sanguinea*
Chinese Houses see *Collinsia heterophylla*
Chinese Jade see *Crassula arborescens*
Chinese Lantern see *Physalis alkekengi*, *Sandersonia aurantiaca*

Chionanthus kee-on-*ǎnth*-us *Oleaceae*. From Gk. *chion* (snow) and *anthos* (a flower) referring to the white flowers. Deciduous shrubs or trees.
  retusus re-*tew*-sus. Notched at the tip (the leaves). Chinese Fringe Tree. China.
  virginicus vir-*jin*-i-kus. Of Virginia. E United States.

Chionodoxa kee-on-o-*doks*-a *Liliaceae*. From Gk. *chion* (snow) and *doxa* (glory). The flowers are often borne with snow on the ground. Bulbous perennials.
  albescens ǎl-*bes*-enz. (= *C. nana*). White (the flowers). Crete.
  cretica kray-ti-ka. Of Crete.
  luciliae loo-sil-ee-ie. (= *C. gigantea*). After Lucile Boissier (1822–49) whose husband (see *Colchicum boissieri*) discovered and named it. W Turkey.
  sardensis sar-*den*-sis. Of Sart, W Turkey.

Chives see *Allium schoenoprasum*
  Chinese see *A. tuberosum*

Chlidanthus klid-*ǎnth*-us *Amaryllidaceae*. From Gk. *chlide* (luxury) and *anthos* (a flower). Bulbous perennial.
  fragrans frah-granz. Fragrant. Andes.

Chlorophytum klō-ro-fi-tum *Liliaceae*. From Gk. *chloros* (green) and *phyton* (a plant). Tender herbs. S Africa.
  capense ka-*pen*-see. Of the Cape of Good Hope.
  comosum ko-*mō*-sum. Tufted (the foliage). Spider Plant.
    'Vittatum' vi-*tah*-tum. Striped (the leaves).

Choisya shwūz-ee-a but usually *choy*-zee-a *Rutaceae*. After Jacques Denis Choisy (1799–1859), Swiss botanist. Evergreen shrub.

Choisya (continued)
ternata ter-*nah*-ta. In threes (the leaflets).
Mexican Orange Blossom. Mexico.

Chokeberry see *Aronia*
  Black see *A. melanocarpa*
  Red see *A. arbutifolia*
Christmas Box see *Sarcococca*
Christmas Cheer see *Sedum* × *rubrotinctum*
Christmas Jewels see *Aechmea racineae*
Christmas Pride see *Ruellia macrantha*
Christmas Rose see *Helleborus niger*

Chrysalidocarpus kri-sah-li-dō-*kar*-pus
Palmae. From Gk. *chrysallis* (a chrysalis) and
karpos (a fruit), the fruit is said to resemble a
chrysalis. Tender palm.
  lutescens loo-*tes*-enz. Yellowish, the leaf
  sheaths and stalks. Yellow palm.
  Madagascar.

Chrysanthemum kris-*änth*-e-mum
Compositae. From Gk. *chrysos* (gold) and
anthos (a flower). Annual and perennial herbs.
  alpinum äl-*peen*-um. Alpine. Alps.
  carinatum kä-ri-*nah*-tum. Keeled (the
  involucral bracts). Painted Daisy.
  Morocco.
  coccineum kok-*kin*-ee-um. Scarlet (the
  flowers). Pyrethrum. W Asia.
  coronarium ko-rō-*nah*-ree-um. Used in
  garlands. Crown Daisy. Mediterranean
  region.
  corymbosum see *Tanacetum corymbosum*
  densum see *Tanacetum densum*
  frutescens froo-*tes*-enz. Shrubby. Paris
  Daisy, White Marguerite. Canary Islands.
  haradjanii see *Tanacetum haradjanii*
  hosmariense hos-mah-ree-*en*-see. Of Beni
  Hosmar, near Tetuán, Morocco.
  indicum in-di-kum. Indian. A parent of the
  florist's chrysanthemum. China, Japan.
  maximum hort. see *C.* × *superbum*
  × *morifolium* mo-ri-*fo*-lee-um. *Morus*-
  leaved. Florist's chrysanthemum.
  multicaule mul-tee-*kaw*-lee. Many-
  stemmed. Algeria.
  parthenium see *Tanacetum parthenium*
  segetum se-ge-tum. Of cornfields. Corn
  Marigold. Europe, W Asia.
  × *superbum* soo-*perb*-um. *C. lacustre* × *C.
  maximum.* (= *C. maximum* hort.). Superb.
  Shasta Daisy.
  vulgare see *Tanacetum vulgare*
  weyrichii way-*rik*-ee-ee. After Dr Weyrich,
  a Russian naval surgeon. Japan.

Chrysogonum kris-*o*-go-num *Compositae*.
From Gk. *chrysos* (golden) and *gonu* (a knee)
referring to the yellow flowers and jointed
stem. Herbaceous perennial.
  virginianum vir-jin-ee-*ah*-num. Of Virginia.
  SE United States.

Cicerbita ki-*ker*-bi-ta *Compositae*. The Italian
name for a sow-thistle. Herbaceous
perennials.
  alpina äl-*peen*-a. (= *Lactuca alpina*).
  Alpine. Europe.
  plumieri ploo-mee-*e*-ree. (= *Lactuca
  plumieri*). After Charles Plumier, see
  Plumeria. C Europe.

Cichorium ki-ko-ree-um *Compositae*.
Derived from the Arabic name. Annual,
biennial or perennial herbs.
  endivia en-*di*-vee-a. See below. Endive.
  India.
  intybus in-tew-bus. From *intubus* the L.
  name. Both this and the above were
  derived from Egyptian *tybi* (January) the
  month when it was eaten. Chicory.
  Europe, N Africa, W Asia.

Cider Gum see *Eucalyptus gunnii*
Cigar Plant see *Cuphea ignea*

Cimicifuga kee-mi-ki-*few*-ga *Ranunculaceae*.
From L. *cimex* (a bug) and *fugo* (to repel),
C. foetida has been used as an insect repellent.
Perennial herbs. Bugbane.
  americana a-me-ri-*kah*-na. American. E
  United States.
  dahurica da-*hewr*-ri-ka. Of Dahuria, SE
  Siberia. NE Asia.
  foetida foy-ti-da. Unpleasantly scented. SE
  Europe, N Asia.
  japonica ja-*pon*-i-ka. Of Japan.
  racemosa rä-kay-*mō*-sa. With flowers in
  racemes. Black Snakeroot. EN America.

Cinderella Slippers see *Sinningia regina*
Cineraria cruenta see *Senecio cruentus*
  × **hybrida** see *Senecio* × *hybridus*
  maritima see *Senecio bicolor cineraria*

Cionura kee-on-*ewr*-ra *Asclepiadaceae*. From
Gk. *kion* (a column) and *oura* (a tail)
presumably referring to the stigma.
Deciduous climber.
  erecta e-*rek*-ta. (= *Marsdenia erecta*). Erect.
  SE Europe, W Asia.

Cissus *kis*-us *Vitaceae*. From Gk. *kissos* (ivy). Tender climbers and succulents.

antarctica ăn-*tark*-ti-ka. Of Antarctic regions. Kangaroo Vine. E Australia.

bainesii baynz-ee-ee. After Thomas Baines (1820–75), artist and explorer in S Africa. SW Africa.

capensis see *Rhoicissus capensis*

discolor dis-ko-lor. Two-coloured (the leaves). Rex-begonia Vine. Indonesia.

quadrangularis kwod-rang-gew-*lah*-ris. Four-angled (the stems). S Africa.

rhombifolia rom-bi-*fo*-lee-a. (= *Rhoicissus rhomboidea* hort.). With diamond-shaped leaves. Grape Ivy. C and S America.

sicyoides si-kee-*oi*-deez. Like *Sicyos* (*Cucurbitaceae*). Princess Vine. Tropical America.

striata stree-*ah*-ta. Striped (the stems). Miniature Grape Ivy. Chile, S Brazil.

Cistus *kis*-tus *Cistaceae*. From the Gk. name. Evergreen shrubs. Rock Rose.

× *aguilari* ă-gwi-*lah*-ree. *C. ladanifer* × *C. populifolius*. Of Aguilar, Spain. Spain, Portugal, Morocco.

'Maculatus' măk-ew-*lah*-tus. Spotted (the petals).

albidus ăl-bi-dus. Whitish (the leaves and shoots). SW Europe, N Africa.

× *canescens* kah-*nes*-enz. *C. albidus* × *C. creticus*. Grey-hairy (the leaves and shoots). Algeria.

× *corbariensis* kor-bah-ree-*en*-sis. (*C. populifolius* × *C. salviifolius*). From Corbières, S France.

crispus kris-pus. Finely wavy (the leaves). SW Europe, N Africa.

× *cyprius* kip-ree-us. *C. ladanifer* × *C. laurifolius*. Of Cyprus. W Mediterranean region.

ladanifer la-*dah*-ni-fer. Bearing ladanum (a gum resin used in perfumery). S Europe, N Africa.

laurifolius low-ri-*fo*-lee-us. *Laurus*-leaved. Mediterranean region, SW Europe.

× *lusitanicus* loo-si-*tah*-ni-kus. *C. hirsutus* × *C. ladanifer*. From Portugal. 'Decumbens' day-*kum*-benz. Prostrate.

palhinhae pa-*leen*-ie. After Ruy Telles Palhinha (1871–1957), Portuguese botanist. Algarve.

populifolius pō-pul-i-*fol*-ee-us. *Populus*-leaved. SW Europe.

lasiocalyx lă-see-ō-*kă*-liks. With a woolly calyx.

× *pulverulentus* pul-ve-roo-*len*-tus. *C. albidus* × *C. crispus*. Dusted.

Cistus (continued)
W Mediterranean region.

× *purpureus* pur-*pewr*-ree-us. *C. creticus* × *C. ladanifer*. Purple (the flowers).

× *skanbergii* skăn-*berg*-ee-ee. *C. monspeliensis* × *C. parviflorus*. After Skanberg. Greece.

Citrange see × *Citroncirus webberi*

× *Citrofortunella* kit-rō-for-tew-*nel*-a *Rutaceae*. Intergeneric hybrid, from the names of the parents. *Citrus* × *Fortunella*. Tender evergreen shrub or tree.

mitis mee-tis. *Citrus reticulata* × *Fortunella* sp. (= *Citrus mitis*). Not spiny. Calamondin.

Citron see *Citrus medica*

× *Citroncirus* kit-ron-si-rus *Rutaceae*. Intergeneric hybrid, from the names of the parents. *Citrus* × *Poncirus*. Tender evergreen tree or shrub.

webberi web-a-ree. *Citrus sinensis* × *Poncirus trifoliata*. After H. J. Webber. Citrange.

Citrullus kit-*rul*-us *Cucurbitaceae*. From *Citrus* q.v., referring to the fruit. Tender, annual herb.

lanatus lah-*nah*-tus. Woolly. Water Melon. Tropical and S Africa.

Citrus *kit*-rus *Rutaceae*. The L. name for citron (*C. medica*). Tender, evergreen trees.

aurantiifolia ow-răn-tee-i-*fo*-lee-a. With leaves like *C. aurantium*. Lime. SE Asia.

aurantium ow-răn-tee-um. Orange (the fruit). Seville Orange, Bitter Orange. S Vietnam.

limon lee-mon. The L. name. Lemon. SE Asia.

× *limonia* lee-mō-nee-a. *C. limon* × *C. reticulata*. The L. name for the lemon tree. Mandarin Lime.

maxima mahk-si-ma. Larger (the fruit). Shaddock, Pummelo. Malay Peninsula. Polynesia.

medica med-i-ka. Used in medicine. Citron. India.

mitis see × *Citrofortunella mitis*

× *paradisi* pă-ra-*dee*-see. *C. maxima* × *C. sinensis*. Of paradise. Grapefruit.

reticulata ray-tik-ew-*lah*-ta. Net-veined. Mandarin Orange, Satsuma, Tangerine. SE Asia.

sinensis si-*nen*-sis. Of China. Sweet Orange. SE Asia.

Cladanthus kla-*dănth*-us *Compositae*. From Gk. *klados* (a branch) and *anthos* (a flower), the flower heads are borne at the ends of the shoots. Annual herb.

    arabicus a-*ră*-bi-kus. (= *Anthemis arabica*). Arabian. S Spain, N Africa.

Cladrastis kla-*drăs*-tis *Leguminosae*. From Gk. *klados* (a branch) and *thraustos* (fragile) referring to the brittle shoots. Deciduous trees.

    lutea loo-tee-a. Yellow (the wood). Yellow Wood. SE United States.

    sinensis si-*nen*-sis. Of China.

Clarkia *klark*-ee-a *Onagraceae*. (= *Godetia*). After Captain William Clark (1770–1838). Annual herbs.

    amoena a-*moy*-na. Pleasant. Satin Flower. California.

    concinna kon-*kin*-a. Elegant. Red Ribbons. California.

    pulchella pul-*kel*-la. Pretty. WN America.

    unguiculata un-gwik-ew-*lah*-ta. (= *C. elegans*). Clawed (the petals). California.

Clary see *Salvia sclarea*

Cleistocactus klay-stō-*kăk*-tus *Cactaceae*. From Gk. *kleistos* (closed) and *Cactus* q.v., the flowers are tubular and nearly closed at the mouth.

    baumannii bow-*mahn*-ee-ee. After Baumann, a cactus grower. S America.

    jujuyensis hoo-hoo-ee-*en*-sis. Of Jujuy. N Argentina.

    smaragdiflorus sma-răg-di-*flō*-rus. With emerald flowers, the inner perianth segments are green. Firecracker Cactus. N Argentina.

    straussii strows-ee-ee. After Strauss. Silver Torch. Bolivia, Argentina.

    tominensis tō-mi-*nen*-sis. Of Tomina. Bolivia.

Clematis *klem*-a-tis *Ranunculaceae*. Gk. name for a climbing plant. Herbaceous perennials and climbers.

    alpina ăl-*peen*-a. Alpine. Europe, N Asia.

    armandii ar-*mond*-ee-ee. After Armand David, see *Davidia*. C and W China.

    chrysocoma kris-o-ko-ma. Golden-haired (the young growths). China.

      sericea say-*rik*-ee-a. Silky hairy.

    cirrhosa ki-*rō*-sa. With tendrils, the leaf stalks act as tendrils. Mediterranean region.

Clematis (continued)

    balearica bă-lee-*ah*-ri-ka. Of the Balearic Islands. Balearic Islands, Corsica.

    × *eriostemon* e-ree-ō-*stay*-mon. *C*.

    integrifolia × *C. viticella*. With woolly stamens.

    'Hendersonii' hen-der-*son*-ee-ee. After Henderson who raised it about 1830.

    flammula flăm-ew-la. An old name for this plant from L. *flammula* (a little flame or small banner). S Europe.

    florida flō-ri-da. Flowering. China.

    'Sieboldii' see-*bōld*-ee-ee. After Siebold, from whose nursery it was introduced in 1836, see *Acanthopanax sieboldii*.

    heracleifolia he-ra-klee-i-*fo*-lee-a.

    Heracleum-leaved. C and N China.

      davidiana dă-vid-ee-*ah*-na. After David who introduced it to France in 1863, see *Davidia*.

    integrifolia in-teg-ri-*fo*-lee-a. With entire leaves. E Europe to C Asia.

    × *jackmanii* jăk-*măn*-ee-ee. *C. lanuginosa* × *C. viticella*. After Jackman's nursery where it was raised in 1860.

    × *jouiniana* zhoo-ăn-ee-*ah*-na. *C. heracleifolia davidiana* × *C. vitalba*. After E. Jouin, manager of the Simon-Louis nursery, Metz, France.

    macropetala măk-rō-*pe*-ta-la. With large petals, referring to the large, petal-like staminodes which distinguish it from *C. alpina*. China, Siberia.

    montana mon-*tah*-na. Of mountains. Himalaya, China.

      rubens ru-benz. Red (the flowers).

      wilsonii wil-*son*-ee-ee. After Ernest Wilson who introduced it, see *Magnolia wilsonii*.

    orientalis o-ree-en-*tah*-lis. Eastern. Caucasus to N China.

    recta rek-ta. Erect. S Europe.

    rehderiana ray-da-ree-*ah*-na. After Rehder, see *Rehderodendron*. Nepal to SW China.

    serratifolia se-rah-ti-*fo*-lee-a. With toothed leaves (leaflets). Korea.

    tangutica tăn-*gew*-ti-ka. Of Gansu (previously Kansu). NW China.

    texensis teks-*en*-sis. Of Texas.

    vitalba vee-*tăl*-ba. Literally, white vine. Old Man's Beard, Traveller's Joy. Europe.

    viticella vee-ti-*kel*-la. Diminutive of *Vitis*. S Europe.

Cleome klay-ō-mee *Capparaceae*. Derivation uncertain, possibly from Gk. *kleos* (glory). Tender annual herb.

Cleome (continued)
hassleriana has-la-ree-*ah*-na. (= *C. spinosa*
hort.). After Emile Hassler. Spider
Flower. S America.

Clerodendrum kle-rō-*den*-drum
Verbenaceae. From Gk. *kleros* (chance) and
dendron (a tree) referring to the variable
medicinal properties. Hardy and tender,
trees, shrubs and climbers.
  bungei bung-gee-ee. After Alexander von
  Bunge (1803–90), Russian botanist. China.
  speciosissimum spe-kee-ō-*sis*-i-mum. Most
  showy. Glory Bower. Java.
  thomsoniae tom-*son*-ee-ie. After the wife of
  the Rev. W. C. Thomson, who was in
  Africa 1849–65. W Tropical Africa.
  trichotomum tri-*ko*-to-mum. Branching into
  three. Japan, China.
    fargesii far-*geez*-ee-ee. After Farges who
    introduced it to France, see *Decaisnea
    fargesii*. China.

Clethra *kleth*-ra *Clethraceae.* From Gk.
klethra (alder). Deciduous trees and shrubs.
  alnifolia ăl-ni-*fo*-lee-a. *Alnus*-leaved. Sweet
  Pepper Bush. E N America.
  arborea ar-*bo*-ree-a. Tree-like. Lily of the
  Valley Tree. Madeira.
  barbinervis bar-bi-*ner*-vis. With bearded
  veins. Japan.
  delavayi del-a-*vay*-ee. After Delavay who
  discovered it in 1884, see *Abies delavayi*.
  China.

Cleyera *klay*-a-ra *Theaceae.* After Andreas
Cleyer, 17th century doctor with the East
India Co. Evergreen shrub.
  japonica ja-*pon*-i-ka. Of Japan. Himalaya,
  E Asia.
  'Tricolor' *tri*-ko-lor. (= *C. fortunei*).
  Three-coloured (the leaves).

Clianthus klee-*ănth*-us *Leguminosae.* From
Gk. *kleos* (glory) and *anthos* (a flower).
Tender sub-shrubs.
  formosus for-*mō*-sus. (= *C. speciosus*).
  Beautiful. Sturt's Desert Pea.
  W Australia.
  puniceus pew-*ni*-kee-us. Reddish-purple
  (the flowers). Glory Pea. New Zealand.

Cliff Brake, Green see *Pellaea viridis*
  Purple see *P. atropurpurea*

Clivia *klie*-vee-a *Amaryllidaceae.* After Lady
Charlotte Florentina Clive, Duchess of

Clivia (continued)
Northumberland (died 1868), granddaughter
of Robert Clive. Tender evergreen perennial
herbs. S Africa.
  × *cyrtanthiflora* kur-tănth-i-*flō*-ra. *C.
  miniata* × *C. nobilis*. Cyrtanthus-flowered.
  miniata min-ee-*ah*-ta. Cinnabar-red (the
  flowers). Kaffir Lily.
  nobilis nō-bi-lis. Notable.

Clock Vine see *Thunbergia grandiflora*
Clover, White see *Trifolium repens*

Cobaea kō-*bie*-a *Polemoniaceae.* After
Bernardo Cobo (1572–1659), Spanish
missionary in Mexico and Peru. Annual
climber.
  scandens skăn-denz. Climbing. Cathedral
  Bells. Mexico.

Cobnut see *Corylus avellana*

Cochlioda kok-lee-ō-da *Orchidaceae.* From
Gk. *kochlos* (a snail-shell) referring to the
shell-like calluses on the lip of the type
species. Greenhouse orchids.
  densiflora dens-i-*flō*-ra. Densely flowered.
  Peru, Bolivia.
  rosea ro-see-a. Rose-coloured (the flowers).
  Peru.

Coconut see *Cocos nucifera*
Cockscomb see *Celosia cristata*
Cockspur Thorn see *Crataegus crus-galli*

Cocos *kō*-kos *Palmae.* From Portuguese *coco*
(a grinning face) from the appearance of the
fruit. Tender palms.
  nucifera new-*ki*-fe-ra. Nut-bearing.
  Coconut. Tropics.
  weddelliana see *Microcoelum weddellianum*

Codiaeum kō-dee-*ie*-um *Euphorbiaceae.*
From the native name. Tender evergreen
shrub.
  variegatum vă-ree-a-*gah*-tum. Variegated.
    pictum pik-tum. Painted. Croton. SE
    Asia.

Codonopsis kō-dōn-*op*-sis *Campanulaceae.*
From Gk. *kodon* (a bell) and -*opsis* indicating
resemblance, referring to the bell-shaped
corolla. Annual and perennial climbers.
  clematidea klem-a-*tid*-ee-a. Like *Clematis*.
  W Asia.
  convolvulacea kon-vol-vew-*lah*-kee-a. (= *C.
  vinciflora*). Like *Convolvulus*. Himalaya,
  W China.

Codonopsis (continued)
ovata ō-*vah*-ta. Ovate (the leaves).
Himalaya, S China.

Coelogyne koy-*lo*-gin-ee but commonly see-ō-*gie*-nee *Orchidaceae*. From Gk. *koilos* (hollow) and *gyne* (female) referring to the hollowed stigma. Greenhouse orchids.
asperata ă-spe-*rah*-ta. Roughened. SE Asia.
barbata bar-*bah*-ta. Bearded (the central lobe of the lip). Himalaya.
Burfordiense bur-ford-ee-*en*-see. *C. asperata* × *C. pandurata*. Of Burford Lodge, Dorking.
cristata kris-tah-ta. Crested (the lip). Himalaya.
elata ay-*lah*-ta. Tall. Himalaya.
flaccida flăk-ki-da. Drooping (the racemes). Himalaya.
massangeana ma-son-zhee-*ah*-na. After M. de Massange, a 19th-century orchid grower.
mooreana mor-ree-*ah*-na. After F. W. Moore who supplied the type specimen. S Vietnam.
ochracea ok-*rah*-kee-a. Ochre-yellow (the lip). Himalaya, Burma.
pandurata păn-dew-*rah*-ta. Fiddle-shaped (the lip). Black Orchid. Malaya, Borneo.
speciosa spe-kee-ō-sa. Showy. Java.

Coffea kof-ee-a *Rubiaceae*. From *kahwah*, the Arabic name. Tender, evergreen shrub.
arabica a-ră-bi-ka. Arabian. Arabian Coffee Plant. Tropical Africa.

Coix kō-iks *Gramineae*. The Gk. name for a similar plant. Annual grass.
lacryma-jobi lă-kri-ma-yō-bee. Job's tears, from the tear-shaped, grey-white seeds. Job's Tears. SE Asia.

Colchicum kol-ki-kum *Liliaceae*. Of Colchis, W Asia where they were said to grow. Cormous perennials.
agrippinum ăg-ri-*peen*-um. After Agrippina. Cult.
autumnale ow-tum-*nah*-lee. Of autumn (flowering). Autumn Crocus. Europe.
boissieri bwŭ-see-e-ree. After Pierre Edmund Boissier (1810–85), Genevese botanist. S Greece.
bornmuelleri born-*moo*-la-ree. After Joseph Bornmüller (1862–1942) who collected in the Balkans and W Asia. Turkey.
byzantinum bi-zan-*tee*-num. Of Istanbul (Byzantium). Cult.
cilicium ki-*li*-kee-um. Of Cilicia (S Turkey).

Colchicum (continued)
luteum loo-tee-um. Yellow (the flowers). C Asia, NW India.
speciosum spe-kee-ō-sum. Showy. Caucasus, N Turkey, Iran.

Coleus ko-lee-us *Labiatae*. From Gk. *koleos* (a sheath) referring to the stamens which are united into a tube. Tender perennial herbs.
blumei bloom-ee-ee. After Carl Ludwig von Blume (1796–1862). Flame Nettle. Java.
× *hybridus* hib-ri-dus. Hybrid. Many of the commonly grown forms belong here.
thyrsoideus thur-*soi*-dee-us. Staff-like (the inflorescence). C Africa.

Colletia ko-lay-tee-a *Rhamnaceae*. After Philibert Collet (1643–1718), French botanist. More or less leafless, spiny shrubs.
armata ar-*mah*-ta. Spiny. S Chile.
paradoxa pă-ra-*doks*-a. (= *C. cruciata*). Unusual. Uruguay.

Collinia elegans see *Chamaedorea elegans*

Collinsia ko-*linz*-ee-a *Scrophulariaceae*. After Zaccheus Collins (1764–1831), Philadelphia botanist. Annual herbs.
grandiflora grăn-di-*flō*-ra. Large-flowered. Blue Lips. W N America.
heterophylla he-te-rō-*fil*-la. With variable leaves. Chinese Houses. California.
verna ver-na. Of spring (flowering). Blue-eyed Mary. E United States.

Colombia Buttercup see *Oncidium cheirophorum*

Colquhounia ka-*hoon*-ee-a *Labiatae*. After Sir Robert Colquhoun (died 1838), who collected in the Himalaya. Semi-hardy shrub.
coccinea kok-*kin*-ee-a. Scarlet (the flowers). Himalaya, SW China.

Columbine see *Aquilegia*

Columnea ko-*lum*-nee-a *Gesneriaceae*. After Fabius Columna (1567–1640). Tender, evergreen climbers. Costa Rica.
gloriosa glō-ree-ō-sa. Glorious.
linearis lin-ee-*ah*-ris. Linear (the leaves).
microphylla mik-rō-*fil*-la. With small leaves.

Colutea ko-*loo*-tee-a *Leguminosae*. From *kolutea* the Gk. name. Deciduous shrubs and trees.
arborescens ar-bo-*res*-enz. Tree-like.

Colutea (continued)
Bladder Senna. Mediterranean region, SE Europe.
× *media* me-dee-a. *C. arborescens* × *C. orientalis*. Intermediate (between the parents).
orientalis o-ree-en-*tah*-lis. Eastern. W Asia.

Comfrey see *Symphytum*
Russian see *S.* × *uplandicum*

Commelina kom-e-*leen*-a *Commelinaceae*. After two Dutch botanists, Johan (1629–92) and Caspar (1667–1731) Commelin. Semi-hardy and tender perennials. Day Flower.
coelestis koy-*les*-tis. Sky-blue (the flowers). Mexico.
erecta e-*rek*-ta. Erect. E and S United States.
tuberosa tew-be-*rō*-sa. Tuberous. Mexico.

Comptonia komp-*ton*-ee-a *Myricaceae*. After Henry Compton, Bishop of London (1632–1713). Deciduous, suckering shrub.
peregrina pe-re-*gree*-na. Foreign. Sweet Fern. E N America.

Cone Flower see *Rudbeckia*

Conophytum kon-o-*fit*-um *Aizoaceae*. From Gk. *konos* (a cone) and *phyton* (a plant) referring to the conical shape of the plant body. S Africa.
albescens ăl-*bes*-enz. White.
bilobum bi-*lō*-bum. Two-lobed.
frutescens froo-*tes*-enz. Shrubby.
gratum grah-tum. Pleasing.
minutum mi-*new*-tum. Very small.
obcordellum ob-kor-*del*-um. Shaped like a small heart.
pearsonii peer-*son*-ee-ee. After Prof. Pearson.

Consolida kon-*so*-li-da *Ranunculaceae*. From L. *consolida* (to make whole) referring to medicinal properties. Annual herbs. Larkspur.
ambigua ăm-*big*-ew-a. (= *Delphinium ajacis*). Doubtful. Mediterranean region.
regalis ray-*gah*-lis. (= *Delphinium consolida*). Royal. SE Europe, Turkey.

Convallaria kon-va-*lah*-ree-a *Liliaceae*. From L. *convallis* (a valley). Herbaceous perennial.
majalis mah-*yah*-lis. Flowering in May. Lily of the Valley. Europe.

Convolvulus kon-*vol*-vew-lus *Convolvulaceae*. From L. *convolva* (to twine around). Annual and perennial herbs and shrubs.
althaeoides ahl-thie-*oi*-deez. Like *Althaea*. S Europe.
cneorum nee-*o*-rum. From *kneorum*, Gk. name for a dwarf, olive-like shrub. S Europe.
sabatius sa-*bah*-tee-us. (= *C. mauritanicus*). Of Savona (Sabbatia), N Italy. Italy, Sicily, N Africa.
tricolor tri-ko-lor. Three-coloured (the flowers). S Europe.

Coprosma ko-*pros*-ma *Rubiaceae*. From Gk. *kopros* (dung) and *osme* (smell) referring to the odour of the foliage. Evergreen, tender and hardy shrubs. New Zealand.
lucida loo-ki-da. Glossy (the leaves).
petriei pet-ree-ee. After Donald Petrie (1846–1925), Scottish inspector of schools and amateur naturalist in New Zealand.
repens ree-penz. Creeping.

Coral Berry see *Ardisia crispa*
Coralberry see *Aechmea fulgens*
Coral Drops see *Bessera elegans*
Coral Plant see *Berberidopsis corallina*, *Russellia equisetiformis*
Coral Tree see *Erythrina crista-galli*
Coral Vine see *Antigonon leptopus*

Cordyline kor-*di*-li-nee *Agavaceae*. From Gk. *kordyle* (a club) referring to the large, fleshy roots of some species. Tender and semi-hardy, evergreen trees and shrubs.
australis ow-*strah*-lis. Southern. Cabbage Tree. New Zealand.
indivisa in-dee-*vee*-sa. Undivided, it is usually single-stemmed. New Zealand.
rubra rub-ra. Red (? the flowers). Cult.
stricta strik-ta. Upright. New South Wales.
terminalis ter-min-*ah*-lis. Terminal (the inflorescence). Good Luck Plant. E Asia.

Coreopsis ko-ree-*op*-sis *Compositae*. From Gk. *koris* (a bug) and *-opsis* indicating resemblance, the seeds look like ticks. Annual and perennial herbs. Tickseed.
auriculata ow-rik-ew-*lah*-ta. With basal lobes (the leaves). SE United States.
basalis ba-*sah*-lis. (= *C. drummondii*). Basal, perhaps referring to the red bases of the ray florets. S United States.
grandiflora grăn-di-*flō*-ra. Large-flowered. SE United States.
lanceolata lăn-kee-ō-*lah*-ta. Lanceolate (the

Coreopsis (continued)
leaves). SE United States.
tinctoria tink-*to*-ree-a. Used in dyeing.
C and W United States.
verticillata ver-ti-ki-*lah*-ta. Whorled, the
leaves are finely cut, appearing whorled.
SE United States.

Coriandrum ko-ree-*ăn*-drum *Umbelliferae*.
From *koriandron* the Gk. name. Annual
herb.
sativum sa-*tee*-vum. Cultivated. Coriander.
N Africa, W Asia.

Coriaria ko-ree-*ah*-ree-a *Coriariaceae*. From
L. *corium* (leather), some species are used in
tanning. Deciduous shrubs.
japonica ja-*pon*-i-ka. Of Japan.
terminalis ter-mi-*nah*-lis. Terminal (the
racemes). Himalaya, China.
xanthocarpa zănth-ō-*kar*-pa. With yellow
fruits.

Corkscrew Rush see *Juncus effusus* 'Spiralis'
Corn see *Zea mays*
Corn Cockle see *Agrostemma githago*
Corn Lily see *Ixia*
Corn Marigold see *Chrysanthemum segetum*
Corn Salad see *Valerianella locusta*
Cornelian Cherry see *Cornus mas*
Cornflower see *Centaurea cyanus*

Cornus *kor*-nus *Cornaceae*. The L. name for
C. mas. Deciduous shrubs and trees.
Dogwood.
alba ăl-ba. White (the fruit). Siberia,
China.
alternifolia ăl-ter-ni-*fo*-lee-a. With alternate
leaves. EN America.
amomum a-*mō*-mum. Gk. name of a spice
plant.
canadensis kăn-a-*den*-sis. Of Canada or NE
North America. N America, E Asia.
capitata kăp-i-*tah*-ta. In a dense head (the
flowers). Himalaya, China.
florida flō-ri-da. Flowering. Flowering
Dogwood. E United States.
kousa koo-sa. The Japanese name. Japan,
China.
chinensis chin-*en*-sis. Of China.
macrophylla măk-ro-*fil*-la. With large
leaves. Himalaya.
mas mahs. Male, an epithet used to
distinguish a robust species from a more
delicate one which was regarded as female.
Cornelian Cherry. Europe.
nuttallii nū-*tahl*-ee-ee. After Thomas
Nuttall (1786–1839). Pacific Dogwood.

Cornus (continued)
W N America.
sanguinea săng-*gwin*-ee-a. Red (autumn
colour). Common Dogwood. Europe.
stolonifera sto-lō-*ni*-fe-ra. Bearing stolons.
N America.
'Flaviramea' flah-vi-*rahm*-ee-a. With
yellow shoots.

Corokia ko-rō-kee-a *Cornaceae*. From the
Maori name *korokia*. Evergreen shrubs. New
Zealand.
cotoneaster ko-tōn-ee-*ă*-ster. The intricately
branched habit is reminiscent of some
cotoneasters.
× *virgata* vir-*gah*-ta. *C. buddleioides* × *C.
cotoneaster*. Twiggy.

Coronilla ko-rō-*nil*-la *Leguminosae*.
Diminutive of L. *corona* (a crown) referring
to the arrangement of the flowers. Hardy and
semi-hardy shrubs.
emeroides e-me-*roi*-des. Like *C. emerus*. SE
Europe. Syria.
emerus e-*me*-rus. From the Italian name
emero. Scorpion Senna. C and S Europe.
glauca glow-ka. Glaucous (the leaves).
S Europe.
valentina văl-en-*teen*-a. Of Valencia, Spain.
S Europe.

Correa *ko*-ree-a *Rutaceae*. After José
Francesco Correa de Serra (1751–1823),
Portuguese botanist. Tender and semi-hardy
evergreen shrubs. SE Australia.
alba ăl-ba. White (the flowers).
reflexa re-*fleks*-a. (= *C. speciosa*). Reflexed
(the corolla lobes).

Cortaderia kor-ta-*de*-ree-a *Gramineae*. From
the Argentinian name. Perennial grasses.
jubata yoo-*bah*-ta. Like a mane (the
inflorescence). S America.
richardii ree-*shard*-ee-ee. After Achille
Richard (1794–1852), French botanist.
New Zealand.
selloana sel-ō-*ah*-na. (= *C. argentea*). After
Sellow, see *Feijoa sellowiana*. Pampas
Grass. S America.

Corydalis ko-*ri*-da-lis *Fumariaceae*. The Gk.
name for a lark, the spur of the flower
resembles that of a lark. Perennial herbs.
bulbosa bul-*bō*-sa. Bulbous (the tuber).
Europe, Asia.
cashmeriana kăsh-me-ree-*ah*-na. Of
Kashmir. Himalaya.
cheilanthifolia kay-lănth-i-*fo*-lee-a. With

Corydalis (continued)
leaves like *Cheilanthus*. China.
lutea loo-tee-a. Yellow (the flowers).
Europe.
nobilis nō-bi-lis. Notable. C Asia.
solida so-li-da. Solid (the tuber). Europe.

Corylopsis ko-ril-*op*-sis *Hamamelidaceae*.
From *Corylus* q.v. and Gk. *-opsis* indicating
resemblance, the habit and leaves are similar
to *Corylus*. Deciduous trees and shrubs.
glabrescens gla-*bres*-enz. Nearly glabrous
(the leaves). Japan, Korea.
pauciflora paw-si-*flō*-ra. Few-flowered (the
spikes). Japan, Taiwan.
sinensis si-*nen*-sis. Of China. China, Tibet.
spicata spee-*kah*-ta. With flowers in spikes.
Japan.
willmottiae wil-*mot*-ee-ie. After Ellen
Willmott, see *Ceratostigma willmottianum*.

Corylus *ko*-ril-us *Corylaceae*. From *korylos*
the Gk. name. Deciduous shrubs and trees.
avellana ă-ve-*lah*-na. Of Avella Vecchia,
S Italy. Hazel, Cobnut. Europe, N Africa,
W Asia.
   'Contorta' kon-*tor*-ta. Twisted (the
   shoots). Harry Lauder's Walking Stick.
colurna ko-*lurn*-a. The classical name.
Turkish Hazel. SE Europe, W Asia.
maxima mahk-si-ma. Larger. Filbert.
S Europe.

Cosmos *kos*-mos *Compositae*. From Gk.
kosmos (beautiful). Annual herbs.
bipinnatus bi-pin-*ah*-tus. With bi-pinnate
leaves. Mexico.
sulphureus sul-*fewr*-ree-us. Sulphur-yellow
(the flowers). Mexico.

Costmary see *Balsamita major*

Cotinus *ko*-ti-nus *Anacardiaceae*. From
kotinos Gk. name for the olive. Deciduous
shrubs or trees.
coggygria ko-*gig*-ree-a. (= *Rhus cotinus*).
From *kokkugia* the Gk. name. Smoke
Tree. C and S Europe, Himalaya, China.
   'Foliis Purpureis' *fo*-lee-is pur-*pewr*-ree-
   is. With purple leaves.
obovatus ob-ō-*vah*-tus. (= *Rhus cotinoides*).
Obovate (the leaves). SE United States.

Cotoneaster ko-tŏn-ee-ă-ster *Rosaceae*.
From L. *cotoneum* (quince) and *-aster*
(resembling somewhat) from the similarity of
the leaves of some species. Deciduous and
evergreen shrubs.

Cotoneaster (continued)
adpressus ăd-*pres*-us. Pressed against,
referring to its low habit. Himalaya,
China.
   praecox prie-koks. Early (fruit ripening).
affinis a-*fee*-nis. Related to (another
species). Himalaya, W China.
   bacillaris bă-ki-*lah*-ris. Staff-like. The
   branches are used to make walking-
   sticks in the Himalaya.
bullatus bul-*lah*-tus. With impressed veins
on the leaves. W China, Tibet.
   floribundus flō-ri-*bun*-dus. Profusely
   flowering.
congestus con-*ges*-tus. Congested (the habit).
Himalaya, W China.
conspicuus kon-*spik*-ew-us. Conspicuous
(the fruit). Tibet.
'Cornubia' kor-*new*-bee-a. L. name for
Cornwall.
dammeri dăm-a-ree. After Dammer.
C China.
distichus dis-ti-kus. Two-ranked (the
leaves). Himalaya.
divaricatus di-vah-ri-*kah*-tus. With
spreading branches. China.
'Exburiensis' eks-ba-ree-*en*-sis. *C. frigidus*
× *C. salicifolius*. From Exbury, where it
was raised.
franchetii fron-*shay*-tee-ee. After Adrien
Franchet (1834–1900), French botanist.
Tibet, W China.
   sternianus stern-ee-*ah*-nus. After Sir
   Frederick Stern. Burma.
frigidus fri-gi-dus. Growing in cold regions.
Himalaya, W China.
horizontalis ho-ri-zon-*tah*-lis. Horizontal
(the habit). China.
'Hybridus Pendulus' *hib*-ri-dus *pen*-dew-
lus. A hybrid of pendulous habit.
lacteus lăk-tee-us. Milky (the flowers).
China.
microphyllus mik-rō-*fil*-lus. Small-leaved.
Himalaya, SW China.
   cochleatus kok-lee-*ah*-tus. Shell-like (the
   leaves).
   thymifolius tiem-i-*fo*-lee-us. *Thymus*-
   leaved.
multiflorus mul-tee-*flō*-rus. Many-flowered.
Caucasus to China.
pannosus pă-*nō*-sus. Felt-like (the young
shoots and the undersides of the leaves).
China.
prostratus pros-*trah*-tus. Prostrate.
Himalaya, SW China.
'Rothschildianus' roths-chield-ee-*ah*-nus.
After Baron Rothschild.
salicifolius să-lik-i-*fo*-lee-us. *Salix*-leaved.

Cotoneaster (continued)
China.
floccosus flok-ō-sus. Woolly (the
underside of the young leaves).
rugosus roog-ō-sus. Wrinkled (the
leaves).
simonsii sie-*monz*-ee-ee. After Mr Simons
who introduced it. Himalaya.
× *watereri* waw-ta-ra-ree. *C. frigidus* × *C.*
henryanus. After Waterer's who raised it.

Cotton, Levant see *Gossypium herbaceum*
Tree see *G. arboreum*
Cotton Rose see *Hibiscus mutabilis*

Cotula kot-ew-la *Compositae*. From Gk.
kotula (a small cup), the base of the leaves
often clasp the stem making a small cup.
Annual and perennial herbs.
atrata ah-*trah*-ta. Black (the flowers). New
Zealand.
barbata bar-*bah*-ta. Bearded (the stems).
S Africa.
coronopifolia ko-rō-no-pi-*fo*-lee-a.
Coronopus-leaved. Brass Buttons. S Africa.
potentillina po-ten-til-*een*-a. Like *Potentilla*.
New Zealand.
squalida skwah-li-da. Dirty. New Zealand.

Cotyledon ko-ti-*lay*-don *Crassulaceae*. From
Gk. *kotyle* (a small cup) referring to the cup-
shaped leaves of some species. Tender
succulents.
barbeyi bar-bee-ee. After William Barbey
(1842–1914), Geneva botanist. E Africa,
Arabia.
ladysmithensis lay-dee-smith-*en*-sis. From
Ladysmith. S Africa.
orbiculata or-bik-ew-*lah*-ta. Orbicular (the
leaves). S Africa.
paniculata pan-ik-ew-*lah*-ta. With flowers
in panicles. S Africa.
reticulata ray-tik-ew-*lah*-ta. Net-veined.
S Africa.
simplicifolia see *Chiastophyllum*
oppositifolium
teretifolia te-reet-i-*fo*-lee-a. With cylindrical
leaves. S Africa.
undulata un-dew-*lah*-ta. Wavy-edged (the
leaves). Silver Crown. S Africa.

Couve Tronchuda see *Brassica oleracea*
Tronchuda
Cow Herb see *Vaccaria pyramidata*
Cowberry see *Vaccinium vitis-idaea*
Cowslip see *Primula veris*
Crab see *Malus*
Siberian see *M. baccata*

Crambe kram-bay *Cruciferae*. Gk. name for
cabbage from the similar leaves. Perennial
herbs.
cordifolia kor-di-*fo*-lee-a. With heart-
shaped leaves. Caucasus.
maritima ma-*ri*-ti-ma. Growing near the sea.
Sea Kale. W Europe to W Asia.

Cranberry see *Oxycoccus*
American see *O. macrocarpus*
Small see *O. palustris*
Crane Lily see *Strelitzia reginae*
Cranesbill see *Geranium*
Bloody see *G. sanguineum*
Dusky see *G. phaeum*
Meadow see *G. pratense*
Wood see *G. sylvaticum*
Crape Myrtle see *Lagerstroemia indica*

Crassula krăs-ew-la *Crassulaceae*. From L.
crassus (thick) referring to the fleshy leaves.
Tender succulents. S Africa.
arborescens ar-bo-res-enz. Tree-like.
Chinese Jade.
argentea ar-gen-tee-a. (= *C. portulaca*).
Silvery. Jade Plant.
cooperi koo-pa-ree. After Thomas Cooper
(1815–1913).
falcata făl-*kah*-ta. (= *Rochea falcata*).
Sickle-shaped (the leaves). Propeller
Plant.
lactea lăk-tee-a. Milky (the flowers).
Tailor's Patch.
lycopodioides lik-ō-pō-dee-*oi*-deez. Like
Lycopodium.
milfordiae mil-*ford*-ee-ie. After Mrs Helen
A. Milford (died 1940) who collected in
S Africa.
perforata per-fo-*rah*-ta. Perforated, the
paired leaves are joined at the base and
thus perforated by the stem.
portulaca see *C. argentea*
rupestris roo-pes-tris. Growing on rocks.
Buttons on a String.
sarcocaulis sar-kō-*kaw*-lis. Fleshy-
stemmed.
schmidtii shmit-ee-ee. After E. Schmidt.
socialis so-kee-*ah*-lis. Growing in colonies.
teres te-res. Cylindrical.

Crataegus kra-*tie*-gus *Rosaceae*. The Gk.
name from *kratos* (strength) referring to the
hard wood. Deciduous trees. Hawthorn.
crus-galli kroos-gǎ-lee. A cock's spur,
referring to the long thorns. Cockspur
Thorn. E N America.
laciniata la-kin-ee-*ah*-ta. Deeply cut (the

Crataegus (continued)
leaves). W Asia.
laevigata lie-vi-*gah*-ta. (= *C. oxyacantha*).
Smooth (the leaves). Midland Hawthorn.
Europe.
× *lavallei* la-*vahl*-ee-ee. After Pierre
Alphonse Martin Lavallée (1836–84).
'Carrierei' kă-ree-*e*-ree-ee. After Elie
Abel Carrière (1816–96), French
botanist and horticulturist.
monogyna mon-*o*-gi-na. With one pistil.
Europe, N Africa, W Asia.
prunifolia proon-i-*fo*-lee-a. *Prunus*-leaved.
?N America.
tanacetifolia tă-na-set-i-*fo*-lee-a. *Tanacetum*-
leaved. W Asia.

Cream Cups see *Platystemon californicus*
Creeping Charlie see *Pilea nummularia*
Creeping Jenny see *Lysimachia nummularia*
Creeping Wintergreen see *Gaultheria procumbens*
Creeping Zinnia see *Sanvitalia procumbens*

Crepis kre-pis *Compositae*. From Gk. *krepis*
(a boot). Annual and perennial herbs.
aurea ow-ree-a. Golden (the flowers). SE
Europe.
incana in-*kah*-na. Grey-hairy (the leaves).
S Greece.
rubra rub-ra. Red (the flowers). SE Europe

Crinodendron krin-ō-*den*-dron
Elaeocarpaceae. From Gk. *krinon* (a lily) and
dendron (a tree). Evergreen, semi-hardy
shrub.
hookerianum huk-a-ree-*ah*-num. After
Hooker. Chile.

Crinum kree-num *Amaryllidaceae*. From Gk.
krinon (a lily). Semi-hardy, bulbous
perennials.
bulbispermum bul-bee-*sperm*-um. (= *C.
capense*). With bulbous seeds. S Africa.
moorei mor-ree-ee. After Moore. Natal.
natans nă-tănz. Floating. W Africa.
× *powellii* powl-ee-ee. *C. bulbispermum* ×
C. moorei. After C. Baden-Powell who
raised it about 1885.

Crocosmia krō-*kos*-mee-a *Iridaceae*. From
Gk. *krokos* (saffron) and *osme* (smell), the
dried flowers smell of saffron. Cormous
perennials.
× *crocosmiiflora* krō-kos-mee-i-*flō*-ra. *C.
aurea* × *C. pottsii*. With flowers like
Crocosmia, it was originally described in

Crocosmia (continued)
another genus. Montbretia.
masoniorum may-son-ee-*or*-rum. After
Canon G. E. and Miss Mason, collectors
of S African plants. S Africa.
paniculata pa-nik-ew-*lah*-ta. (= *Antholyza
paniculata*. *Curtonus paniculatus*). With
flowers in panicles. S Africa.
pottsii pots-ee-ee. After George Harrington
Potts (1830–1907), who introduced it in
1877. S Africa.

Crocus krō-kus *Iridaceae*. From Gk. *krokos*
(saffron). Cormous herbs.
asturicus see *C. serotinus salzmannii*
ancyrensis ăn-ki-*ren*-sis. Of Ankara.
W Turkey.
angustifolius ăn-gus-ti-*fo*-lee-us. (= *C.
susianus*). Narrow-leaved. Crimea.
aureus see *C. flavus*.
balansae ba-*lahn*-zie. After Benedict
Balansa (1825–91), French botanist who
collected in W Asia. W Asia.
banaticus ba-*nah*-ti-kus. Of Banat (N
Romania). E Europe.
biflorus bi-*flō*-rus. Two-flowered. Scotch
Crocus. S Europe, W Asia.
cancellatus kăn-ke-*lah*-tus. Latticed (the
corm tunic). Greece, Turkey.
candidus kăn-di-dus. White (the flowers).
W Turkey.
chrysanthus kris-ănth-us. Golden-flowered.
SE Europe, S Turkey.
clusii see *C. serotinus clusii*
corsicus kor-si-kus. Of Corsica.
dalmaticus dăl-mă-ti-kus. Of Dalmatia.
Yugoslavia, Albania.
etruscus e-*troos*-kus. Of Tuscany. W Italy.
flavus flah-vus. (= *C. aureus*). Yellow (the
flowers). SE Europe, W Asia.
fleischeri flie-sha-ree. After M. Fleischer
(1861–1930). W and S Turkey.
imperati im-pe-*rah*-tee. After Ferrante
Imperato (1550–1625), Naples
apothecary. W Italy.
suaveolens swah-vee-ō-lenz. (= *C.
suaveolens*). Sweet-scented.
korolkowii ko-rol-*kov*-ee-ee. After General
Nikolai Iwanovitsch Korolkow (born
1837). C Asia.
kotschyanus kot-shee-*ah*-nus. After
Theodor Kotschy (1813–66), Austrian
botanist. W Asia.
laevigatus lie-vi-*gah*-tus. Smooth (the corm
tunic). S Greece.
longiflorus long-gi-*flō*-rus. Long-flowered.
SW Italy, Sicily, Malta.
medius me-dee-us. Intermediate. SE

Crocus (continued)
France, NW Italy.
minimus min-i-mus. Smaller. Corsica,
Sardinia.
niveus niv-ee-us. Snow white. S Greece.
nudiflorus new-di-*flō*-rus. Flowering before
the leaves emerge. SW Europe.
ochroleucus ok-ro-*loo*-kus. Yellowish-white.
W Asia.
olivieri o-liv-ee-*e*-ree. After Guillaume
Antoine Olivier (1756–1814), French
naturalist. SE Europe, Turkey.
pulchellus pul-*kel*-lus. Pretty. Balkans,
W Turkey.
reticulatus ray-tik-ew-*lah*-tus. Net-veined
(the corm tunic). SE Europe.
sativus sa-*tee*-vus. Cultivated. Saffron. Cult.
serotinus se-*ro*-ti-nus. Late flowering.
Portugal.
    clusii clooz-ee-ee. (= *C. clusii*). After
    Clusius, see *Gentiana clusii*. Portugal,
    Spain.
    salzmannii sahlts-*mahn*-ee-ee. (= *C.
    asturicus*). After Philip Salzmann
    (1781–1851), French botanist. Spain.
sieberi see-ba-ree. After France William
Sieber (1789–1844), a plant collector. SE
Europe.
speciosus spe-kee-*ō*-sus. Showy. W Asia.
suaveolens see *C. imperati suaveolens*
susianus see *C. angustifolius*
tommasinianus tom-ma-see-nee-*ah*-nus. After
Muzio Giuseppe Spirito de Tommasini
(1794–1879), Italian botanist. E Europe.
vernus ver-nus. Of spring (flowering).
Dutch Crocus. Europe.
versicolor ver-*si*-ko-lor. Variously coloured.
Maritime Alps.

Crocus, Autumn see *Colchicum autumnale*
Cross Vine see *Bignonia capreolata*

Crossandra kros-*ăn*-dra *Acanthaceae*. From
Gk. *krossos* (a fringe) and *aner* (male)
referring to the fringed anthers. Tender,
evergreen shrubs.
    infundibuliformis in-fun-dib-ew-lee-*form*-is.
    (= *C. undulifolia*). Trumpet-shaped (the
    flowers). Firecracker Flower. S India, Sri
    Lanka.
    nilotica ni-*lo*-ti-ka. Of the Nile Valley.
    Tropical Africa.

Croton see *Codiaeum variegatum pictum*
Crowberry see *Empetrum*
Crown Daisy see *Chrysanthemum coronarium*
Crown of Thorns see *Euphorbia milii*
Cruel Plant see *Araujia sericofera*

Cryophytum crystallinum see
Mesembryanthemum crystallinum

Cryptanthus krip-*tănth*-us *Bromeliaceae*.
From Gk. *krypto* (to hide) and *anthos* (a flower)
referring to the hidden flowers. Tender
perennials. Earth Star. Brazil.
    acaulis a-*kaw*-lis. Stemless. Starfish Plant.
    bivittatus bi-vi-*tah*-tus. With two stripes
    (the leaves).
    bromelioides brom-ee-lee-*oi*-deez. Like
    Bromelia. Rainbow Star.
    fosterianus fos-ta-ree-*ah*-nus. After Milford
    Bateman Foster (born 1888) a collector
    and grower of bromeliads.
    zonalis zō-*nah*-lis. Banded (the leaves).
    Zebra Plant.

Cryptocoryne krip-tō-*ko*-ri-nee *Araceae*.
From Gk. *krypto* (to hide) and *coryne* (a club).
The spadix (i.e. the club) is hidden by the
spathe. Aquarium plants. Water Trumpet.
    aponogetifolia a-pon-ō-gay-ti-*fo*-lee-a.
    Aponogeton-leaved.
    balansae ba-*lahn*-zie. After Balansa, see
    Crocus balansae. SE Asia.
    beckettii be-*ket*-ee-ee. After Thomas W.
    Naylor Beckett (1839–1906), a coffee
    planter who collected in Sri Lanka. Sri
    Lanka.
    blassii blahs-ee-ee. After A. Blass, German
    horticulturist. Thailand.
    ciliata ki-lee-*ah*-ta. Fringed with hairs (the
    spathe). SE Asia.
    griffithii gri-*fith*-ee-ee. After Griffith, see
    Ceratostigma griffithii. Malay Peninsula.
    nevilii ne-*vil*-ee-ee. After H. Nevill who
    collected the type specimen. Sri Lanka.
    petchii pech-ee-ee. After Petch. Sri Lanka.
    undulata un-dew-*lah*-ta. Wavy-edged (the
    leaves).
    willisii wil-*is*-ee-ee. After Willis. Sri Lanka.

Cryptogramma krip-tō-*grăm*-ma
Cryptogrammaceae. From Gk. *krypto* (to hide)
and *gramma* (a line), the sori which form a
line around the margins of the pinnules are
hidden by the rolled leaf margins. Fern.
    crispa kris-pa. Crisped (the fronds). Parsley
    Fern. Europe, W Asia.

Cryptomeria krip-to-*me*-ree-a *Taxodiaceae*.
From Gk. *krypto* (to hide) and *meris* (a part)
referring to the concealed parts of the flowers.
    japonica ja-pon-i-ka. Of Japan. Japan,
    China.
        'Elegans' ay-le-gahnz. Elegant.

Cryptomeria (continued)
'Lobbii' *lob*-ee-ee. After its introducer,
Thomas Lobb (1820–94), who collected
for Veitch in SE Asia.
'Vilmoriniana' vil-mo-rin-ee-*ah*-na. After
M. de Vilmorin, French nurseryman,
in whose garden it was found.

Ctenanthe ten-*änth*-ee *Marantaceae*. From
Gk. *kteinos* (a comb) and *anthos* (a flower)
from the arrangement of the bracts. Tender
perennials. Brazil.
lubbersiana lub-erz-ee-*ah*-na. After C.
Lubbers, head gardener at the Brussels'
Botanic Garden at the end of the 19th
century.
oppenheimiana o-pan-hiem-ee-*ah*-na. After
Edouard Oppenheim. Never-never Plant.
Brazil.

Cuckoo Flower see *Cardamine pratensis*
Cuckoo Pint see *Arum maculatum*
Cucumber see *Cucumis sativus*
Cucumber Tree see *Magnolia acuminata*

Cucumis *kew*-kew-mis *Cucurbitaceae*. The
L. name for cucumber. Tender annuals.
melo may-lo. L. name for an apple-shaped
melon. Melon. W Africa.
sativus sa-*tee*-vus. Cultivated. Cucumber,
Gherkin. S Asia.

Cucurbita kew-*kur*-bi-ta *Cucurbitaceae*. L.
name for a gourd. Annual herbs.
maxima mahk-si-ma. Largest. Pumpkin.
Cult.
pepo pe-pō. L. name for a large pumpkin
or marrow. Marrow. Cult.

Cunninghamia kūn-ing-*häm*-ee-a
Taxodiaceae. After James Cunningham, a
surgeon with the East India Co. who found
the following in 1701. Evergreen conifer.
lanceolata län-kee-ō-*lah*-ta. Lanceolate (the
leaves). China.

Cup Flower see *Nierembergia*

Cuphea *kew*-fee-a *Lythraceae*. From Gk.
kyphos (curved) referring to the curved seed
capsule. Tender herbs and shrubs.
cyanea see-*ah*-nee-a. Blue (the petals).
Mexico.
hyssopifolia hi-sōp-i-*fo*-lee-a. *Hyssopus*-
leaved. Mexico, Guatemala.
ignea ig-nee-a. Glowing (the red calyx) or
on fire (the flowers resemble a lighted
cigar). Cigar Plant. Mexico, Jamaica.

Cupid's Dart see *Catananche caerulea*

× **Cupressocyparis** kew-pres-ō-*kew*-pa-ris
Cupressaceae. Intergeneric hybrid, from the
names of the parents. *Chamaecyparis* ×
Cupressus. Evergreen conifer.
leylandii lay-*länd*-ee-ee. *Chamaecyparis
nootkatensis* × *Cupressus macrocarpa*. After
C. J. Leyland who grew some of the first-
raised trees at Haggerston Hall. Leyland
Cypress.

Cupressus kew-*pres*-us *Cupressaceae*. The L.
name for *C. sempervirens*. Evergreen
conifers. Cypress.
cashmeriana kăsh-me-ree-*ah*-na. Of
Kashmir. Origin uncertain.
glabra glă-bra. Smooth (the bark). Arizona.
lusitanica loo-si-*tah*-ni-ka. Of Portugal,
from where it was introduced to England.
Cedar of Goa. ? Mexico.
macrocarpa măk-rō-*kar*-pa. Large-fruited.
Monterey Cypress. California.
sempervirens sem-per-*vi*-renz. Evergreen.
Italian Cypress. Cult.

Currant, Black see *Ribes nigrum*
  Buffalo see *R. odoratum*
  Flowering see *R. sanguineum*
  Mountain see *R. alpinum*
  Red see *R. rubrum*
Curry Plant see *Helichrysum italicum serotinum*
Curtonus paniculatus see *Crocosmia
paniculata*
Custard Apple see *Annona cherimola, A.
reticulata*
Cut-leaved Bramble see *Rubus laciniatus*

Cyananthus kee-a-*nănth*-us *Campanulaceae*.
From Gk. *kyanos* (blue) and *anthos* (a
flower). Perennial herbs.
lobatus lō-*bah*-tus. Lobed (the leaves).
Himalaya, Assam, W China.
microphyllus mik-rō-*fil*-lus. Small-leaved.
Himalaya, Nepal.
sherriffii she-*rif*-ee-ee. After George Sherriff
(1898–1967), who collected in the
Himalaya. Himalaya.

Cyanotis kee-a-*nō*-tis *Commelinaceae*. From
Gk. *kyanos* (blue) and *ous* (an ear) referring
to the ear-like, blue petals. Tender perennial
herbs.
kewensis kew-*en*-sis. Of Kew. Teddy Bear
Plant. India.
somaliensis so-mah-lee-*en*-sis. Of Somalia.
Pussy Ears. Tropical Africa.

Cyathodes kee-a-*thō*-deez *Epacridaceae*.
From Gk. *kyathodes* (a small cup) referring
to the cup-shaped disc under the ovary.
Evergreen, heath-like shrub.
  colensoi ko-*len*-zo-ee. After the Rev.
  William Colenso (1811–99), printer,
  missionary and botanist in New Zealand,
  who collected the type specimen. New
  Zealand.

Cycas *see*-kas *Cycadaceae*. Gk. name for a
palm, which this resembles in habit and leaf.
Evergreen, tender, palm-like plant.
  revoluta re-vo-*loo*-ta. Revolute (the margins
  of the leaflets). Sago Palm. Ryuku Islands,
  S Japan.

Cyclamen *sik*-la-men, classically *koo*-kla-men
Primulaceae. The Gk. name. Cormous herbs.
Sowbread.
  cilicium ki-*li*-kee-um. Of Cilicia, S Turkey.
  Turkey.
  coum *kō*-um. (= *C. atkinsii, C. orbiculatum*).
  Of Kos. SE Europe, W Asia.
  creticum *kray*-ti-kum. Of Crete.
  cyprium *kip*-ree-um. Of Cyprus.
  graecum *grie*-kum. of Greece. Greece,
  Turkey, Cyprus.
  hederifolium he-de-ri-*fo*-lee-um. (= *C.
  neapolitanum*). *Hedera*-leaved. SE Europe,
  Turkey.
  persicum *per*-si-kum. Of Iran (Persia).
  Florist's Cyclamen. E Mediterranean
  region, W Asia.
  purpurascens pur-pew-*răs*-enz. (= *C.
  europaeum*). Purplish (the flowers).
  Europe, Caucasus.
  repandum re-*păn*-dum. Wavy-margined (the
  leaves). Mediterranean region.
  rohlfsianum rolfs-ee-*ah*-num. After Rohlfs
  who first collected it. Libya.

Cydonia si-*dō*-nee-a or classically koo-*dō*-
nee-a *Rosaceae*. L. name, from Cydonia, NE
Crete, now Khania. Deciduous tree.
  japonica see *Chaenomeles japonica*
  lagenaria see *Chaenomeles speciosa*
  maulei see *Chaenomeles japonica*
  oblonga ob-*long*-ga. Oblong (the leaves).
  Quince. W and C Asia.
  speciosa see *Chaenomeles speciosa*.

Cymbalaria sim-ba-*lah*-ree-a
Scrophulariaceae. From Gk. *kymbalon* (a
cymbal) referring to the shape of the leaves.
Herbaceous perennials.

Cymbalaria (continued)
  aequitriloba ie-kwee-tri-*lō*-ba. With three
  equal lobes (the leaves). S Europe.
  hepaticifolia he-pă-ti-ki-*fo*-lee-a. *Hepatica*-
  leaved. Corsica.
  muralis mew-*rah*-lis. (= *Linaria
  cymbalaria*). Growing on walls. Ivy-leaved
  Toadflax. Europe.
  pallida pă-li-da. Pale (the flowers). Italy.

Cymbidium kim-*bid*-ee-um *Orchidaceae*.
From Gk. *kymbe* (a boat), referring to the
hollowed lip. Greenhouse orchids.
  Alexanderi ă-leks-*ahn*-da-ree. After H. G.
  Alexander (1875–1972), orchid hybridist.
  aloifolium a-lō-ee-*fo*-lee-um. *Aloe*-leaved.
  Himalaya, SE Asia.
  canaliculatum kă-na-lik-ew-*lah*-tum.
  Channelled (the leaves). Queensland.
  Coningsbyanum kon-ingz-bee-*ah*-num. *C.
  grandiflorum* × *C. insigne*. After the raiser
  Arthur Coningsby (c. 1888–1966), an
  orchid grower with Sander's.
  dayanum see *C. eburneum dayanum*
  devonianum de-vō-nee-*ah*-num. Of Devon.
  Himalaya.
  eburneum e-*burn*-ee-um. Ivory white.
  Himalaya, SE Asia.
    dayanum day-*ah*-num. (= *C. dayanum*).
    After John Day (1824–88), amateur
    orchid grower.
  elegans *ay*-le-gahnz. Elegant. N India.
  ensifolium ayn-si-*fo*-lee-um. With sword-
  shaped leaves. E Asia.
  erythrostylum e-rith-ro-*stil*-um. With a red
  column. Vietnam.
  finlaysonianum fin-lay-son-ee-*ah*-num. After
  George Finlayson (died 1823), a naturalist
  with the East India Co. SE Asia.
  giganteum gi-*găn*-tee-um. Very large.
  Himalaya to SE Asia.
  grandiflorum grăn-di-*flō*-rum. With large
  flowers. Himalaya, China.
  insigne in-*sig*-nee. Remarkable. Indochina.
  lowianum lō-ee-*ah*-num. After Sir Hugh
  Low (1824–1905), a diplomat in SE Asia
  and an authority on orchids. Burma.
  Pauwelsii pow-*elz*-ee-ee. *C. insigne* × *C.
  lowianum concolor*. After Pauwels who
  collected orchids for Sander's.
  pendulum *pen*-dew-lum. Pendulous.
  Himalaya, SE Asia.
  pumilum pew-mi-lum. Dwarf. China.
  tigrinum ti-*gree*-num. Striped like a tiger.
  SE Asia.
  tracyanum tray-see-*ah*-num. After Henry
  Amos Tracy (c. 1850–1910), orchid and
  bulb nurseryman. SE Asia.

Cynara si-*nah*-ra or classically koo-*nah*-ra
Compositae. The L. name. Perennial herbs.
  cardunculus kar-dun-*kew*-lus. Diminutive o
  Carduus, a genus of thistles. Cardoon. SW
  Europe.
  scolymus sko-li-mus. The L. name for
  Scolymus hispanicus. Cult.

Cynoglossum si-no-*glos*-um *Boraginaceae*.
From Gk. *kyon* (a dog) and *glossum* (a
tongue) referring to the leaves.
  amabile a-*mah*-bi-lee. Beautiful.
  E Himalaya, W China.
  nervosum ner-*vō*-sum. Distinctly veined
  (the leaves). Himalaya.
  officinale o-fi-ki-*nah*-lee. Sold as a herb.
  Hound's Tongue. Europe.

Cyperus si-*pe*-rus or classically koo-*pe*-rus
Cyperaceae. Gk. name for a sedge. Tender
and hardy, grass-like plants. Sedge.
  albostriatus ăl-bō-stree-*ah*-tus. White-
  striped, referring to the pale leaf veins.
  S Africa.
  involucratus in-vo-loo-*krah*-tus. (= *C.
  alternifolius* hort.). With an involucre,
  referring to the large, leaf-like bracts borne
  under the inflorescence. Africa.
  longus long-gus. Long, the tall stems.
  Galingale. Europe to India.
  papyrus pa-*pi*-rus. Gk. name for the paper
  made from this plant. Papyrus. C Africa,
  Nile Valley.

Cypress see *Cupressus*
  Hinoki see *Chamaecyparis obtusa*
  Italian see *Cupressus sempervirens*
  Lawson see *Chamaecyparis lawsoniana*
  Leyland see × *Cupressocyparis leylandii*
  Monterey see *Cupressus macrocarpa*
  Nootka see *Chamaecyparis nootkatensis*
  Sawara see *Chamaecyparis pisifera*
Cypress Spurge see *Euphorbia cyparissias*
Cypress Vine see *Ipomaea quamoclit*

Cypripedium kip-ree-*pee*-dee-um
Orchidaceae. From Gk. *kypris* (Venus) and
pedilon (a slipper) referring to the shape of
the flowers. Hardy orchids. Lady's Slipper
Orchid. For the tender orchids sometimes
listed here see *Paphiopedilum*.
  acaule a-*kaw*-lee. Stemless. E United
  States.
  arietinum ă-ree-e-*tee*-num. Like a ram's-
  head. Ram's Head Lady's Slipper Orchid.
  EN America.
  calceolus kăl-*kee*-o-lus. A small shoe.

Cypripedium (continued)
  Europe, Asia.
  candidum kăn-di-dum. White. E United
  States.
  reginae ray-*geen*-ie. Of the Queen. E N
  America.

Cyrtanthus kur-*tănth*-us *Amaryllidaceae*.
From Gk. *kyrtos* (arched) and *anthos* (a flower`)
referring to the curved perianth tube. Tender
and semi-hardy bulbous herbs. S Africa.
  angustifolius ăn-gus-ti-*fo*-lee-us. Narrow-
  leaved.
  mackenii ma-*ken*-ee-ee. After Mark John
  M'ken (1823–72), a Scotsman who became
  curator of the Natal Botanic Garden and
  collected in S Africa.
  obrienii ō-*brie*-an-ee-ee. After Mr Jas.
  O'Brien who imported it.
  ochroleucus ok-rō-*loo*-kus. Yellowish-white.
  parviflorus par-vi-*flō*-rus. Small-flowered.
  sanguineus sang-*win*-ee-us. Blood-red (the
  flowers).

Cyrtomium falcatum see *Polystichum
falcatum*

Cystopteris kis-*top*-te-ris *Athyriaceae*. From
Gk. *kystos* (a bladder) and *pteris* (a fern)
referring to the bladder-like indusium.
Bladder Fern.
  bulbifera bul-*bi*-fe-ra. Bulb-bearing, the
  small, bulb-like growths on the fronds
  from which new plants grow. Berry
  Bladder Fern. N America.
  dickieana dik-ee-*ah*-na. After Prof. George
  Dickie (1812–82), of Aberdeen. Arctic
  Bladder Fern. N hemisphere.
  fragilis fră-gi-lis. (= *C. alpina*). Brittle (the
  stalks). Brittle Bladder Fern. Widely
  distributed.
  montana mon-*tah*-na. Of mountains.
  Mountain Bladder Fern. N hemisphere.

Cytisus *si*-ti-sus *Leguminosae*. From *kytisos*,
Gk. name for these or similar shrubs.
Deciduous and evergreen shrubs. Broom.
  battandieri ba-ton-dee-*e*-ree. After Jules
  Aimé Battandier, French botanist who
  studied Algerian plants. Pineapple-scented
  Broom. Morocco.
  × *beanii* been-ee-ee. *C. ardoinii* × *C.
  purgans*. After William Jackson Bean
  (1863–1947), Kew botanist and authority
  on hardy, woody plants.
  'Burkwoodii' burk-*wud*-ee-ee. After Albert
  Burkwood of Burkwood and Skipwith.

Cytisus (continued)
decumbens day-*kum*-benz. Prostrate.
S Europe.
× *kewensis* kew-*en*-sis. *C. ardoinii* × *C.
multiflorus.* Of Kew where it was raised in
1891.
monspessulanus mon-spès-ew-*lah*-nus. Of
Montpelier. Montpelier Broom. S Europe,
N Africa, Syria.
multiflorus mul-tee-*flō*-rus. Many-flowered.
White Spanish Broom. Spain, Portugal.
nigricans nig-ri-känz. Blackish, the flowers
turn black when dried. C and S Europe.
× *praecox* prie-koks. *C. multiflorus* × *C.
purgans.* Early (flowering).
procumbens prō-*kum*-benz. Prostrate. SE
Europe.
purgans pur-ganz. Purging. SW Europe.
purpureus pur-*pewr*-ree-us. Purple (the
flowers). Alps, SE Europe.
racemosus hort. see *C.* × *spachianus*
scoparius skō-*pah*-ree-us. Broom-like.
  'Andreanus' on-dray-*ah*-nus. After
  André who discovered it in Normandy,
  see *Anthurium andreanum.*
× *spachianus* spah-chee-*ah*-nus. *C.
canariensis* × *C. stenopetalus.* (= *C.
racemosus* hort. *Genista fragrans* hort.).
After Edouard Spach (1801–79), French
botanist.
supranubias soo-pra-*new*-bee-as. (=
Spartocytisus nubigenus). From above the
clouds. Canary Islands.
× *versicolor* ver-si-ko-lor. *C. hirsutus* × *C.
purpureus.* Variously coloured (the
flowers).
  'Hillieri' *hil*-ee-a-ree. After Hillier's who
  raised it.

D

Daboecia dă-bō-*ee*-kee-a *Ericaceae.* After St.
Dabeoc. Evergreen, heath-like shrubs.
azorica a-*zo*-ri-ka. Of the Azores.
cantabrica kän-*tă*-bri-ka. Of Cantabria,
N Spain. W Europe.
  'Atropurpurea' ah-trō-pur-*pewr*-ree-a.
  Deep purple.
  'Bicolor' *bi*-ko-lor. Two-coloured.
  'Praegerae' *pray*-ga-rie. After Mrs
  Praeger who found it.
× *scotica* sko-ti-ka. *D. azorica* × *D.
cantabrica.* Of Scotland where it was raised.

Dacrydium da-*krid*-ee-um *Podocarpaceae.*
From Gk. *dacrydion* (a small tear), referring to

Dacrydium (continued)
the exuded resin drops. Evergreen conifer.
franklinii frank-*lin*-ee-ee. After Sir John
Franklin, Arctic explorer (1786–1847).
Huon Pine. Tasmania.

Dactylorhiza dăk-til-ō-*ree*-za *Orchidaceae.*
From Gk. *dactylos* (a finger) and *rhiza* (a root)
referring to the finger-like tubers. Hardy
orchids.
elata ay-*lah*-ta. (= *Orchis elata*). Tall.
N Africa.
foliosa fo-lee-*ō*-sa. (= *Orchis foliosa, Orchis
maderensis*). Leafy. Madeira.
fuchsii fuks-ee-ee. After Fuchs. Common
Spotted Orchid. Europe.
incarnata in-kar-*nah*-ta. (= *Orchis
incarnata*). Pink (the flowers). Meadow
Orchid. Europe.
majalis mah-*yah*-lis. Flowering in May.
Europe.
  praetermissa prie-ter-*mis*-a. (=
  *Dactylorhiza praetermissa. Orchis
  praetermissa*). Overlooked. NW Europe.

Daffodil, Hoop-petticoat see *Narcissus
bulbocodium*
  Pheasant's Eye see *N. poeticus recurvus*
  Wild see *N. pseudonarcissus*

Dahlia dah-lee-a *Compositae.* After Dr
Anders Dahl (1751–89), Swedish botanist.
Tender and semi-hardy tuberous perennials.
Mexico.
coccinea kok-*kin*-ee-a. Scarlet (the ray
flowers).
pinnata pin-*ah*-ta. Pinnate (the leaves).

Daisy see *Bellis perennis*
Daisy Bush see *Olearia*
Dame's Violet see *Hesperis matronalis*

Danae dă-na-ay *Liliaceae.* After Danae of
Gk. mythology. Evergreen shrub.
racemosa ră-kay-*mō*-sa. (= *Ruscus
racemosus*). With flowers in racemes.
Alexandrian Laurel. SW Asia.

Daphne dăf-nay *Thymelaeaceae.* The Gk.
name for *Laurus nobilis.* Deciduous and
evergreen shrubs.
arbuscula ar-*bus*-kew-la. Like a dwarf tree.
E Czechoslovakia.
bholua bo-loo-a. From the native name,
Bholu Swa. Himalaya.
blagayana blă-gay-*ah*-na. After Count
Blagay who discovered it in 1837. SE
Europe.

Daphne (continued)
× *burkwoodii* burk-*wud*-ee-ee. *D. caucasica*
× *D. cneorum*. After Albert Burkwood
who raised it.
cneorum nee-*o*-rum. Gk. name for an olive-
like shrub. Garland Flower. Europe.
genkwa genk-wa. The Japanese version of
the Chinese name, Genk'wa. China.
laureola low-*ree*-o-la. L. name for a little
laurel crown. Spurge Laurel. S and W
Europe, N Africa.
× *mantensiana* măn-tenz-ee-*ah*-na. *D.* ×
burkwoodii × *D. retusa*. After Manten's
nursery. British Columbia.
mezereum me-*ze*-ree-um. From *mezereon* the
L. name. Mezereon. Europe, Siberia.
odora o-*dō*-ra. Fragrant. China.
'Aureo-marginata' *ow*-ree-ō-mar-gi-*nah*-
ta. Gold-margined (the leaves).
petraea pe-*trie*-a. Growing on rocks.
N Italy.
retusa re-*tew*-sa. With a notched apex (the
leaves). E Himalaya, W China.
tangutica tăn-*gew*-ti-ka. Of Gansu (Kansu).
W China.

Daphniphyllum dăf-nee-*fil*-lum
Daphniphyllaceae. From Gk. *daphne* (laurel)
and *phyllon* (a leaf) from the resemblance of
the leaves to those of *Laurus nobilis*.
Evergreen shrub.
macropodum ma-*kro*-po-dum. With a large
stalk (the leaves). Japan.

Darlingtonia dar-ling-*ton*-ee-a
Sarraceniaceae. After Dr William Darlington
(1782–1863). Carnivorous herb.
californica kăl-i-*forn*-i-ka. Californian.
California Pitcher Plant. California,
Oregon.

Date Plum see *Diospyros lotus*

Datura da-*tewr*-ra *Solanaceae*. From a native
name. Tender herbs and shrubs.
arborea ar-*bo*-ree-a. (= *D. cornigera*). Tree-
like. Angel's Trumpet. S America.
ceratocaula ke-ra-tō-*kaw*-la. With a horn-
like stem. Mexico.
inoxia in-*oks*-ee-a. (= *D. meteloides*). Not
spiny. SW United States, Mexico.
metel me-tel. The native name. SW China.
sanguinea san-*gwin*-ee-a. Blood-red (the
corolla). S America.
suaveolens swah-*vee*-o-lenz. Sweetly
scented. Brazil.

Daucus dow-kus *Umbelliferae*. The L. name
Biennial herb.
carota ka-*rot*-a. From *karoton* the Gk. name
Wild Carrot. Europe, Asia.
sativus sa-*tee*-vus. Cultivated. Carrot.

Davallia da-*vahl*-ee-a *Davalliaceae*. After
Edmond Davall (1763–1798), Swiss botanist.
Evergreen, tender ferns.
canariensis ka-nah-ree-*en*-sis. Of the Canary
Islands. Deer's-foot Fern. Canary Islands,
Portugal, Spain, Madeira.
fejeensis fee-jee-*en*-sis. Of Fiji. Rabbit's-foot
Fern.
mariesii ma-*reez*-ee-ee. After Charles Maries
(c. 1851–1902), who collected for Veitch
in China and Japan. Squirrel's-foot Fern,
Ball Fern. E Asia.
trichomanoides trik-ō-mahn-*oi*-deez. Like
Trichomanes. Squirrel's-foot Fern.
Malaysia.

Davidia da-*vid*-ee-a *Davidiaceae*. After Abbé
Armand David (1826–1900), French
missionary and plant collector in China who
discovered the following in 1869. Deciduous
tree.
involucrata in-vo-loo-*krah*-ta. With an
involucre, referring to the showy bracts.
Dove Tree, Handkerchief Tree. China.
vilmoriniana vil-mo-rin-ee-*ah*-na. After
Maurice Vilmorin, French nurseryman
who first raised it.

Dawn Redwood see *Metasequoia*
glyptostroboides
Day Flower see *Commelina*
Day Lily see *Hemerocallis*
Dead Nettle see *Lamium*
  Giant see *L. orvala*
  Spotted see *L. maculatum*

Decaisnea de-*kayz*-nee-a *Lardizabalaceae*.
After Joseph Decaisne (1807–82), French
botanist and director of the Jardin des
Plantes, Paris. Deciduous shrub.
fargesii far-*geez*-ee-ee. After Père Paul
Guillaume Farges (1844–1912), French
missionary who introduced it to France in
1895. W China.

Decumaria dek-ew-*mah*-ree-a
Hydrangeaceae. From L. *decimus* (ten), the
parts of the flower are in tens. Woody
climbers.
barbara bar-ba-ra. Foreign, it was originally

Decumaria (continued)
thought to be introduced. SE United
States.
sinensis sin-*en*-sis. Of China.

Delphinium del-*fin*-ee-um *Ranunculaceae*.
From the Gk. name, from *delphis* (a dolphin)
referring to the shape of the flowers.
Perennial herbs.
ajacis see *Consolida ambigua*
cardinale kar-di-nah-lee. Scarlet (the
flowers). California.
consolida see *Consolida regalis*
elatum ay-*lah*-tum. Tall. Europe, Siberia.
exaltatum eks-al-*tah*-tum. Very tall. SE
United States.
grandiflorum grän-di-*flō*-rum. Large-
flowered. Nepal to China.
× *magnificum* mag-*ni*-fi-kum. *D.*
cheilanthifolium × *D. grandiflorum.* (*D.
formosum* hort.). Magnificent.
nudicaule new-di-*kaw*-lee. With a bare
stem. California, Oregon.
× *ruysii* ries-ee-ee. *D. elatum* × *D.
nudicaule.* After B. Ruys, Dutch
nurseryman.
tatsienense tăt-see-en-*en*-see. Of Tatsienlu
(now K'ang-ling), Sichuan, W China.
zalil zah-lil. The native name. Afghanistan.

Dendrobium den-*drō*-bee-um *Orchidaceae*.
From Gk. *dendron* (a tree) and *bios* (life)
referring to their epiphytic habit. Greenhouse
orchids.
aphyllum a-*fil*-lum. (= *D. pierardii*).
Leafless. Himalaya, SE Asia.
aureum see *D. heterocarpum*
bigibbum bi-*gib*-um. With two swellings (at
the base of the lip). N Australia.
brymerianum brie-ma-ree-*ah*-num. After
W. E. Brymer. SE Asia.
densiflorum dens-i-*flō*-rum. Densely-
flowered. Himalaya, Assam, Burma.
fimbriatum fim-bree-*ah*-tum. Fringed (the
lip). Himalaya, India, Burma.
heterocarpum he-te-ro-*kar*-pum. (= *D.
aureum*). With variable fruit. Nepal,
Sikkim, Burma.
infundibulum in-fun-*dib*-ew-lum. Funnel-
shaped (the flowers). Burma.
kingianum king-ee-*ah*-num. After Captain
King. Australia.
longicornu long-gi-*kor*-noo. With a long
horn. Himalaya, Assam, Burma.
moschatum mos-*kah*-tum. Musk-scented.
Himalaya, India, Burma.
nobile *nō*-bi-lee. Notable. Himalaya, SE
Asia.

Dendrobium (continued)
pierardii see *D. aphyllum*
primulinum preem-ew-*leen*-um. Primrose-
coloured. Himalaya, N Burma.
victoriae-reginae vik-*tor*-ree-ie-ray-*geen*-ie.
After Queen Victoria. Philippines.
williamsonii wil-yam-*son*-ee-ee. After Mr
W. J. Williamson who discovered it in
1868. Himalaya.

Dendrochilum den-drō-*keel*-um *Orchidaceae*.
From Gk. *dendron* (a tree) and *cheilos* (a lip),
referring to the epiphytic habit and
conspicuous lip. Greenhouse orchids.
Philippines.
cobbianum cob-ee-*ah*-num. After Walter
Cobb (c. 1836–1922) of Sydenham who
first flowered it.
filiforme fee-lee-*form*-ee. Thread-like.
glumaceum gloo-*mah*-kee-um. With dry,
chaffy bracts.

Dennstaedtia den-*stet*-ee-a *Dennstaedtiaceae*.
After August Wilhelm Dennstedt
(1776–1826). German botanist. Fern.
punctilobula punk-tee-*lob*-ew-la. With
dotted lobules. Hay-scented Fern. E N
America.

Dentaria see *Cardamine*

Dendromecon den-drō-*may*-kon
Papaveraceae. From Gk. *dendron* (a tree) and
mecon (a poppy). Semi-hardy evergreen
shrub.
rigida ri-gi-da. Rigid (the leaves).
California.

Deodar see *Cedrus deodara*
Desert Privet see *Peperomia magnoliifolia*

Desfontainea des-(or day-)fon-*tay*-nee-a
Potaliaceae. After René Louiche
Desfontaines (1753–1833), French botanist.
Semi-hardy, evergreen shrub.
spinosa spee-*nō*-sa. Spiny (the leaves).
Andes.

Desmodium des-*mō*-dee-um *Leguminosae*.
From Gk. *desmos* (a chain) referring to the
jointed stamen. Annual herb.
motorium mō-*to*-ree-um. (= *D. gyrans*).
Moving (the leaflets). Telegraph Plant.
Tropical Asia.

Deutzia doytz-ee-a *Philadelphaceae*. After
Johann van der Deutz (1743–88), a patron
of Thunberg. Deciduous shrubs.

Deutzia (continued
compacta com-*păk*-ta. Compact (the
inflorescence). China.
× *elegantissima* ay-le-gan-*tis*-i-ma. *D.*
purpurascens × *D. sieboldiana*. Most
elegant.
gracilis gră-ki-lis. Graceful. Japan.
longifolia long-gi-*fo*-lee-a. With long leaves.
W China.
× *magnifica* magh-*ni*-fi-ka. *D. scabra* × *D.*
vilmoriniae. Magnificent.
pulchra pul-kra. Pretty. Philippines,
Taiwan.
× *rosea* ros-ee-a. *D. gracilis* × *D.*
purpurascens. Rose-coloured (the flowers).
scabra skă-bra. Rough (the leaves). Japan,
China.
setchuanensis sech-wahn-*en*-sis. Of Sichuan
(Setchwan). China.
  corymbiflora ko-rim-bee-*flō*-ra. With
  flowers in corymbs.

Devil Flower see *Tacca chantrieri*
Devil's Backbone see *Kalanchoe*
daigremontiana
Devil's Claw see *Physoplexis comosa*
Devil's Fig see *Argemone mexicana*
Devil's Ivy see *Epipremnum aureum*
Devil's Paintbrush see *Hieracium aurantiacum*
Devil's Tongue see *Amorphophallus rivieri*

Dianella dee-a-*nel*-la *Liliaceae*. Diminutive
of Diana, goddess of the chase. Rhizomatous
perennials. Flax Lily.
caerulea kie-*ru*-lee-a. Deep blue. New
South Wales.
intermedia in-ter-*me*-dee-a. Intermediate.
New Zealand.
tasmanica tăz-*măn*-i-ka. Of Tasmania.

Dianthus dee-*ănth*-us *Caryophyllaceae*. From
Gk. *Di* (of Zeus or Jove) and *anthos* (a
flower). Annual, biennial and perennial
herbs. Pink.
× *allwoodii* awl-*wud*-ee-ee. *D. caryophyllus*
× *D. plumarius*. After M. C. W. Allwood
(c. 1879–1958), nurseryman specialising in
Dianthus.
alpinus ăl-*peen*-us. Alpine. Alps.
arenarius ă-ray-*nah*-ree-us. Growing in
sandy places. S Sweden.
× *arvenensis* ar-ven-*en*-sis. *D.*
monspessulanus × *D. seguieri*. Of the
Auvergne. France.
barbatus bar-*bah*-tus. Bearded (the petals).
Sweet William. S and E Europe.
caesius see *D. gratianopolitanus*
× *calalpinus* kăl-al-*peen*-us. *D. alpinus* × *D.*

Dianthus (continued)
callizonus. From the names of the parents.
callizonus kă-lee-*zōn*-us. Beautifully zoned
(the petals). S Carpathians.
carthusianorum kar-thew-zee-a-*nor*-rum. Of
the monks of the Carthusian Monastery
nr. Grenoble. SW and C Europe.
caryophyllus kă-ree-ō-*fil*-lus. Smelling of
walnut leaves. Carnation, Clove Pink.
Mediterranean region.
chinensis chin-*en*-sis. Of China. Indian
Pink.
deltoides del-*toi*-deez. Shaped like the Gk.
letter delta (Δ) (the petals). Europe.
glacialis glă-kee-*ah*-lis. Growing near
glaciers. Glacier Pink. E Alps,
Carpathians.
gratianopolitanus grah-tee-ah-nō-po-li-*tah*-
nus. Of Grenoble. Cheddar Pink. Europe.
haematocalyx hie-mah-tō-*kă*-liks. With a
blood-red calyx. Yugoslavia, Greece.
knappii nahp-ee-ee. After Joseph Armin
Knapp (1843–99), botanist of Vienna.
W Yugoslavia.
myrtinervius mur-tee-*ner*-vee-us. Veined like
Myrtus. Greece, Yugoslavia.
neglectus see *D. pavonius*
noeanus see *D. petraeus noeanus*
pavonius pah-*vō*-nee-us. (= *D. neglectus*).
From L. *pavo* (a peacock) referring to the
brilliantly coloured flowers. Alps.
petraeus pe-*trie*-us. Growing on rocks. SE
Europe.
  noeanus nō-ee-*ah*-nus. (= *D. noeanus*).
  After Friedrich Wilhelm Noë. Bulgaria.
plumarius ploo-*mah*-ree-us. Plumed (the
fringed petals). E and C Europe.
superbus soo-*perb*-us. Superb. Fringed
Pink. Europe to Japan.

Diascia dee-ă-skee-a *Scrophulariaceae*. From
Gk. *di* (two) and *askos* (a sac), the flowers
have two spurs. Perennial herbs. S Africa.
barberiae bar-*be*-ree-ie. After Mrs Barber.
Twinspur.
cordata kor-*dah*-ta. Heart-shaped (the
leaves).
rigescens ri-*ges*-enz. Somewhat rigid.

Dicentra di-*ken*-tra *Fumariaceae*. From Gk.
di (two) and *kentron* (a spur), the flowers
have two spurs. Perennial herbs.
cucullaria kuk-ew-*lah*-ree-a. Hood-like (the
flowers). Dutchman's Breeches. E N
America.
eximia eks-*i*-mee-a. Distinguished.
E United States.
formosa for-*mō*-sa. Beautiful. W N
America.

Dicentra (continued)
spectabilis spek-*tah*-bi-lis. Spectacular.
Bleeding Heart. Japan.

Dichelostemma di-kel-ō-*stem*-a
maryllidaceae. From Gk. *dicha* (bifid) and
emma (a garland) referring to the stamen
ppendages. Cormous herbs.
ida-maia ee-da-*mah*-ya. (= *Brodiaea ida-
maia*). After Ida May Burke of California.
Floral Firecracker. Oregon, California.
pulchellum pul-*kel*-lum. Pretty. W N
America.

Dicksonia dik-*son*-ee-a *Dicksoniaceae*. After
ames Dickson (1738–1822), Scottish
urseryman and naturalist. Tender Tree
erns.
antarctica ăn-*tark*-ti-ka. Of Antarctic
regions. Woolly Tree Fern. Australia,
Tasmania.
fibrosa fi-*brō*-sa. Fibrous (the trunk).
Golden Tree Fern. New Zealand.
squarrosa skwah-*rō*-sa. With the parts
spreading at right angles. New Zealand.

Dictamnus dik-*tăm*-nus *Rutaceae*. Gk. name
or an origanum. Perennial herb.
albus ăl-bus. (= *D. fraxinella*). White, the
flowers vary from white to purple.
Dittany, Burning Bush. S Europe to
N China.

Didiscus caeruleus see *Trachymene caerulea*

Dieffenbachia dee-fan-*bahk*-ee-a *Araceae*.
After J. F. Dieffenbach (1790–1863).
Tender perennials. Dumb Cane.
amoena a-*moyn*-a. Pleasant. S America.
× *bausei bowz*-ee-ee. *D. maculata* × *D.
weirii*. After Christian Frederick Bause (c.
1839–95), German nurseryman with the
R.H.S. at Chiswick, later with Veitch.
bowmannii bow-*măn*-ee-ee. After its
introducer, David Bowman (1838–68),
who collected for Veitch in S America.
Brazil.
'Exotica' eks-*o*-ti-ka. Exotic. Costa Rica.
imperialis im-pe-ree-*ah*-lis. Showy. Peru.
maculata măk-ew-*lah*-ta. (= *D. picta*).
Spotted (the leaves). C and S America.
oerstedii ur-*sted*-ee-ee. After Anders Oersted
who collected in S America in the 19th
century. Mexico, Costa Rica.

Dierama dee-e-*rah*-ma *Iridaceae*. From Gk.
dierama (a funnel) referring to the shape of
the flowers. Cormous perennial.

Dierama (continued)
pulcherrimum pul-*ke*-ri-mum. Very pretty.
Angel's Fishing Rod, Wand Flower.
S Africa.

Diervilla dee-er-*vil*-la *Caprifoliaceae*. After
Dr N. Dierville, a French surgeon who
introduced the following to Europe about
1700. Deciduous shrub.
lonicera lon-i-*se*-ra. From *Lonicera* q.v., a
related genus. E N America.

Digitalis di-gi-*tah*-lis *Scrophulariaceae*. From
L. *digitus* (a finger) referring to the finger-
like flowers. Biennial and perennial herbs.
dubia *dub*-ee-a. Dubious (that it belongs to
the genus), not typical. Balearic Islands.
ferruginea fe-roo-*gin*-ee-a. Rusty (markings
on the flowers). SE Europe, W Asia.
grandiflora grăn-di-*flō*-ra. Large-flowered.
E and C Europe, W Asia.
lanata lah-*nah*-ta. Woolly (the racemes).
E Europe.
lutea loo-tee-a. Yellow. W Europe.
× *mertonensis* mer-ton-*en*-sis. *D. grandiflora*
× *D. purpurea*. Of Merton.
purpurea pur-*pewr*-ree-a. Purple. Foxglove.
Europe.

Dill see *Anethum graveolens*

Dimorphotheca di-mor-fô-*thee*-ka
Compositae. From Gk. *dis* (twice), *morphe*
(shape) and *theka* (a fruit) referring to the
different kinds of fruit produced by the ray
and disc flowers. Tender annuals. S Africa.
barberiae see *Osteospermum jucundum*
ecklonis see *Osteospermum ecklonis*
pluvialis ploo-vee-*ah*-lis. (= *D. annua*). Of
rain, it flowers after the rains.
sinuata sin-ew-*ah*-ta. (= *D. aurantiaca* hort.
D. calendulacea). Wavy-edged (the leaves).
Star of the Veldt.

Dionaea dee-on-*ie*-a *Droseraceae*. Gk. name
for Venus. Carnivorous herb.
muscipula mus-*kip*-ew-la. Fly-catching.
Venus's Fly Trap. N and S Carolina
(bogs).

Dioscorea dee-os-*ko*-ree-a *Dioscoreaceae*.
After Pedanios Dioscorides, 1st century Gk.
herbalist. Tender Climbers. Yam.
discolor dis-ko-lor. Two-coloured (the
leaves). Ornamental Yam. Ecuador.
elephantipes e-le-*făn*-ti-pays. Like an
elephant's foot (the half-exposed tuber).
Elephant's Foot, Tortoise Plant. S Africa.

Diospyros dee-*os*-pi-ros *Ebenaceae*. From Gk. *dios* (divine) and *pyros* (wheat) referring to the edible fruits. Deciduous trees.
  kaki kah-ki. The Japanese name. Chinese Persimmon, Kaki. China.
  lotus lō-tus. Gk. *lotos*, which was applied to many plants. Date Plum. China.
  virginiana vir-jin-ee-*ah*-na. Of Virginia. Persimmon. SE United States.

Dipelta di-*pel*-ta *Caprifoliaceae*. From Gk. *di* (two) and *pelta* (a shield) referring to the conspicuous bracts which enclose the fruit. Deciduous shrubs.
  floribunda flō-ri-*bun*-da. Profusely flowering. C and W China.
  ventricosa ven-tri-*kō*-sa. Swollen on one side (the base of the corolla). W China.
  yunnanensis yoo-nan-*en*-sis. Of Yunnan, China.

Diplacus aurantiacus see *Mimulus aurantiacus*
  puniceus see *M. aurantiacus puniceus*
Dipladenia see *Mandevilla*

Dipsacus dip-sa-kus *Dipsacaceae*. The Gk. name, from *dipsa* (thirst), water collects in the cup formed by the leaf bases. Biennial herb.
  fullonum fu-lō-num. Of fullers, the prickly fruiting heads of *D. sativus* were used by fullers to tease cloth. Teasel. Europe, Asia.

Disanthus dis-*ănth*-us *Hamamelidaceae*. From Gk. *dis* (twice) and *anthos* (a flower), the flowers are borne in pairs. Deciduous shrub.
  cercidifolius ker-ki-di-*fo*-lee-us. *Cercis*-leaved. Japan.

Disporum di-spo-rum *Liliaceae*. From Gk. *dis* (two) and *spora* (a seed), each chamber of the ovary contains two seeds. Perennial herb. Fairy Bells.
  sessile se-si-lee. Stalkless (the leaves). Japan, China.

Distylium di-*stil*-ee-um *Hamamelidaceae*. From Gk. *dis* (two) and *stylos* (a style), the flowers have two styles. Evergreen Shrub.
  racemosum ră-kay-*mō*-sum. With flowers in racemes. S Japan.

Dittany see *Dictamnus albus*

Dizygotheca di-zi-go-*thee*-ka *Araliaceae*. From Gk. *dis* (twice), *zygos* (a yoke) and *theka* (a case) the anthers have four lobes, twice the normal number. Tender, evergreen shrub.
  elegantissima ay-le-găn-*tis*-i-ma. (= *Aralia elegantissima*). Most elegant. Finger Aralia, False Aralia. New Caledonia, Polynesia.

Dodekatheon dō-dek-a-*thee*-on *Primulaceae*. From Gk. *dodeka* (twelve) and *thios* (god). Perennial herbs. Shooting Star.
  frigidum fri-gi-dum. Of cold regions. NW North America, NE Asia.
  hendersonii hen-der-*son*-ee-ee. After Louis Fourniquet Henderson (1853–1942). WN America.
  jeffreyi jef-ree-ee. After Jeffrey, see *Pinus jeffreyi*.
  meadia mee-dee-a. After Richard Mead (1673–1754), English physician and botanical patron. E United States.

Dog's-tooth Violet see *Erythronium dens-canis*
Dogwood see *Cornus*
  Common see *C. sanguinea*
  Flowering see *C. florida*
  Pacific see *C. nuttallii*

Dombeya dom-bee-a *Sterculiaceae*. After Joseph Dombey (1742–94), French botanist who collected in S America. Tender, evergreen shrubs.
  burgessiae bur-*jes*-ee-ie. (= *D. mastersii*). After Miss Burgess of Birkenhead. E and S Africa.
  × *cayeuxii* kie-yuz-ee-ee. *D. burgessiae* × *D. wallichii*. After M. Henri Cayeux of the Lisbon Botanic Garden who raised it in 1895.

Doritis do-*ree*-tis *Orchidaceae*. From Gk. *dory* (a spear) referring to the spear-shaped lip. Greenhouse orchid.
  pulcherrima pul-*ke*-ri-ma. (= *Phalaenopsis esmeralda*). Very pretty. SE Asia.

Doronicum do-*ron*-i-kum *Compositae*. Derivation obscure. Perennial herbs. Leopard's Bane.
  austriacum ow-stree-*ah*-cum. Austrian. Europe.
  columnae ko-*lum*-nie. (= *D. caucasicum*. *D. cordatum*). After Fabius Columna (1567–1640). SE Europe, W Asia.
  pardalianches par-da-lee-*ahn*-kees. Gk. name for a poisonous plant, originally

Doronicum (continued)
thought to be this one, meaning literally,
strangling leopards. Europe.
plantagineum plahn-ta-*gin*-ee-um. *Plantago*-
like. Europe.

Dorotheanthus do-ro-thee-*ănth*-us
Aizoaceae. Dr Martin Heinrich Schwantes
named this genus after his mother Dorothea.
Succulent annuals. S Africa.
bellidiformis bel-id-ee-*form*-is. (=
Mesembryanthemum criniflorum). *Bellis*-
like. Livingstone Daisy.
tricolor tri-ko-lor. (= *Mesembryanthemum
tricolor*). Three-coloured (the flowers).

Dorycnium do-*rik*-nee-um *Leguminosae*.
From *doryknion*, Gk. name for a *Convolvulus*.
Sub-shrub.
hirsutum hir-*soo*-tum. Hairy. S Europe.

Douglas Fir see *Pseudotsuga menziesii*
Douglasia vitaliana see *Vitaliana primuliflora*
Dove Tree see *Davidia involucrata*

Draba drah-ba *Cruciferae*. From *drabe*, Gk.
name for a related plant. Perennial herbs.
aizoides ie-zō-*ee*-deez. Like *Aizoon*.
Europe.
aizoon see *D. lasiocarpa*
alpina ăl-*peen*-a. Alpine. Europe.
bruniifolia brun-ee-i-*fo*-lee-a. With leaves
like *Brunia*. Caucasus.
bryoides see *D. rigida*.
dedeana dee-dee-*ah*-na. After Dede. N and
E Spain.
imbricata see *D. rigida imbricata*
lasiocarpa lă-see-ō-*kar*-pa. Woolly-fruited.
E Europe.
mollisima mol-*lis*-i-ma. Very soft. Caucasus.
polytricha po-*li*-tri-ka. With many hairs.
Armenia.
repens see *D. sibirica*
rigida ri-gi-da. (= *D. bryoides*). Rigid (the
leaves). Caucasus.
   imbricata im-bri-*kah*-ta. (= *D. imbricata*).
   Overlapping (the leaves).
× *salomonii* sah-lo-*mon*-ee-ee. *D. bruniifolia*
× *D. dedeana*. After Salomon.
sibirica si-*bi*-ri-ka. (= *D. repens*). Of
Siberia. Siberia, Caucasus, Greenland.

Dracaena dra-*kie*-na *Agavaceae*. From Gk.
drakɔina (a dragon). It has also been
suggested that it was named after Sir Francis
Drake. Tender, evergreen shrubs and trees.
deremensis de-rem-*en*-sis. Of Derema,

Dracaena (continued)
Tanzania. Tropical Africa.
'Bausei' *bowz*-ee-ee. After Bause, see
Dieffenbachia × *bausei*.
draco drǎ-ko. A dragon. Dragon tree.
Canary Islands.
fragrans frah-granz. Fragrant (the flowers).
Upper Guinea.
'Lindenii' lin-*den*-ee-ee. After Linden,
Belgian nurseryman.
'Massangeana' ma-*son*-zhee-ah-na. After
M. de Massange.
goldieana gōl-dee-*ah*-na. After the Rev.
Hugh Goldie, American missionary in
W Africa in the late 19th century. Upper
Guinea.
hookeriana huk-a-ree-*ah*-na. After Hooker.
S Africa.
marginata mar-gi-*nah*-ta. Margined (the
leaves). Madagascar.
reflexa re-*fleks*-a. (= *Pleomele reflexa*).
Reflexed (the inflorescence). Madagascar,
Mauritius.
sanderiana sahn-da-ree-*ah*-na. After Henry
Sander, founder of the famous Sander's
Nursery of St Albans and Bruges.
Cameroons.
surculosa sur-kew-*lō*-sa. (= *D. godseffiana*).
Suckering. Tropical W Africa.

Dracocephalum drǎ-kō-*kef*-a-lum *Labiatae*.
From Gk. *draco* (a dragon) and *cephale* (a
head) referring to the shape of the flowers.
Herbaceous perennials.
forrestii fo-*rest*-ee-ee. After George Forrest,
see *Abies delavayi forrestii*. W China.
grandiflorum grǎn-di-*flō*-rum. Large-
flowered. W Siberia.
hemsleyanum hemz-lee-*ah*-num. After
William Botting Hemsley (1843–1924).
Tibet.
ruyschiana riesh-ee-*ah*-na. After Frederick
Ruysch (1638–1731), Prof. of botany at
Amsterdam. Siberia, N China.

Dracunculus dra-*kun*-kew-lus *Araceae*. L.
name for another plant, meaning a small
dragon. Tuberous perennials. Mediterranean
region.
muscivorus musk-*i*-vo-rus. (= *Helicodiceros
muscivorum*). Fly-eating.
vulgaris vul-*gah*-ris. Common. Dragon
Arum.

Dragon Arum see *Dracunculus vulgaris*
Dragon Tree see *Dracaena draco*

Drimys *drim*-is *Winteraceae*. The Gk. word for acrid, from the taste of the bark. Evergreen shrubs and trees.
  lanceolata lǎn-kee-ō-*lah*-ta. (= *D. aromatica*). Lanceolate (the leaves). Mountain Pepper. SE Australia, Tasmania.
  winteri win-ta-ree. After Captain William Winter who sailed with Frances Drake and collected bark for medicinal and culinary purposes. Winter's Bark. C and S Chile.

Dropwort see *Filipendula vulgaris*

Drosera dro-se-ra *Droseraceae*. From Gk. *droseros* (dewy), referring to the appearance of the leaves. Carnivorous, perennial herbs. Sundew.
  binata bi-*nah*-ta. Paired, the leaf is forked into two segments. SE Australia, New Zealand.
  capensis ka-*pen*-sis. Of the Cape of Good Hope.
  filiformis fee-lee-*form*-is. Thread-like (the leaves). SE United States.

Drunkard's Dream see *Hatiora salicornioides*

Dryas dree-as *Rosaceae*. Gk. *dryas*, from Dryades, daughters of Zeus and originally nymphs of the oak, the leaves somewhat resemble oak leaves. Evergreen shrubs.
  drummondii drū-*mond*-ee-ee. After its discoverer Thomas Drummond (1780–1835). Arctic N America.
  octopetala ok-tō-*pe*-ta-la. Eight-petalled. Mountain Avens. Arctic and alpine N hemisphere.
  × *suendermannii* soon-der-*mahn*-ee-ee. *D. drummondii* × *D. octopetala*. After Franz Sündermann (1864–1946). Arctic N America.

Dryopteris dree-*op*-te-ris *Aspidiaceae*. The Gk. name from *drys* (oak) and *pteris* (a fern). Buckler Fern.
  aemula ie-mew-la. Imitating. Hay-scented Buckler Fern. W Europe.
  cristata kris-*tah*-ta. Crested. Crested Buckler Fern. N America, Europe, Asia.
  dilatata dil-a-*tah*-ta. (= *D. austriaca*). Expanded (the fronds). Broad Buckler Fern. N America, Europe, Asia.
  erythrosora e-rith-rō-*so*-ra. With red sori. China, Japan.
  filix-mas fil-iks-mahs. Literally male fern, to distinguish it from the more delicate Lady Fern. Male Fern. Europe, N America.

Dryopteris (continued)
  goldieana gōl-dee-*ah*-na. After its discoverer, John Goldie (1793–1886). Giant Wood Fern. E N America.

Duchesnea dew-*shez*-nee-a *Rosaceae*. After Antoine Nicolas Duchesne (1747–1827), French horticulturist. Perennial herb.
  indica in-di-ka. (= *Fragaria indica*). Indian. Mock Strawberry. Afghanistan, Himalaya, E Asia.

Dumb Cane see *Dieffenbachia*
Dusty Miller see *Artemisia stelleriana*
Dutchman's Breeches see *Dicentra cucullaris*
Dutchman's Pipe see *Aristolochia macrophylla*
Dyer's Greenweed see *Genista tinctoria*

E

Earth Star see *Cryptanthus*
Easter Lily see *Lilium longiflorum*
Eastern Red Cedar see *Juniperus virginiana*

Eccremocarpus e-krem-ō-*kar*-pus *Bignoniaceae*. From Gk. *ekkremus* (hanging) and *karpos* (a fruit) referring to the hanging pods. Semi-hardy woody climber or annual.
  scaber skǎ-ber. Rough. Chile.

Echeveria e-kee-*ve*-ree-a *Crassulaceae*. After Athanasio Echeverriay Godoy, 18th-century botanical artist. Tender succulents. Mexico.
  affinis a-*feen*-is. Related to.
  agavoides a-gahv-*oi*-deez. Like *Agave*.
  carnicolor kar-*ni*-ko-lor. Flesh-coloured (the flowers).
  derenbergii de-ran-*berg*-ee-ee. After J. Derenberg (1873–1928).
  elegans ay-le-gahnz. Elegant.
  gibbiflora gib-bi-*flō*-ra. With the flowers swollen on one side.
  glauca see *E. secunda glauca*
  harmsii harmz-ee-ee. After Dr Hermann Harms (1870–1942), German botanist.
  leucotricha loo-ko-tri-ka. White-haired.
  multicaulis mul-tee-*kaw*-lis. Many-stemmed.
  nodulosa nō-dew-*lō*-sa. With nodules (papillae on the leaves).
  pulvinata pul-vee-*nah*-ta. Cushion-like.
  runyonii rūn-*yon*-ee-ee. After Robert Runyon.
  secunda se-*kun*-da. With flowers on one side of the stalk.

Echeveria (continued)

glauca glow-ka. (= *E. glauca*). Glaucous (the leaves).

setosa say-*tō*-sa. Bristly-hairy (the stems).

shaviana shah-vee-*ah*-na. After the Missouri Botanic Garden (Shaw's Garden, after the founder, Henry Shaw 1800–99) whose staff collected it.

Echinacea e-kee-*nah*-kee-a *Compositae*. From Gk. *echinos* (a hedgehog) referring to the prickly receptacle scales. Perennial herb.

purpurea pur-*pewr*-ree-a. (= *Rudbeckia purpurea*). Purple (the flowers). E United States.

Echinocactus e-keen-ō-*kǎk*-tus *Cactaceae*. From Gk. *echinos* (a hedgehog) and *Cactus* q.v.

grusonii gru-*son*-ee-ee. After Herman Gruson who built up a large collection of cacti. Golden Barrel Cactus. Mexico.

horizonthalonius ho-ri-zon-tha-*lō*-nee-us. Referring to the horizontally-held areoles. SW United States, Mexico.

ingens ing-gaynz. Huge. Mexico.

Echinocereus e-keen-ō-*kay*-ree-us *Cactaceae*. From Gk. *echinos* (a hedgehog) and *Cereus* q.v.

engelmannii eng-gal-*mahn*-ee-ee. After Georg Engelmann (1809–44), German physician and botanist. SW United States, N Mexico.

enneacanthus en-ee-a-*kǎnth*-us. 9-spined (the areoles). SW United States, N Mexico.

stramineus strah-*min*-ee-us. Straw-coloured (the spines)

fitchii fich-ee-ee. After W. R. Fitch. Texas.

knippelianus ni-pel-ee-*ah*-nus. After Karl Knippel, German cactus dealer. Mexico.

pectinatus pek-tin-*ah*-tus. Comb-like. SW United States, N Mexico.

rigidissimus ri-gi-*dis*-i-mus. Very rigid. S Arizona, N Mexico.

pentalophus pen-ta-*lof*-us. With five crests. S Texas.

pulchellus pul-*kel*-us. Pretty. Mexico.

viridiflorus vi-ri-di-*flō*-rus. With green flowers. SW United States.

Echinochloa e-keen-*o*-klō-a *Gramineae*. From Gk. *echinos* (a hedgehog) and *chloe* (a grass). Annual grass.

crus-galli kroos-*gǎ*-lee. A cock's spur. Asia.

Echinodorus e-keen-*o*-do-rus *Alismataceae*. From Gk. *echinos* (a hedgehog) and *doros* (a

Echinodorus (continued)

bag) referring to the clustered, spiny fruits. Aquarium plants. Burhead.

berteroi bert-a-*rō*-ee. After Carlo Giuseppe Bertero (1789–1831), Italian physician. S United States, W Indies.

cordifolius kor-di-*fo*-lee-us. With heart-shaped leaves. S United States.

horizontalis hoi-ri-zon-*tah*-lis. Horizontal (the leaves). Amazon.

longistylis long-*gi*-sti-lis. With a long style. Brazil.

magdalenensis mǎg-da-layn-*en*-sis. Of the Rio Magdalena, Colombia.

paniculatus pan-ik-ew-*lah*-tus. With flowers in panicles. S America.

quadricostatus kwod-ree-kos-*tah*-tus. Four-ribbed. Tropical S America.

tenellus te-*nel*-us. Dainty. S America.

Echinofossulocactus see *Stenocactus*

Echinomastus macdowellii see *Neolloydia macdowellii*

Echinops e-kee-nops *Compositae*. From Gk. *echinos* (a hedgehog) and *ops* (appearance). Perennial herbs. Globe Thistle.

bannaticus ba-*nǎ*-ti-kus. Of Banat, N Romania. SE Europe, W Asia.

humilis hu-mi-lis. Low growing. W Asia.

ritro rit-rō. A S European name. E Europe, W Asia.

ruthenicus roo-*then*-i-kus. (= *E. ruthenicus*). Of Ruthenia, SW Russia. E Europe.

sphaerocephalus sfie-rō-*kef*-a-lus. With a spherical head. Europe, W Asia.

Echinopsis e-kee-*nop*-sis *Cactaceae*. From Gk. *echinos* (a hedgehog) and *-opsis* (appearance). Sea Urchin Cactus.

aurea ow-ree-a. Golden (the flowers). Argentina.

eyriesii ie-*reez*-ee-ee. After Alexander Eyries who introduced it from Uruguay in 1830. S America.

ferox fe-rōks. (= *Lobivia ferox*). Spiny. Bolivia.

leucantha loo-*kǎnth*-a. White-flowered. W Argentina.

multiplex mul-tee-pleks. With many stems. S Brazil.

rhodotricha rod-*o*-tri-ka. Red-haired. Paraguay, N Argentina.

Echium e-*kee*-um *Boraginaceae*. From *echion* the Gk. name. Biennial herbs.

plantagineum plahn-ta-*gin*-ee-um. (= *E.*

Echium (continued)
lycopsis). Like *Plantago*. Europe,
Caucasus.
russicum ru-si-kum. (= *E. rubrum*).
Russian. E Europe, W Asia.
vulgare vul-*gah*-ree. Common. Viper's
Bugloss. Europe, Asia.

Edgeworthia ej-*werth*-ee-a *Thymelaeaceae*.
After Michael Pakenham Edgeworth
(1812–81), amateur botanist and plant
collector with the E India Co. Semi-hardy,
deciduous shrub.
chrysantha kris-*ănth*-a. (= *E. papyrifera*).
With golden flowers. China.

Edraianthus ed-rie-*ănth*-us *Campanulaceae*.
From Gk. *hedraios* (sitting) and *anthos* (a
flower) referring to the sessile flowers.
Perennial herbs.
dalmaticus dăl-*mă*-ti-kus. (= *E. caudatus*).
Of Dalmatia.
graminifolius grah-min-i-*fol*-ee-us. Grass-
leaved. SE Europe.
pumilio pew-*mil*-lee-ō. Dwarf. Yugoslavia.
serpyllifolius ser-pi-li-*fo*-lee-us. Thyme-
leaved. Yugoslavia.

Egeria ay-*ge*-ree-a *Hydrocharitaceae*. After a
Roman goddess of water. Aquatic perennial
herb.
densa den-sa. (= *Elodea densa*). Dense.
S America.

Egg Plant see *Solanum melongena*
Eglantine see *Rosa eglanteria*
Egyptian Star Cluster see *Pentas lanceolata*

Eichhornia iek-*horn*-ee-a *Pontederiaceae*.
After J. A. F. Eichhorn (1779–1856), an
eminent Prussian. Tender aquatic perennial.
crassipes krăs-i-pays. With a thick stalk (the
leaves). Water Hyacinth. Tropical
America.

Elaeagnus e-lee-*ăg*-nus *Elaeagnaceae*. A Gk.
name originally applied to a willow, from
helodes (growing in marshes) and *hagnos*
(pure), referring to the white fruit masses (of
the willow). Deciduous and evergreen shrubs.
angustifolia ăn-gust-i-*fo*-lee-a. Narrow-
leaved. W Asia.
commutata kom-ew-*tah*-ta. Changeable.
Silver Berry. N America.
× *ebbingei* e-*bing*-gee-ee. *E. macrophylla* ×
E. pungens. After J. W. E. Ebbinge of
Boskoop.
glabra glă-bra. Glabrous. China, Japan.

Elaeagnus (continued)
macrophylla măk-rō-*fil*-la. Large-leaved.
Korea, Japan.
pungens pung-genz. Spiny. Japan.
'Maculata' măk-ew-*lah*-ta. Blotched (the
leaves).
umbellata um-be-*lah*-ta. The flowers are in
umbel-like clusters. Himalaya, China,
Japan.

Elder see *Sambucus*
Common see *S. nigra*
Red-berried see *S. racemosa*
Elecampane see *Inula helenium*

Eleocharis e-lee-o-ka-ris *Cyperaceae*. From
Gk. *helodes* (growing in marshes) and *charis*
(grace). Aquatic perennials.
acicularis a-kik-ew-*lah*-ris. Needle-like (the
stems). Hair Grass. N temperate regions.
dulcis dul-kis. Sweet (the edible tuber).
Chinese Water Chestnut. Tropical Asia.
W Africa.

Elephant's Ears see *Caladium*
Elephant's Foot see *Dioscorea elephantipes*

Elettaria e-le-*tah*-ree-a *Zingiberaceae*. From
elettari the Indian name. Tender herb.
cardamomum kar-da-*mō*-mum. A Gk. name.
India.

Elfin Herb see *Cuphea hyssopifolia*
Elk's Horns see *Rhombophyllum nelii*

Elliottia e-lee-o-tee-a *Ericaceae*. After
Stephen Elliott (1771–1830) who discovered
the following. Deciduous shrub.
racemosa ră-kay-*mō*-sa. With flowers in
racemes. SE United States.

Elm see *Ulmus*
Belgian see *U.* 'Belgica'
Camperdown see *U.* 'Camperdownii'
Cornish see *U. angustifolia cornubiensis*
Dutch see *U.* 'Hollandica'
English see *U. procera*
Exeter see *U.* 'Exoniensis'
Goodyer's see *U. angustifolia*
Jersey see *U.* 'Sarniensis'
Smooth see *U. carpinifolia*
Wheatley see *U.* 'Sarniensis'
Wych see *U. glabra*

Elodea e-lō-dee-a *Hydrocharitaceae*. From
Gk. *helodes* (growing in marshes). Aquatic
herbs.
canadensis kăn-a-*den*-sis. Of Canada or NE

Elodea (continued)
North America. N America.
crispa see *Lagarosiphon major*
densa see *Egeria densa*

Elsholtzia el-*sholtz*-ee-a *Labiatae*. After
Johann Sigismund Elsholtz (1623–88).
Deciduous shrub.
stauntonii stawn-*ton*-ee-ee. After Sir George
Staunton (1737–1801). N China.

Elymus e-li-mus *Gramineae*. From *elymos* the
Gk. name for millet. Perennial Grass.
arenarius ă-ray-*nah*-ree-us. Growing in
sandy places, it is a coloniser of sand
dunes. Lyme Grass. Europe, Asia.

Embothrium em-*both*-ree-um *Proteaceae*.
From Gk. *en* (in) and *bothrion* (a small pit),
the anthers are borne in cup-shaped pits on
the perianth segments. Evergreen shrub or
tree.
coccineum kok-*kin*-ee-um. Scarlet (the
flowers). Fire Bush. Chile, Argentina.

Emilia em-*ee*-lee-a *Compositae*. Derivation
obscure, presumably commemorative.
Annual herb.
javanica ja-vă-ni-ka. Of Java. Tassel
Flower. Tropics.

Emmenopterys e-men-*op*-te-ris *Rubiaceae*.
From Gk. *emmenes* (enduring) and *pteryx* (a
wing) one lobe of the calyx enlarges into a
conspicuous, leaf-like wing. It has not
flowered in this country. Deciduous tree.
henryi hen-ree-ee. After Augustine Henry,
see *Illicium henryi*. C and SW China.

Empetrum em-pe-trum *Empetraceae*.
empetron the Gk. name, from *en* (on) and
petros (a rock) referring to its habitat.
Evergreen shrub. Crowberry.
nigrum nig-rum. Black (the fruit).
N hemisphere.

Endymion see *Hyacinthoides*
Endive see *Cichorium endivia*

Enkianthus eng-kee-*ănth*-us *Ericaceae*. From
Gk. *enkyos* (pregnant) and *anthos* (a flower),
in *E. quinqueflorus* each flower appears to bear
another inside it. Deciduous shrubs.
campanulatus kăm-păn-ew-*lah*-tus. Bell-
shaped (the corolla). Japan.
cernuus ker-new-us. Nodding (the racemes).
Japan.
rubens roo-benz. Red (the flowers).

Enkianthus (continued)
chinensis chin-*en*-sis. Of China. W. China,
Upper Burma.
perulatus pe-ru-*lah*-tus. With conspicuous
bud scales. Japan.

Epidendrum e-pi-*den*-drum *Orchidaceae*.
From Gk. *epi* (upon) and *dendron* (a tree)
referring to their epiphytic habit. Greenhouse
orchids.
brassavolae bra-*sah*-vo-lie. After *Brassavola*
q.v. C America.
ciliare ki-lee-*ah*-ree. Edged with hairs (the
fringed lip). Tropical America.
cochleatum kok-lee-*ah*-tum. Shell-like (the
lip). Tropical America.
difforme di-*form*-ee. Of unusual shape.
Florida, Tropical America.
endresii en-*dres*-ee-ee. After Senor A. R.
Endres (died 1877) who collected in Costa
Rica in 1870. Costa Rica.
fragrans frah-granz. Fragrant. Mexico to
S America.
ibaguense ee-ba-*gen*-see. From Ibagué,
Colombia. C and S America.
mariae mă-ree-ie. After Mrs Mary Östlund
who grew Mexican orchids. Mexico.
nocturnum nok-*tur*-num. Night flowering.
Tropical America.
parkinsonianum par-kin-son-ee-*ah*-num.
After John Parkinson (1772–1847).
Consul-General in Mexico who sent plants
to Kew. C America.
polybulbon po-lee-*bul*-bon. With many
bulbs (pseudobulbs). C America,
W Indies.
prismatocarpum pris-măt-ō-*kar*-pum. With a
prism-shaped fruit. C America.
radiatum ră-dee-*ah*-tum. Radiating, the
purple, radial lines on the lip. C America,
Mexico.
vitellinum vi-te-*leen*-um. Egg-yolk yellow.
Mexico, Guatemala.

Epigaea e-pi-*gie*-a *Ericaceae*. From Gk. *epi*
(on) and *gaia* (the earth) referring to their
creeping habit. Evergreen shrubs.
asiatica ah-see-*ah*-ti-ka. Asian. Japan.
repens ree-penz. Creeping. Trailing
Arbutus. E N America.

Epilobium e-pi-*lo*-bee-um *Onagraceae*. From
Gk. *epi* (upon) and *lobos* (a pod), the corolla
is borne on the end of the ovary. Perennial
herbs. Willow Herb.
chlorifolium klō-ri-*fo*-lee-um. With leaves
like *Chlora* (now *Blackstonia*). New
Zealand.

Epilobium (continued)

kaikourense kie-ku-*ren*-see. Of Kaikoura, New Zealand.

dodonaei dō-do-*nie*-ee. (= *E. rosmarinifolium*). After Rembert Dodoens (1517–85), Flemish physician and herbalist. Europe.

fleischeri flie-sha-ree. After M. Fleischer (1861–1930). Alps.

glabellum gla-*bel*-um. Nearly glabrous. New Zealand.

latifolium lah-tee-*fo*-lee-um. Broad-leaved. N America, Europe, Asia.

Epimedium e-pi-*may*-dee-um *Berberidaceae*. From *epimedion*, Gk. name for another plant. Perennial herbs. Barrenwort.

alpinum ăl-*peen*-um. Alpine. S Europe.

× *cantabrigiense* kăn-ta-brig-ee-*en*-se. *E. alpinum* × *E. pubigerum*. Of Cambridge where it was found by W. T. Stearn and R. Thoday in 1950.

grandiflorum grănd-i-*flō*-rum. Large-flowered. E Asia.

perralderianum pe-ral-de-ree-*ah*-num. After Henri René le Tourneux de la Perraudière (1831–61), French naturalist. Algeria.

pubigerum pew-*bi*-ge-rum. Hairy. Balkans, Caucasus, W Asia.

× *rubrum rub*-rum. *E. alpinum* × *E. grandiflorum*. Red (the sepals).

× *versicolor* ver-si-ko-lor. *E. grandiflorum* × *E. pinnatum colchicum*. Variously coloured (the flowers).

× *warleyense* wor-lee-*en*-see. *E. alpinum* × *E. pinnatum colchicum*. Of Warley Place, see *Ceratostigma willmottianum*.

× *youngianum* yung-ee-*ah*-num. *E. diphyllum* × *E. grandiflorum*. After Young. 'Niveum' *ni*-vee-um. Snow-white (the flowers).

× **Epiphronitis** e-pi-*frō*-ni-tis *Orchidaceae*. Intergeneric hybrid, from the names of the parents. *Epidendrum* × *Sophronitis*. Greenhouse orchids.

Epiphyllum e-pi-*fil*-lum *Cactaceae*. From Gk. *epi* (upon) and *phyllon* (a leaf), the flattened, green stems which bear the flowers, resemble leaves.

ackermannii see *Nopalxochia ackermannii*

anguliger ang-*gew*-li-ger. Hooked (the stems). S Mexico.

caudatum kaw-*dah*-tum. Prolonged into a slender tail (the shoots). S Mexico.

chrysocardium kris-ō-*kar*-dee-um. With a golden heart (the yellow filaments in the

Epiphyllum (continued)

centre of the flower). S Mexico.

crenatum kray-*nah*-tum. With shallow, rounded teeth (on the shoots). C America.

Epipremnum e-pi-*prem*-num *Araceae*. From Gk. *epi* (upon) and *premnum* (a tree stump) referring to the epiphytic habit. Tender, evergreen climber.

aureum ow-ree-um. (= *Scindapsus aureus*). Golden (the variegated leaves). Devil's Ivy. Solomon Islands.

Episcia e-*pis*-kee-a *Gesneriaceae*. From Gk. *episkios* (shaded), they grow in shady places. Tender herbs.

cupreata kew-pree-*ah*-ta. Coppery (the leaves). Flame Violet. N S America.

dianthiflora dee-ănth-i-*flō*-ra. *Dianthus*-flowered. Lace Flower Vine. Mexico.

lilacina li-la-*keen*-a. Lilac (the flowers). C America.

reptans rep-tănz. Creeping. N S America.

Eragrostis e-ra-*gros*-tis *Gramineae*. From Gk. *eros* (love) and *agrostis* (a grass). Annual grasses.

amabilis a-*mah*-bi-lis. Beautiful. Japanese Love Grass. SE Asia.

elegans ay-le-gahnz. Elegant. Love Grass. Brazil.

tef tef. The native name. NE Africa.

Eranthis e-*răn*-this *Ranunculaceae*. From Gk. *er* (spring) and *anthos* (a flower), referring to the early flowers. Tuberous herbs.

cilicia ki-*li*-kee-a. Of Cilicia, S Turkey. W Asia.

hyemalis hee-e-*mah*-lis. Of winter (flowering). Winter Aconite. S Europe.

× *tubergenii* tew-ber-*gen*-ee-ee. *E. cilicia* × *E. hyemalis*. After the Dutch bulb nursery, van Tubergen.

Eremurus e-ray-*mew*-rus *Liliaceae*. From Gk. *eremia* (a desert) and *oura* (a tail) referring to their habitat and the shape of the inflorescence. Perennial herbs. Foxtail Lily.

elwesii el-*wez*-ee-ee. After H. J. Elwes, see *Galanthus elwesii*. Cult.

himalaicus hi-ma-*lah*-i-kus. Of the Himalaya.

olgae ol-gie. After Olga Fedtschenko (1845–1921). N Iran, C Asia.

robustus rō-*bust*-us. Robust. C Asia.

stenophyllus sten-ō-*fil*-lus. (= *E. bungei*). Narrow-leaved. C Asia, Iran.

Erica e-*ree*-ka *Ericaceae.* The classical name, probably for *E. arborea.* Hardy and tender, evergreen shrubs. Heath.
  arborea ar-*bo*-ree-a. Tree-like. Tree Heath. S Europe, N and E Africa, W Asia.
    alpina ăl-*peen*-a. Alpine. Spain.
  australis ow-*strah*-lis. Southern. Spanish Heath. Spain, Portugal.
  canaliculata kăn-ah-lik-ew-*lah*-ta. Channelled (the leaves). S Africa.
  carnea see *E. herbacea*
  cinerea ki-*ne*-ree-a. Grey. Bell Heather. W Europe.
  × *darleyensis* dar-lee-*en*-sis. *E. erigena* × *E. herbacea*. Of Darley Dale, Derbyshire, where it was raised.
  erigena e-ri-*gen*-a. (= *E. mediterranea*). Irish. W Europe.
  gracilis gră-ki-lis. Graceful. S Africa.
  herbacea her-*bah*-kee-a. (= *E. carnea*). Herbaceous. C and E Europe (mountains).
  hyemalis hee-e-*mah*-lis. Of winter (flowering). ? S Africa.
  mediterranea see *E. erigena*
  terminalis ter-mi-*nah*-lis. Terminal (the flowers). SW Europe.
  tetralix *tet*-ra-liks. *tetralice*, an Athenian name for *Erica.* Cross-leaved Heath. N and W Europe.
  vagans vă-gănz. Wandering, the spreading habit. SW Europe, Cornwall.
  ventricosa ven-tri-*kō*-sa. Swollen on one side (the corolla). S Africa.
  × *watsonii* wot-*son*-ee-ee. *E. ciliaris* × *E. tetralix*. After H. C. Watson who discovered it near Truro. NW Europe.
  × *williamsii* wil-*yămz*-ee-ee. *E. tetralix* × *E. vagans*. After P. D. Williams who introduced it to cultivation in 1910. Cornwall.

Erigeron e-*ri*-ge-ron *Compositae.* The classical name of a plant, probably groundsel, from Gk. *eri* (early) and *geron* (an old man), referring to the fluffy, white seed heads. Perennial herbs. Fleabane.
  aurantiacus ow-răn-tee-*ah*-kus. Orange (the flowers). Turkestan.
  aureus ow-ree-us. Golden (the flowers). NW United States.
  compositus kom-*po*-si-tus. Compound (the leaves).
  glaucus glow-kus. Glaucous (the leaves). Beach Aster. W United States.
  karvinskianus kar-vin-skee-*ah*-nus. (= *E. mucronatus*). After Wilhelm Friedrich Karwinski von Karwin (1780–1855), who

Erigeron (continued)
    collected in S America. Mexico to Venezuela.
  leiomerus lay-*o*-me-rus. With smooth parts. W United States.
  simplex *sim*-pleks. Simple (the stem, i.e. unbranched). W United States.
  speciosus spek-ee-*ō*-sus. Showy. W N America.
    macranthus ma-*krănth*-us. (= *E. macranthus*). Large-flowered.

Erinacea e-ri-*nah*-kee-a *Leguminosae.* L. for resembling a hedgehog, referring to its habit. Deciduous, spiny shrub.
  anthyllis ăn-*thil*-lis. (= *E. pungens*). Gk. name for the related kidney vetch. Hedgehog Broom. Spain, N Africa.

Erinus e-ri-nus *Scrophulariaceae.* Gk. name for another plant. Perennial herb.
  alpinus ăl-*peen*-us. Alpine. Fairy Foxglove. Europe.

Eriobotrya e-ree-ō-*bot*-ree-a *Rosaceae.* From Gk. *erion* (wool) and *botrys* (a bunch of grapes), referring to the woolly inflorescence. Evergreen, semi-hardy shrub or small tree.
  japonica ja-*pon*-i-ka. Of Japan. Loquat. China, Japan.

Eritrichium e-ri-*trik*-ee-um *Boraginaceae.* From Gk. *erion* (wool) and *thrix* (hair). Perennial herbs.
  canum kah-num. (= *E. rupestre pectinatum*). Grey. Afghanistan, Himalaya, Tibet.
  nanum nah-num. Dwarf. Fairy Forget-me-not, King of the Alps. Alps.
  rupestre roo-*pes*-tree. Growing on rocks. Asia.

Erodium e-rō-dee-um *Geraniaceae.* From Gk. *erodios* (a heron) referring to the shape of the fruits. Perennial herbs. Heron's Bill.
  absinthoides ăb-sinth-*oi*-deez. Like *Artemisia absinthium*. W Asia.
  chrysanthum kris-*ănth*-um. With golden flowers. Greece.
  corsicum kor-si-kum. Of Corsica. Corsica, Sardinia.
  guttatum gu-*tah*-tum. Spotted (the petals). SW Mediterranean region.
  manescavi măn-es-*kah*-vee. After Manescau (died 1875), Italian merchant and naturalist. Pyrenees.
  petraeum pe-*trie*-um. Growing in rocky places. S France.
    glandulosum glăn-dew-*lō*-sum. (= *E.*

82   Ero-Esc

Erodium (continued)
macradenum). Glandular. Pyrenees,
N Spain.
reichardii rie-*kard*-ee-ee. (= *E.*
chamaedryoides). After Reichard. Balearic
Islands.
× *variabile* vă-ree-*ah*-bi-le. *E. corsicum* ×
E. reichardii. Variable.
'Roseum' *ro*-see-um. (= *E.*
chamaedryoides 'Roseum'). Rose-
coloured (the flowers).

Eryngium e-*ring*-gee-um *Umbelliferae.* From
eryggion the Gk. name for *E. campestre.*
Perennial herbs.
agavifolium a-gah-vi-*fo*-lee-um. *Agave*-
leaved. Argentina.
alpinum ăl-*peen*-um. Alpine. Europe.
amethystinum ă-me-*thist*-i-num. Violet (the
flowers). Europe.
bourgatii bour-*găt*-ee-ee. After M. Bourgat
who collected in the Pyrenees.
Mediterranean region.
giganteum gi-*găn*-tee-um. Very large.
Caucasus.
maritimum ma-*ri*-ti-mum. Growing near the
sea. Sea Holly. Europe (coasts).
× *oliverianum* o-li-va-ree-*ah*-num. After
Oliver.
planum *plah*-num. Flat (the leaves).
Europe, Asia.
tripartitum tri-*par*-tee-tum. Three-parted
(the leaves). Cult.

Erysimum e-*ri*-si-mum *Cruciferae.* From
erysimon the Gk. name. Annual, biennial and
perennial herbs.
asperum a-*spe*-rum. Rough (the leaves).
W and C N America.
capitatum kap-i-*tah*-tum. In a dense head
(the flowers). W N America.
helveticum hel-*vay*-ti-kum. Of Switzerland.
Pyrenees to SE Europe.
linifolium leen-i-*fo*-lee-um. *Linum*-leaved.
Spain, Portugal.
murale mew-*rah*-lee. Growing on walls.
Sicily, N Africa.
perofskianum pe-rof-skee-*ah*-num. After V.
A. Perofsky (1794-c. 1857). Afghanistan,
Pakistan.
pulchellum pul-*kel*-um. (= *E. rupestre*).
Pretty. Greece, W Asia.

Erythrina e-rith-*reen*-a *Leguminosae.* From
Gk. *erythros* (red) referring to the colour of
the flowers. Tender shrub or tree.
crista-galli kris-ta-*gă*-lee. Cock's comb.
Coral Tree. S America.

Erythronium e-rith-*ron*-ee-um *Liliaceae.*
From *erythronion* the Gk. name for another
plant. Perennial herbs.
albidum ăl-bi-dum. White (the flowers).
N America.
americanum a-me-ri-*kah*-num. American.
Yellow Adder's Tongue. E N America.
californicum kăl-i-*forn*-i-kum. Of California.
citrinum ki-*tree*-num. Lemon-yellow (the
flowers). W N America.
dens-canis dens-*kă*-nis. A dog's tooth.
Dog's-tooth Violet. S Europe, Turkey.
grandiflorum grănd-i-*flō*-rum. Large-
flowered. W N America.
hendersonii hen-der-*son*-ee-ee. After Louis
Fourniquet Henderson (1853–1942). W N
America.
howellii how-*el*-ee-ee. After Thomas
Howell (1842–1912). W N America.
multiscapoideum mul-tee-ska-*poi*-dee-um.
With many scapes. N California.
oregonum o-ree-*gŏ*-num. Of Oregon. W N
America.
revolutum re-vo-*loo*-tum. Turned back (the
perianth lobes). Trout Lily. W N
America.
tuolumnense too-o-lum-*nen*-see. Of
Tuolumne County, California.
C California.

Erythrorhipsalis e-rith-rō-*rip*-sa-lis
Cactaceae. From Gk. *erythros* (red) and
Rhipsalis q.v. referring to the red fruits.
pilocarpa pi-lō-*kar*-pa. (= *Rhipsalis*
pilocarpa). With a hairy fruit. Brazil.

Escallonia es-ka-*lon*-ee-a *Escalloniaceae.*
After Senor Escallon, a Spanish traveller in
S America. Evergreen shrubs.
bifida bi-fi-da. (= *E. montevidiensis*). Split
into two (the leaf apex). E S America.
'Edinensis' e-din-*en*-sis. Of Edinburgh,
where it was raised.
× *exoniensis* eks-ō-nee-*en*-sis. Of Exeter,
where it was raised by Messrs. Veitch.
'Iveyi' *ie*-vee-ee. *E. bifida* × *E.× exoniensis.*
After Mr Ivey, gardener at Caerhays who
found it.
laevis *lie*-vis. Smooth (the leaves). Brazil.
'Langleyensis' lang-lee-*en*-sis. Of Langley,
where it was raised in 1893.
rubra *ru*-bra. (= *E. punctata*). Red (the
flowers). Chile, Argentina.
macrantha ma-*krănth*-a. (= *E.*
macrantha). Large-flowered. Chiloe.
virgata vir-*gah*-ta. Twiggy. Chile,
Argentina.

Eschscholzia esh-*sholts*-ee-a *Papaveraceae*.
After Johann Friedrich Eschscholz
1793–1831), Russian botanist. Annual herbs.
caespitosa kie-spi-*tō*-sa. Tufted. California.
californica kăl-i-*forn*-i-ka. Of California.
Californian Poppy.

Espostoa es-*po*-stō-a *Cactaceae*. After
Nicolas Esposto, a botanist in Lima, Peru.
lanata lah-*nah*-ta. Woolly. Peru, Ecuador.

Eucalyptus ew-ka-*lip*-tus *Myrtaceae*. From
Gk. *eu* (well) and *kalypto* (to cover) referring
to the calyx which forms a lid over the flowers
in bud. Hardy to tender evergreen trees.
citriodora kit-ree-o-*dō*-ra. Lemon-scented.
Lemon-scented Gum. Queensland.
coccifera kok-*ki*-fe-ra. Berry-bearing.
Mount Wellington Peppermint,
Tasmanian Snow Gum. Tasmania.
cordata kor-*dah*-ta. With heart-shaped
leaves. Silver Gum. Tasmania.
dalrympleana dăl-rim-plee-*ah*-na. After
Dalrymple. Broad-leaved Kindling Bark.
Tasmania, SE Australia.
glaucescens glow-*kes*-enz. Somewhat
glaucous (the foliage). Tingiringi Gum. SE
Australia.
globulus glob-ew-lus. Like a small globe (the
buds). Tasmanian Blue Gum. Tasmania.
gunnii gŭn-ee-ee. After Ronald Campbell
Gunn (1808–81), a magistrate and botanist
in Tasmania. Tasmania.
niphophila ni-*fo*-fi-la. Snow-loving. Snow
Gum. SE Australia.
parvifolia par-vi-*fo*-lee-a. Small-leaved.
Small-leaved Gum. New South Wales.
pauciflora paw-si-*flō*-ra. Few-flowered.
Cabbage Gum. Tasmania, SE Australia.
perriniana pe-rin-ee-*ah*-na. After George
Samuel Perrin (1849–1900), a forester in
SE Australia. Spinning Gum. SE Australia.
pulverulenta pul-ve-ru-*len*-ta. Dusty,
referring to the glaucous bloom on the
leaves. New South Wales.
urnigera ur-*ni*-ge-ra. Urn-bearing (the fruit
is urn-shaped). Urn Gum. Tasmania.
viminalis vee-mi-*nah*-lis. With osier-like
shoots. Ribbon Gum. Tasmania, SE
Australia.

Eucomis ew-*kom*-is *Liliaceae*. From Gk. *eu*
(good) and *kome* (hair) referring to the
attractive flower heads. Bulbous herbs.
S Africa.
bicolor bi-ko-lor. Two-coloured (the
flowers).

Eucomis (continued)
comosa ko-*mō*-sa. With a tuft (of leafy
bracts at the apex of the raceme).

Eucommia ew-*kom*-ee-a *Eucommiaceae*.
From Gk. *eu* (good) and *kommi* (gum). It is
the only hardy tree that can produce rubber.
Deciduous tree.
ulmoides ul-*moi*-deez. Like *Ulmus*. China.

Eucryphia ew-*krif*-ee-a *Eucryphiaceae*. From
Gk. *eu* (well) and *kryphios* (covered), the
sepals form a cap over the flower bud. Hardy
and semi-hardy trees and shrubs.
cordifolia kor-di-*fo*-lee-a. With heart-
shaped leaves. Chile.
glutinosa gloo-ti-*nō*-sa. Sticky. Chile.
× *intermedia* in-ter-*me*-dee-a. *E. glutinosa* ×
E. lucida. Intermediate (between the
parents).
lucida loo-ki-da. Glossy (the leaves).
Tasmania.
milliganii mil-li-*găn*-ee-ee. After Joseph
Milligan (1807–c. 1883), a Scottish
surgeon who collected in Tasmania.
Tasmania.
× *nymansensis* nie-manz-*en*-sis. *E. cordifolia*
× *E. glutinosa*. Of Nymans, Sussex where
it originated.
'Nymansay' *nie*-manz-ay. Derived from
one of two clones of the above first
exhibited, Nymans A and Nymans B.

Euodia ew-ō-dee-a *Rutaceae*. From Gk.
euodia (a sweet scent) referring to the
fragrant flowers. Deciduous trees.
daniellii dăn-*yel*-ee-ee. After William
Freeman Daniell (1818–65), a British
army surgeon who found it in China.
N China, Korea.
hupehensis hew-pee-*hen*-sis. Of Hubei
(Hupeh). China.

Euonymus ew-*on*-i-mus *Celastraceae*. The L.
name. Deciduous and evergreen trees and
shrubs.
alatus ah-*lah*-tus. Winged (the shoots).
China, Japan.
europaeus oy-rō-*pie*-us. European. Spindle
Tree. Europe, W Asia.
fortunei for-*tewn*-ee-ee. After Robert
Fortune, see *Fortunella*, E Asia.
'Coloratus' ko-lo-*rah*-tus. Coloured, the
leaves colour in winter.
'Kewensis' kew-*en*-sis. Of Kew, to where
it was first introduced.
radicans rah-di-kănz. With rooting stems.
grandiflorus grănd-i-*flō*-rus. Large-flowered.

Euonymus (continued)
Himalaya, W China.
salicifolius så-li-ki-*fo*-lee-us. *Salix*-leaved.
hamiltonianus hå-mil-ton-ee-*ah*-nus. After
Francis Buchanan-Hamilton, Scottish
surgeon and botanist. Himalaya, E Asia.
sieboldianus see-bōld-ee-*ah*-nus. After
Siebold, see *Acanthopanax sieboldianus*.
E Asia.
japonicus ja-*pon*-i-kus. Of Japan.
nanus nah-nus. Dwarf. Caucasus to China.
turkestanicus tur-kes-*tahn*-i-kus. Of
Turkestan. C Asia.
oxyphyllus oks-ee-*fil*-lus. With sharp-
pointed leaves. Japan, Korea.
phellomanus fel-ō-*mahn*-us. With corky
shoots. China.
planipes *plahn*-i-pays. (= *E. sachalinensis*
hort.). With a flat stalk. NE Asia.
wilsonii wil-*son*-ee-ee. After Ernest Wilson,
see *Magnolia wilsonii*.

Eupatorium ew-pa-*to*-ree-um *Compositae*.
The Gk. name, from Eupator, King of
Pontus. Perennial herbs and tender shrubs.
cannabinum kån-a-*been*-um. Like *Cannabis*
(the leaves). Hemp Agrimony. Europe,
N Africa to C Asia.
coelestinum koy-les-*teen*-um. Sky-blue (the
flowers). Mistflower. S and E N America,
W Indies.
ligustrinum lig-us-*tree*-num. (= *E.
micranthum. E. weinmannianum*). Like
Ligustrum. C America.
maculatum måk-ew-*lah*-tum. Spotted (the
leaves). Joe-pye Weed. E N America.
purpureum pur-*pewr*-ree-um. Purple (the
flowers). Joe-pye Weed. E N America.
rugosum roo-gō-sum. (= *E. ageratoides*).
Wrinkled (the leaves). White Snake-root.
E N America.

Euphorbia ew-*for*-bee-a *Euphorbiaceae*. The
classical name after Euphorbus, physician to
Juba, king of Mauritania. Annuals,
perennials, sub-shrubs, tender shrubs and
succulents.
bubalina bew-ba-*leen*-a. Of the African
gazelle. S Africa.
caput-medusae kå-put-may-*dew*-sie.
Medusa's head, to which this plant has
been likened. Medusa's Head. S Africa.
characias ka-rå-kee-ahs. L. name of a plant.
Mediterranean region.
wulfenii wul-*fen*-ee-ee. (= *E. wulfenii*).
After Wulfen, see *Wulfenia*. SE Europe.
cyathophora see-åth-ō-*fo*-ra. (= *E.
heterophylla* hort.). Cup-bearing. Fire on

Euphorbia (continued)
the Mountain, Annual Poinsettia. E United
States, Mexico.
cyparissias kew-pa-*ris*-ee-as. Cypress-like.
Cypress Spurge. Europe.
echinus e-*keen*-us. Spiny. Morocco.
epithymoides e-pi-tiem-*oi*-deez. (= *E.
polychroma*). Like *Epithymum* (= *Cuscuta*).
E Europe.
fulgens *ful*-genz. Shining (the bracts).
Mexico.
grandicornis gránd-i-*kor*-nis. With large
horns (on the shoots). S Africa.
griffithii gri-*fith*-ee-ee. After Griffith, see
Ceratostigma griffithii. Himalaya, Tibet.
hermentiana see *E. trigona*
heterophylla hort. see *E. cyathophora*
lathyris lå-thi-ris. The classical name. Caper
Spurge, Mole Plant. Europe.
mammillaris må-mi-*lah*-ris. Bearing nipples.
Corncob Cactus. S Africa.
marginata mar-gi-*nah*-ta. Margined (the
upper leaves). Snow on the Mountain.
United States.
meloformis may-lō-*form*-is. Melon-shaped.
S Africa.
milii *mil*-ee-ee. Said to be after M. Millius,
Governor of the Isle of Bourbon where it
was grown. Crown of Thorns. Madagascar.
   'Splendens' *splen*-denz. (= *E. splendens*).
   Splendid.
myrsinites mur-sin-*ee*-teez. Like *Myrsine*.
Europe.
obesa o-*bay*-sa. Fat. Baseball Cactus.
S Africa.
palustris pa-*lus*-tris. Growing in marshes.
Europe.
polychroma see *E. epithymoides*
pseudocactus soo-dō-*kåk*-tus. False cactus.
S Africa.
pulcherrima pul-*ke*-ri-ma. Very pretty.
Poinsettia. C America.
resinifera ray-see-*ni*-fe-ra. Resin-bearing.
Morocco.
robbiae rob-ee-ie. After Mary Ann Robb
(1829–1912) who introduced it. Mrs
Robb's Bonnet. W Asia.
seguieriana seg-wee-e-ree-*ah*-na. After Jean
Francis Seguier (1703–84), French
botanist. Europe.
   niciciana nee-cheech-ee-*ah*-na. After
   Nicic. SE Europe.
sikkimensis sik-im-*en*-sis. Of Sikkim.
Himalaya, Tibet.
splendens see *E. milii* 'Splendens'
submammillaris sub-må-mi-*lah*-ris. With
small nipples. S Africa.
trigona tri-gō-na. (= *E. hermentiana*).

Euphorbia (continued)
Three-angled (the stem). SW Africa.
valida vă-li-da. Robust. S Africa.
wulfenii see *E. characias wulfenii*

Eurya ew-ree-a *Theaceae*. From Gk. *euru*
(broad). Evergreen shrub.
japonica ja-*pon*-i-ka. Of Japan. Himalaya,
Japan, SE Asia.

Euryops ew-ree-ops *Compositae*. From Gk.
eu (well) and *ops* (appearance). Evergreen
shrubs.
acraeus a-*krie*-us. (= *E. evansii* hort.).
Growing in high places. Drakensberg
Mountains, S Africa.
pectinatus pek-ti-*nah*-tus. Comb-like (the
pinnate leaves). S Africa.

Evening Primrose see *Oenothera biennis*
Everlasting see *Helichrysum bracteatum*

Exochorda eks-ō-*kor*-da *Rosaceae*. From Gk.
exo (outside) and *chorda* (a cord), referring
to fibres outside the placenta in the ovary.
Deciduous shrubs.
giraldii ji-*răl*-dee-ee. After Giraldi who
introduced it, see *Callicarpa bodinieri
giraldii*. NW China.
wilsonii wil-*son*-ee-ee. After Ernest
Wilson who introduced it in 1907, see
Magnolia wilsonii.
korolkowii ko-rol-*kov*-ee-ee. After
Korolkow, see *Crocus korolkowii*.
Turkestan.
× *macrantha* ma-*krănth*-a. *E. korolkowii* ×
E. racemosa. Large-flowered.
racemosa ră-kay-*mō*-sa. With flowers in
racemes. N China.

Exacum eks-a-kum *Gentianaceae*. From
exacon the Gallic name for *Centaurium*.
Tender annual or biennial.
affine a-*fee*-nee. Related to. Persian Violet.
Socotra.

F

Fabiana fah-bee-*ah*-na *Solanaceae*. After
Archbishop Francisco Fabian y Fuero
(1719–1801). Evergreen, semi-hardy shrub.
imbricata im-bri-*kah*-ta. Closely
overlapping (the leaves).

Fagus fah-gus *Fagaceae*. The L. name.
Deciduous trees. Beech.

Fagus (continued)
orientalis o-ree-en-*tah*-lis. Eastern.
Caucasus, W Asia.
sylvatica sil-*vă*-ti-ka. Of woods. Common
Beech. Europe.
'Asplenifolia' a-splay-ni-*fo*-lee-a. With
leaves like *Asplenium*.
'Pendula' *pen*-dew-la. Pendulous.
Purpurea pur-*pewr*-ree-a. Purple (the
leaves). Copper Beech.
'Riversii' ri-*verz*-ee-ee. After Messrs
Rivers who raised it.
'Rohanii' rō-*hahn*-ee-ee. After Prince
Camille de Rohan on whose estate it was
raised.
'Zlatia' *zlah*-tee-a. Golden (the leaves),
from Serbian *zlatos*.

Fair Maids of France see *Ranunculus
aconitifolius* 'Flore Pleno'
Fairy Bells see *Disporum*
Fairy Forget-me-not see *Eritrichium nanum*
Fairy Foxglove see *Erinus alpinus*
False Acacia see *Robinia pseudacacia*
False African Violet see *Streptocarpus
saxorum*
False Aralia see *Dizygotheca elegantissima*
False Hellebore see *Veratrum*
False Spikenard see *Smilacina racemosa*
Fameflower see *Talinum*
Fanwort see *Cabomba*

× ***Fatshedera*** făts-*he*-de-ra *Araliaceae*.
Intergeneric hybrid, from the names of the
parents. *Fatsia* × *Hedera*. Evergreen, semi-
scandent shrub.
lizei lee-zay-ee. *Fatsia japonica* 'Moseri' ×
Hedera hibernica. After Messrs. Lizé
Frères of Nantes who raised it in 1910.

Fatsia făts-ee-a *Araliaceae*. From a Japanese
name. Evergreen shrub.
japonica ja-*pon*-i-ka. (= *Aralia sieboldii*). Of
Japan. False Castor-oil Plant.

Faucaria fow-*kah*-ree-a *Aizoaceae*. From L.
faux (a gullet), the paired, toothed leaves
resemble open jaws. Tender succulents.
S Africa.
tigrina tig-*reen*-a. Tiger-like. Tiger's Jaws.
tuberculosa tew-ber-kew-*lō*-sa. With
tubercles, the white spots on the leaves.

Feather Grass see *Stipa pennata*

Feijoa fie-*hō*-a *Myrtaceae*. After Don de Silva
Feijo, 19th century Brazilian botanist.
Evergreen semi-hardy shrub.

Feijoa (continued)
  sellowiana se-lō-ee-*ah*-na. After its
  discoverer, Friedrich Sellow (Sello)
  (1789–1831), a German botanist who
  collected in S America. S America.

Felicia fe-*lik*-ee-a *Compositae*. After Felix, a
German official. Annuals and sub-shrubs.
S Africa.
  amelloides ă-mel-*oi*-deez. Like *Aster
  amellus*.
  bergeriana ber-ga-ree-*ah*-na. After Berger.
  Kingfisher Daisy.
  pappei păp-ee-ee. (= *Aster pappei*). After
  Carl Wilhelm Ludwig Pappe (1802–62) a
  German botanist who worked in S Africa.
  rosulata ros-ew-*lah*-ta. (= *Aster natalensis*).
  With leaves in a rosette.

Felt Bush see *Kalanchoe beharensis*
Fennel see *Foeniculum vulgare*
  Florence see *F. vulgare azoricum*
  Giant see *Ferula communis*
Fern, Alpine Lady see *Athyrium distentifolium*
  American Sword see *Polystichum munitum*
  Arctic Bladder see *Cystopteris dickieana*
  Asparagus see *Asparagus setosus*
  Berry Bladder see *Cystopteris bulbifer*
  Bird's Nest see *Asplenium nidus*
  Bladder see *Cystopteris*
  Boston see *Nephrolepis exaltata*
    'Bostoniensis'
  Brazil Tree see *Blechnum brasiliense*
  Brittle Bladder see *Cystopteris fragilis*
  Broad Buckler see *Dryopteris dilatata*
  Buckler see *Dryopteris*
  Button see *Pellaea rotundifolia*
  Christmas see *Polystichum acrostichoides*
  Cinnamon see *Osmunda cinnamonea*
  Crested Buckler see *Dryopteris cristata*
  Crown see *Blechnum discolor*
  Deer's Foot see *Davallia canariensis*
  Elk's Horn see *Platycerium bifurcatum*
  Erect Sword see *Nephrolepis cordifolia*
  Filmy see *Hymenophyllum*
  Floating see *Salvinia auriculata*
  Giant Wood see *Dryopteris goldieana*
  Golden Tree see *Dicksonia fibrosa*
  Hairy Lip see *Cheilanthes lanosa*
  Hammock see *Blechnum occidentale*
  Hard see *Blechnum spicant*
  Hard Shield see *Polystichum aculeatum*
  Hare's Foot see *Polypodium vulgare*
  Hart's Tongue see *Phyllitis scolopendrium*
  Hay-scented see *Dennstaedtia punctilobula*
  Hay-scented Buckler see *Dryopteris aemula*
  Hen and Chicken see *Asplenium bulbiferum*

Fern (continued)
  Holly see *Polystichum acrostichoides*, *P.
  falcatum*
  Interrupted see *Osmunda claytoniana*
  Japanese Painted see *Athyrium
  goeringianum* 'Pictum'
  Lady see *Athyrium filix-femina*
  Lip see *Cheilanthes*
  Maidenhair see *Maidenhair Fern*
  Male see *Dryopteris filix-mas*
  Mountain Bladder see *Cystopteris montana*
  Necklace see *Asplenium bulbiferum*
  Oak see *Gymnocarpium dryopteris*
  Ostrich-feather see *Matteuccia struthiopteris*
  Palm Leaf see *Blechnum capense*
  Parsley see *Cryptogramma crispa*
  Rabbit's Foot see *Davallia fejeensis*
  Rib see *Blechnum brasiliense*
  Royal see *Osmunda regalis*
  Rusty Back see *Ceterach officinarum*
  Sensitive see *Onoclea sensibilis*
  Soft Shield see *Polystichum setiferum*
  Squirrel's Foot see *Davallia mariesii*,
  D. trichomanoides
  Stag's Horn see *Platycerium bifurcatum*
  Sword see *Nephrolepis*
  Tunbridge Filmy see *Hymenophyllum
  tunbrigense*
  Walking see *Camptosorus rhizophyllus*
  Wilson's Filmy see *Hymenophyllum wilsoni*
  Woolly Rock see *Cheilanthes distans*
  Woolly Tree see *Dicksonia antarctica*

Ferocactus fe-rō-*kăk*-tus *Cactaceae*. From
L. *ferox* (savage) referring to the spines, and
Cactus q.v.
  acanthodes a-*kănth*-ō-deez. Spiny. SW
  United States, N Mexico.
  hamatacanthus hah-mah-ta-*kănth*-us. (=
  Hamatocactus hamatacanthus). With
  hooked spines. SW United States,
  N Mexico.
  latispinus lah-tee-*speen*-us. With broad
  spines. Mexico.
  setispinus say-tee-*speen*-us. (= *Hamatocactu
  setispinus*). With bristle-like spines.
  S Texas, N Mexico.
  viridescens vi-ri-*des*-enz. Greenish (the
  flowers). California.
  wislizenii wiz-li-*zen*-ee-ee. After
  Wislizenius. SW United States,
  N Mexico.

Ferula *fe*-ru-la *Umbelliferae*. The L. name.
Perennial herbs.
  communis kom-*ew*-nis. Common. Giant
  Fennel. S Europe, W Asia.

Ferula (continued)
  tingitana ting-gi-*tah*-na. Of Tangier.
  N Africa, W Asia.

Festuca fes-*too*-ka *Gramineae*. L. name for a
grass stalk. Perennial grasses.
  alpina ăl-*peen*-a. Alpine. Alps.
  amethystina ă-me-*this*-ti-na. Violet.
  C Europe.
  glacialis glă-kee-*ah*-lis. Growing in icy
  places. Pyrenees.
  glauca glow-ka. Glaucous (the leaves).
  S France.

Ficus *fee*-kus *Moraceae*. The L. name for *F.
carica*. Deciduous and tender, evergreen
trees and shrubs.
  benghalensis beng-ga-*len*-sis. Of Benghal.
  Banyan Tree. India, Pakistan.
  benjamina ben-ja-*meen*-a. From *benjan*, the
  Indian name. Weeping Fig. Himalaya, SE
  Asia to N Australia.
  carica kah-ri-ka. Of Caria, W Asia.
  Common Fig.
  diversifolia see *F. deltoidea*
  deltoidea del-*toi*-dee-a. (= *F. diversifolia*).
  Deltoid i.e. shaped like the Gk. letter delta
  (Δ) (the leaves). Mistletoe Fig. Malaysia.
  elastica e-*lăs*-ti-ka. Producing elastic.
  Rubber Plant. Himalaya, SE Asia.
  lyrata li-*rah*-ta. Fiddle-shaped (the leaves).
  Tropical Africa.
  pumila pew-mi-la. (= *F. repens*). Dwarf.
  Creeping Fig. E Asia.
  radicans see *F. sagittata*
  religiosa ray-lig-ee-ō-sa. Sacred, a sacred
  tree in India. India, SE Asia.
  repens see *F. pumila*
  retusa re-*tew*-sa. Notched at the apex (the
  leaves). SE Asia.
  rubiginosa roo-bi-gi-*nō*-sa. Rusty (the hairs
  on the underside of the leaf). Rusty Fig.
  New South Wales.
  sagittata săg-i-*tah*-ta. (= *F. radicans*).
  Shaped like an arrow-head (the leaves).
  E Asia.

Fiddler's Trumpets see *Sarracenia
leucophylla*
Fig, Common see *Ficus carica*
  Creeping see *F. pumila*
  Mistletoe see *F. deltoidea*
  Rusty see *F. rubiginosa*
  Weeping see *F. benjamina*
Filbert see *Corylus maxima*

Filipendula fi-li-*pen*-dew-la *Rosaceae*. From
L. *filum* (a thread) and *pendulus* (hanging)

Filipendula (continued)
referring to the threads connecting the root
tubers. Perennial herbs.
  camtschatica kămt-*shă*-ti-ka. Of
  Kamtchatka. NE Asia.
  palmata pahl-*mah*-ta. Lobed like a hand
  (the leaves). NE Asia.
  purpurea pur-*pewr*-ree-a. Purple (the
  flowers). Japan.
    'Nana' *nah*-na. (= *F. digitata* 'Nana').
    Dwarf.
  rubra rub-ra. Red (the flowers). Queen of
  the Prairie. E United States.
  ulmaria ul-*mah*-ree-a. *Ulmus*-like (the
  leaflets). Meadowsweet. Europe, Asia.
  vulgaris vul-*gah*-ris. Common. Dropwort.
  Europe, N Africa, W Asia, Siberia.

Finger Aralia see *Dizygotheca elegantissima*
Finger-nail Plant see *Neoregelia spectabilis*
Finocchio see *Foeniculum vulgare azoricum*
Fir see *Abies*
  Alpine see *A. lasiocarpa*
  Balsam see *A. balsamea*
  Caucasian see *A. nordmanniana*
  European Silver see *A. alba*
  Flaky see *A. squamata*
  Giant see *A. grandis*
  Greek see *A. cephalonica*
  Himalayan see *A. spectabilis*
  Korean see *A. koreana*
  Nikko see *A. homolepis*
  Noble see *A. procera*
  Pacific Silver see *A. amabilis*
  Red see *A. magnifica*
  Red Silver see *A. amabilis*
  Santa Lucia see *A. bracteata*
  White see *A. concolor*
Fire Bush see *Embothrium coccineum*
Fire-on-the-Mountain see *Euphorbia
cyathophora*
Firecracker Flower see *Crossandra
infundibuliformis*
Firecracker Vine see *Manettia inflata*
Firethorn see *Pyracantha*
Firewheel Tree see *Stenocarpus sinuatus*

Fittonia fi-*ton*-ee-a *Acanthaceae*. After
Elizabeth and Sarah Mary Fitton. Tender
perennial herbs. NE South America.
  gigantea gi-*găn*-tee-a. Very large.
  verschaffeltii vair-sha-*felt*-ee-ee. After M.
  Verschaffelt, a 19th century Belgian
  nurseryman.
    argyroneura ar-gi-ro-*newr*-ra. Silver-
    veined. Mosaic Plant.

Fitzroya fitz-*roy*-a *Cupressaceae*. After Captain Robert Fitzroy (1805–65), commander of The Beagle during Darwin's voyage. Evergreen conifer.
  cupressoides kew-pres-*oi*-deez *Cupressus*-like. Chile, Argentina.

Five Fingers see *Syngonium auritum*
Flame Creeper see *Tropaeolum speciosum*
Flame Nettle see *Coleus blumei*
Flame of the Woods see *Ixora coccinea*
Flame Plant see *Anthurium scherzerianum*
Flame Violet see *Episcia cupreata*
Flaming Sword see *Vriesia splendens*
Flamingo Flower see *Anthurium scherzerianum*
Flax see *Linum usitatissimum*
  Golden see *L. flavum*
  Tree see *L. arboreum*
  Yellow see *Reinwardtia indica*
Fleabane see *Erigeron*
Floral Firecracker see *Dichelostemma idamaia*
Floss Flower see *Ageratum conyzoides*
Flower of an Hour see *Hibiscus trionum*
Flower of the Western Wind see *Zephyranthes candida*
Flowering Rush see *Butomus umbellatus*
Foam Flower see *Tiarella cordifolia*

Foeniculum fee-*nik*-ew-lum *Umbelliferae*. The L. name. Perennial herbs.
  vulgare vul-*gah*-ree. Common. Fennel. S Europe.
    azoricum a-*zo*-ri-kum. Of the Azores. Florence Fennel, Finocchio.

Fokienia fo-kee-*en*-ee-a *Cupressaceae*. From Fukien (now Fujien), China where it grows. Semi-hardy, evergreen conifer.
  hodginsii hoj-*inz*-ee-ee. After Captain A. Hodgins who discovered it.

Fontinalis fon-ti-*nah*-lis *Fontinalaceae*. From L. *fontinalis* (of springs or fountains) referring to its habitat. Aquatic moss.
  antipyretica ăn-ti-pi-*ret*-i-ka. Against fire, it was packed around chimneys in wooden houses to prevent fire. N hemisphere.

Forest Lily see *Veltheimia viridiflora*
Forget-me-not see *Myosotis*

Forsythia for-*sieth*-ee-a *Oleaceae*. After William Forsyth (1737–1804) a Scottish gardener who became superintendent of the Royal Garden of Kensington Palace.

Forsythia (continued)
  Deciduous shrubs.
  × *intermedia* in-ter-*med*-ee-a. *F. suspensa* × *F. viridissima*. Intermediate (between the parents).
  ovata ō-*vah*-ta. Ovate (the leaves). Korean Forsythia. Korea.
  suspensa sus-*pens*-a. Hanging (the flowers). Golden Bell. China.
  viridissima vi-ri-*di*-si-ma. Most green (the shoots). China.
    'Bronxensis' bronks-*en*-sis. Of the Bronx. It was grown at the New York Botanic Garden.

Fortunella for-tew-*nel*-a *Rutaceae*. After Robert Fortune (1812–80) who collected in China. He introduced the tea plant from China into India.
  japonica ja-*pon*-i-ka. Of Japan. Kumquat. ? S China.

Fothergilla fo-tha-*gil*-a *Hamamelidaceae*. After Dr John Fothergill (1712–80), English physician who grew American plants. Deciduous shrubs. SE United States.
  gardenii gar-*den*-ee-ee. After Dr Garden who discovered it, see *Gardenia*.
  major mah-yor. (= *F. monticola*). Larger.

Fountain Grass see *Pennisetum setaceum*
Fountain Plant see *Amaranthus tricolor* 'Salicifolius'
Four o'clock Plant see *Mirabilis jalapa*
Foxglove see *Digitalis purpurea*
Foxtail Grass see *Alopecurus pratensis*
Foxtail Lily see *Eremurus*

Fragaria fra-*gah*-ree-a *Rosaceae*. From *fraga* the L. name, referring to the scent of the fruit. Perennial herbs. Strawberry.
  × *ananassa* ăn-a-*năs*-a. *F. chiloensis* × *F. virginiana*. From *Ananas* q.v. it was originally referred to as the pine or pineapple strawberry. Garden Strawberry.
  indica see *Duchesnea indica*
  moschata mos-*kah*-ta. Musk-scented. Hautbois Strawberry. Europe, W Asia.
  vesca ves-ka. Little. Alpine Strawberry. Europe, Asia, EN America.

Francoa frang-*kō*-a *Saxifragaceae*. After Francisco Franco a 16th-century Spanish physician. Semi-hardy perennial herb.
  sonchifolia son-ki-*fo*-lee-a. *Sonchus*-leaved. Bridal Wreath. Chile.

Frangipani see *Plumeria rubra*

Frankenia frang-*ken*-ee-a *Frankeniaceae*.
After Johan Frankenius (1590–1661).
Evergreen sub-shrub.
  thymifolia tiem-i-*fo*-lee-a. *Thymus*-leaved.
  N Africa.

Franklinia frank-*lin*-ee-a *Theaceae*. After
Benjamin Franklin (1706–90), the American
statesman. Deciduous tree.
  alatamaha a-lah-ta-*mah*-ha. (*Gordonia
  alatamaha*). Of the Altamaha River,
  Georgia, United States, near which it grew
  before extinction.

Fraxinus *fraks*-i-nus *Oleaceae*. The L. name.
Deciduous trees. Ash.
  americana a-me-ri-*kah*-na. American.
  White Ash. E N America.
  angustifolia ang-gus-ti-*fo*-lee-a. Narrow-
  leaved. Narrow-leaved Ash.
  W Mediterranean region, N Africa.
    oxycarpa oks-i-*kar*-pa. (= *F. oxycarpa*).
    With pointed fruits. E Mediterranean
    region, W Asia.
  excelsior eks-*kel*-see-or. Taller. Common
  Ash. Europe, Caucasus.
    'Jaspidea' yas-*pid*-ee-a. Jasper-like (the
    yellow shoots).
  ornus or-nus. L. name for the mountain ash.
  Manna Ash. S Europe, W Asia.
  velutina vel-ew-*teen*-a. Velvety (the leaves
  and shoots of some forms). Arizona Ash.
  SW United States, Mexico.

Freckle Face see *Hypoestes phyllostachya*

Freesia *freez*-ee-a *Iridaceae*. After Friedrich
Heinrich Theodor Freese (died 1876), a
German physician. Tender or semi-hardy
cormous perennials. S Africa.
  alba *ăl*-ba. (= *F. refracta alba*). White (the
  flowers).
  × *hybrida* hib-ri-da. Hybrid. Common
  Freesia.
  refracta re-*frăk*-ta. Broken.

Fremontia see *Fremontodendron*

Fremontodendron free-mont-ō-*den*-dron
Sterculiaceae. (*Fremontia*). After Major-
General John Charles Fremont (1813–90),
explorer and plant collector in the W United
States who discovered *F. californicum*.
  californicum kăl-i-*forn*-i-kum. Of California.
  mexicanum meks-i-*kah*-num. Mexican.
  S California, Mexico.

French Bean see *Phaseolus vulgaris*
French Honeysuckle see *Hedysarum
coronarium*
Friendship Plant see *Billbergia nutans, Pilea
involucrata*
Fringe Tree see *Chionanthus virginicus*
  Chinese see *C. retusus*

Fritillaria fri-ti-*lah*-ree-a *Liliaceae*. From L.
fritillus (a dicebox) referring to the chequered
flowers. Bulbous perennials. Fritillary.
  acmopetala ăk-mō-*pe*-ta-la. With anvil-
  shaped petals. W Asia.
  assyriaca hort. see *F. uva-vulpis*
  bithynica bi-*thin*-i-ka. (= *F. schliemanii*). Of
  Bithynia (NW Turkey).
  camschatkensis kăm-shăt-*ken*-sis. Of
  Kamtchatka. WN America, N Japan.
  caucasica kaw-*kăs*-i-ka. Of the Caucasus.
  Turkey, Iran, Caucasus.
  davisii day-*vis*-ee-ee. After Peter Hadland
  Davis (born 1918). Greece.
  gracilis see *F. messanensis gracilis*
  graeca grie-ka. Of Greece. S Greece.
    thessala thes-a-la. Of Thessalia,
    N Greece. S Europe.
  imperialis im-pe-ree-*ah*-lis. Showy. Crown
  Imperial. Turkey to Kashmir.
  involucrata in-vo-loo-*krah*-ta. With an
  involucre, the upper three leaves are
  whorled. S France, NW Italy.
  lanceolata lăn-kee-ō-*lah*-ta. Lanceolate (the
  leaves). W N America.
  latifolia lah-tee-*fo*-lee-a. Broad-leaved.
  Caucasus, NE Turkey.
  lusitanica loo-si-*tah*-ni-ka. Of Portugal.
  Spain, Portugal.
  meleagris mel-ee-*ah*-gris. Spotted (the
  flowers). Europe.
  messanensis mes-an-*en*-sis. Of Messina,
  Sicily. S Europe, N Africa.
    gracilis gră-ki-lis. (= *F. gracilis*).
    Graceful. Yugoslavia, Albania.
  pallidiflora pa-li-di-*flō*-ra. Pale-flowered.
  C Asia.
  persica *per*-si-ka. Of Iran (Persia). W Asia.
  pluriflora ploo-ri-*flō*-ra. Many-flowered.
  California.
  pontica *pon*-ti-ka. Of Pontus (N Turkey).
  SE Europe, NW Turkey.
  pudica pud-*ee*-ka. Modest. W N America.
  pyrenaica pi-ray-*nah*-i-ka. Of the Pyrenees.
  Pyrenees, NW Spain.
  raddeana rah-dee-*ah*-na. After Gustav
  Ferdinand Richard Radde (1831–1903) of
  Gdansk. Iran, C Asia.
  recurva re-*kur*-va. Curved back (the

Fritillaria (continued)
   perianth lobes). W United States.
   roylei royl-ee-ee. After John Forbes Royle
   (1798–1858), a surgeon with the E India
   Co. who collected in India and the
   Himalaya. Himalaya.
   ruthenica roo-*then*-i-ka. Of Ruthenia, SW
   Russia. SE Europe, W Asia.
   schliemanii see *F. bithynica*
   uva-vulpis oo-va-*vul*-pis. Fox's grape, a
   translation of the Kurdish name *tarai
   raiwa*. W Asia.
   verticillata ver-ti-ki-*lah*-ta. Whorled (the
   leaves). C Asia, China.

Frog's Bit see *Hydrocharis morsus-ranae*

Fuchsia fuks-ee-a *Onagraceae*. After
Leonhart Fuchs (1501–66), a German
physician and herbalist. Deciduous and
evergreen, semi-hardy and tender shrubs.
   'Corallina' ko-ra-*leen*-a. Coral red (the
   flowers).
   excorticata eks-kor-ti-*kah*-ta. With peeling
   bark. New Zealand.
   magellanica mă-ge-*lăn*-i-ka. From the
   region of the Magellan Straits. S Chile,
   S Argentina.
      gracilis gră-ki-lis. Graceful.
      molinae mo-*leen*-ie. (= *F. magellanica*
      'Alba'). After Juan Ignacio Molina
      (1740–1829).
      'Versicolor' ver-*si*-ko-lor. Variously
      coloured (the flowers).
   procumbens prŏ-*kum*-benz. Prostrate.
   Trailing Fuchsia. New Zealand.
   'Riccartonii' ri-kar-*ton*-ee-ee. Of Riccarton,
   Scotland where it was raised.

Furze see *Ulex*

G

Gagea gayj-ee-a *Liliaceae*. After Sir Thomas
Gage (1761–1820), Suffolk botanist whose
grandfather is commemorated in the
greengage. Bulbous perennials.
   fistulosa fist-u-*lō*-sa. With hollow leaves.
   Europe, Caucasus.
   lutea loo-tee-a. Yellow (the flowers).
   Europe.

Gaillardia gay-*lard*-ee-a *Compositae*. After
Gaillard de Charentonneau, 18th-century
French magistrate and botanical patron.
Blanket Flower.

Gaillardia (continued)
   × *grandiflora* grănd-i-*flō*-ra. *G. aristata* ×
   G. pulchella. (= *G. aristata* hort.). Large-
   flowered.
   pulchella pul-*kel*-a. Pretty. United States,
   Mexico.

Galanthus ga-*lănth*-us *Amaryllidaceae*. From
Gk. *gala* (milk) and *anthos* (a flower)
referring to the colour of the flowers. Bulbous
herbs. Snowdrop.
   allenii a-*len*-ee-ee. After James Allen, a
   snowdrop grower who found it. Possibly
   a hybrid.
   byzantinus bi-zan-*teen*-us. Of Istanbul
   (Byzantium). NW Turkey.
   caucasicus kaw-*kă*-si-cus. Of the Caucasus.
   elwesii el-*wez*-ee-ee. After H. J. Elwes
   (1846–1922), English sportsman and
   naturalist who introduced it. SE Europe,
   W Turkey.
   fosteri *fos*-ta-ree. After Sir Michael Foster
   (1836–1907). Turkey, Lebanon.
   ikariae i-*kah*-ree-ie. Of Ikaria, an Aegean
   island. Caucasus, N Turkey.
   nivalis ni-*vah*-lis. Of the snow. Common
   Snowdrop. Europe.
   platyphyllus plăt-ee-*fil*-lus. (= *G. latifolius*).
   Broad-leaved. Caucasus.
   plicatus pli-*kah*-tus. Pleated (the leaves).
   E Europe.

Galax gă-lăx *Diapensiaceae*. From Gk. *gala*
(milk) referring to the white flowers.
Perennial herb.
   urceolata ur-kee-ō-*lah*-ta. (= *G. aphylla*).
   Urn-shaped. E United States.

Galega ga-*lee*-ga *Leguminosae*. From Gk.
gala (milk) it was fed to goats to improve
milk flow. Perennial herbs.
   officinalis o-fi-ki-*nah*-lis. Sold as a herb.
   Goat's Rue. Europe, W Asia.
   orientalis o-ree-en-*tah*-lis. Eastern.
   Caucasus.

Galeobdolon gă-lee-*ob*-do-lon *Labiatae*. L.
name for a nettle-like plant, from *galeo* (to
cover with a helmet) and *dolon* (a fly's sting).
Perennial herb.
   luteum loo-tee-um. (= *Lamiastrum
   galeobdolon*. *Lamium galeobdolon*). Yellow
   (the flowers). Europe, W Asia.

Galingale see *Cyperus longus*

Galium gă-lee-um *Rubiaceae*. From Gk. *gala* (milk). *G. verum* (lady's bedstraw) has been used to curdle milk. Perennial herb.
   odoratum o-dō-*rah*-tum. (= *Asperula odorata*). Scented. Sweet Woodruff. Europe, N Africa, W Asia.

Galtonia gawl-*ton*-ee-a *Liliaceae*. After Sir Frances Galton (1812–1911). Bulbous herbs. S Africa.
   candicans kăn-di-kănz. (= *Hyacinthus candicans*). White (the flowers). Summer Hyacinth.
   princeps *pring*-keps. Most distinguished.

Gardener's Garters see *Phalaris arundinacea* 'Picta'

Gardenia gar-*den*-ee-a *Rubiaceae*. After Dr Alexander Garden (1730–91), a Scottish physician and botanist who lived in S Carolina. Tender, evergreen shrub.
   jasminoides yăs-min-*oi*-deez. (= *G. florida*). *Jasminum*-like (the fragrant flowers). China.

Garland Flower see *Daphne cneorum*, *Hedychium coronarium*
Garlic see *Allium sativum*

Garrya gă-ree-a *Garryaceae*. After Nicholas Garry of the Hudson's Bay Co. Evergreen shrubs or small trees.
   elliptica e-*lip*-ti-ka. Elliptic (the leaves). W United States
   × *thuretii* thu-*ray*-tee-ee. *G. elliptica* × *G. fadyenii*. After Gustave Thuret who raised it in about 1862.

Gasteria găs-*te*-ree-a *Liliaceae*. From Gk. *gaster* (a belly) referring to the swollen base of the corolla tube. Tender succulents. S Africa.
   brevifolia brev-i-*fo*-lee-a. Short-leaved.
   disticha dis-ti-ka. (= *G. lingua*). Two-ranked (the leaves).
   liliputiana li-lee-put-ee-*ah*-na. Of Lilliput i.e. very small.
   maculata măk-ew-*lah*-ta. Spotted (the leaves).
   marmorata mar-mo-*rah*-ta. Marbled (the leaves).
   trigona tri-*gō*-na. Three-angled.
   verrucosa ve-roo-*kō*-sa. Warty (the leaves). Wart Gasteria.

× *Gaulnettya* gawl-*net*-ee-a *Ericaceae*. Intergeneric hybrid, from the names of the

× *Gaulnettya* (continued)
parents. *Gaultheria* × *Pernettya*. Evergreen shrub.
   wisleyensis hort. wiz-lee-en-sis. *Gaultheria shallon* × *Pernettya mucronata*. Of Wisley where it arose.

Gaultheria gawl-*the*-ree-a *Ericaceae*. After Dr Gaulthier (c. 1708–58), a Canadian botanist and physician.
   procumbens prō-*kum*-benz. Prostrate. Creeping Wintergreen. E N America.
   shallon shă-lon. The native name. Sabal, Shallon. W N America.

Gay Feather see *Liatris*

Gazania ga-*zah*-nee-a *Compositae*. After Theodore of Gaza (1398–1478). Semi-hardy perennials.
   rigens ri-gayns. (= *G. splendens*). Rigid. Treasure Flower. Cult.
      leucolaena loo-ko-*lie*-na. (= *G. uniflora*). White-cloaked, the white-hairy leaves. S Africa.

Gean see *Prunus avium*

Gelsemium gel-*sem*-ee-um *Loganiaceae*. From *gelsomino* the Italian for jasmine. Semi-hardy, evergreen, twining shrub.
   sempervirens sem-per-*vi*-rens. Evergreen. Yellow Jessamine. S United States, C America.

Genista ge-*nis*-ta *Leguminosae*. The L. name. Deciduous or nearly leafless shrubs. Broom.
   aetnensis iet-*nen*-sis. Of Mt Etna, Sicily. Mt Etna Broom. Sicily, Sardinia.
   delphinensis del-fin-*en*-sis. Of Dauphiné, France. S France.
   fragrans hort. see *Cytisus* × *spachianus*
   hispanica hi-*spah*-ni-ka. Spanish. Spanish Gorse. SW Europe.
   januensis yăn-ew-*en*-sis. Of Genoa. Genoa Broom. Italy, SW Europe.
   lydia li-dee-a. Of Lydia (W Turkey). SE Europe, W Asia.
   pilosa pi-*lō*-sa. With long, soft hairs. Europe.
   sylvestris sil-*ves*-tris. Of woods. Dalmatian Broom. SE Europe.
   tenera te-ne-ra. Tender, delicate. Madeira, Tenerife.
   tinctoria tink-*to*-ree-a. Used in dyeing. Dyer's Greenweed. Europe to Siberia.

Gentiana gen-tee-*ah*-na *Gentianaceae*. After Gentius, a king of Illyria in the 2nd century B.C., who is said to have discovered the medicinal properties of *G. lutea*. Perennial herbs. Gentian.

acaulis a-*kaw*-lis. (= *G. excisa*). Without a stem. Trumpet Gentian. S Europe.

andrewsii ăn-*drooz*-ee-ee. After H. C. Andrews. NE North America.

angustifolia ang-gus-ti-*fo*-lee-a. Narrow-leaved. Alps.

asclepiadea a-sklay-pee-*ah*-dee-a. Like *Asclepias*. Willow Gentian. Europe, W Asia.

bellidifolia bel-li-di-*fo*-lee-a. *Bellis*-leaved. New Zealand.

cachemerica cash-*me*-ri-ka. Of Kashmir.

clusii klooz-ee-ee. After Charles de l'Ecluse (Carolus Clusius) (1526–1609), Flemish botanist. Alps.

dinarica di-*nah*-ri-ka. Of the Dinaric Alps. SW Yugoslavia, Albania.

excisa see *G. acaulis*

farreri fă-ra-ree. After Farrer, see *Viburnum farreri*. NW China.

gracilipes gra-*kil*-i-pays. Slender-stalked (the flowers). NW China.

hexaphylla heks-a-*fil*-la. Six-leaved, the leaves are in whorls of six. Tibet.

lagodechiana lă-gō-dek-ee-*ah*-na. Of Lagodechi. Caucasus.

lutea loo-tee-a. Yellow (the flowers). Yellow Gentian. Europe.

ornata or-*nah*-ta. Decorative. Himalaya, Tibet.

pneumonanthe new-mon-*anth*-ee. Lung flower, from supposed medicinal properties. Marsh Gentian. Europe, Caucasus.

punctata punk-*tah*-ta. Spotted (the flowers). Europe.

pyrenaica pi-ray-*nah*-i-ka. Of the Pyrenees. Europe, Asia.

saxosa saks-ō-sa. Growing in rocky places. New Zealand.

septemfida sep-*tem*-fi-da. Divided into seven (the corolla). W Asia.

sino-ornata see-nō-or-*nah*-ta. The Chinese *G. ornata*. W China, Tibet.

veitchiorum veech-ee-*o*-rum. After the Veitch nursery. W China.

verna ver-na. Of spring (flowering). Spring Gentian, Star Gentian. Europe, Asia.

Geogenanthus gay-o-gen-*anth*-us *Commelinaceae*. From Gk. *ge* (earth) *genea* (birthplace) and *anthos* (a flower), the flowers

Geogenanthus (continued) are borne almost at ground level. Tender perennial herbs.

undatus un-*dah*-tus. Wavy-edged (the leaves). Seersucker Plant. Brazil, Peru.

Geranium ge-ră-nee-um *Geraniaceae*. From *geranion* the Gk. name, from *geranos* (a crane) referring to the beak-like fruits. Perennial herbs. Cranesbill. For the common bedding geranium, see *Pelargonium*.

armenum see *G. psilostemon*

candicans hort. see *G. lambertii*

cinereum ki-*ne*-ree-um. Grey (the leaves). Pyrenees.

   subcaulescens sub-kawl-*es*-enz. With a short stem. E Europe.

dalmaticum dăl-*mat*-i-kum. Of Dalmatia.

endressii en-*dres*-ee-ee. After P. A. C. Endress (1806–31) who collected in the Pyrenees. SW France, N Spain.

farreri fă-ra-ree. After Farrer, see *Viburnum farreri*. China.

grandiflorum hort. see *G. himalayense*

himalayense hi-mah-lay-*en*-see. (= *G. grandiflorum* hort.). Of the Himalaya. C Asia, Himalaya.

ibericum hort. see *G.* × *magnificum*

lambertii lăm-*bert*-ee-ee. (= *G. candicans* hort.). After Lambert. Himalaya, Tibet.

macrorrhizum măk-rō-*ree*-zum. With a large root. E and SE Europe.

× *magnificum* mahg-*ni*-fi-kum. *G. ibericum* × *G. platypetalum*. (= *G. ibericum* hort. *G. platypetalum* hort.). Splendid.

malviflorum măl-vi-*flō*-rum. *Malva*-flowered. S Spain.

nodosum nō-*dō*-sum. With conspicuous nodes. S Europe.

phaeum *fie*-um. Dusky. Mourning Widow, Dusky Cranesbill. Europe.

platypetalum hort. see *G.* × *magnificum*

pratense prah-*tayn*-see. Of meadows. Europe, C Asia, Himalaya.

procurrens prō-*ku*-renz. Spreading. Himalaya, Assam.

psilostemon see-*lo*-ste-mon. (= *G. armenum*). With glabrous stamens. Armenia.

renardii re-*nar*-dee-ee. After Charles Claude Renard (1809–86). Caucasus.

sanguineum sang-*gwin*-ee-um. Bloody (the flowers). Bloody Cranesbill. Europe, Caucasus.

   lancastriense lang-kăs-tree-*en*-see. Of Lancaster.

sylvaticum sil-*vă*-ti-kum. Of woods. Wood Cranesbill. Europe, N Asia.

Geranium (continued)
versicolor ver-*si*-ko-lor. Variously coloured.
S and SE Europe.
wallichianum wo-lik-ee-*ah*-num. After
Wallich who introduced it in 1819, see
Pinus wallichiana. Afghanistan, Himalaya.

Gerbera *ger*-ba-ra *Compositae*. After
Traugott Gerber (died 1743), a German
naturalist. Tender perennial.
jamesonii jaym-*son*-ee-ee. After Jameson.
Barberton Daisy, Transvaal Daisy.
Transvaal.

German Ivy see *Senecio mikanioides*
Germander see *Teucrium*
  Shrubby see *T. fruticans*
  Wall see *T. chamaedrys*

Gesneria ges-*ne*-ree-a *Gesneriaceae*. After
Conrad von Gessner (1516–65), Swiss
naturalist. Tender perennial herbs.
cuneifolia kew-nee-i-*fo*-lee-a. With leaves
tapered to the base. Cuba, Puerto Rica.

Geum *gay*-um *Rosaceae*. The L. name.
Perennial herbs.
× *borisii* bo-*ris*-ee-ee. *G. bulgaricum* × *G. reptans*. After Boris.
bulgaricum bul-*gah*-ri-kum. Of Bulgaria. SE
Europe.
chiloense ki-lō-*en*-se. Of Chiloe.
rivale ree-*vah*-lee. Growing by streams.
Water Avens. Europe, Asia, N America.
montanum mon-*tah*-num. Of mountains.
S Europe.
reptans rep-tănz. Creeping. Europe.

Gevuina ga-*veen*-a *Proteaceae*. The native
name. Semi-hardy, evergreen shrub or tree.
avellana ă-ve-*lah*-na. Like *Corylus avellana*
(the nuts). Chilean Hazel. Chile.

Gherkin see *Cucumis sativus*
Ghost Plant see *Graptopetalum paraguayense*
Giant Caladium see *Alocasia cuprea*
Giant Elephant's Ear see *Alocasia macrorrhiza*
Giant Hogweed see *Heracleum mantegazzianum*
Giant Reed see *Arundo donax*

Gilia gi-*lee*-a *Polemoniaceae*. After Filippo
Luigi Gilii (1756–1821), Italian astronomer.
Annual herbs.
achilleifolia a-ki-lee-i-*fo*-lee-a. *Achillea*-
leaved. S California, N Mexico.
capitata kă-pi-*tah*-ta. In a dense head (the

Gilia (continued)
flowers). Blue Thimble Flower. W N
America.
lutea see *Linanthus androsaceus luteus*
rubra see *Ipomopsis rubra*
tricolor tri-*ko*-lor. Three-coloured (the
flowers). Bird's Eyes. California.

Gillenia gi-*len*-ee-a *Rosaceae*. After Arnold
Gille (Gillenius), 17th-century German
botanist. Perennial herb.
trifoliata tri-fo-lee-*ah*-ta. With three leaves
(leaflets). Indian Physic. E N America.

Ginger see *Zingiber officinale*
  Wild see *Asarum*
Ginger Lily see *Hedychium*
  Scarlet see *H. coccineum*

Ginkgo gink-gō *Ginkgoaceae*. From Japanese
ginkyo (silver apricot), originally from old
Chinese *ngin-ghang*. Deciduous tree.
biloba bi-lō-ba. Two-lobed (the leaves).
Maidenhair Tree. China.

Gladdon see *Iris foetidissima*

Gladiolus gla-*dee*-o-lus *Iridaceae*. L. name
for a small sword, referring to the leaves.
Cormous perennials.
blandus see *G. carneus*
byzantinus bi-zan-*teen*-us. Of Istanbul
(Byzantium). S Europe, N Africa.
cardinalis kar-di-*nah*-lis. Scarlet (the
flowers). S Africa.
carneus kar-nee-us. (= *G. blandus*). Flesh-
coloured (the flowers). S Africa.
× *colvillii* kol-*vil*-ee-ee. *G. cardinalis* × *G. tristis*. (= *G. nanus* hort.). After James
Colvill (1777–1832), who grew S African
bulbs at Chelsea.
cuspidatus see *G. undulatus*
× *gandavensis* găn-da-*ven*-sis. Of Ghent.
grandis see *G. liliaceus*
× *hortulanus* hort-ew-*lah*-nus. Of gardens.
illyricus i-*li*-ri-kus. Of Illyria. Europe,
Caucasus.
imbricatus im-bri-*kah*-tus. Overlapping.
E Europe, W Asia.
liliaceus lee-lee-*ah*-kee-us. (= *G. grandis*).
Like *Lilium*. S Africa.
nanus hort. see *G.* × *colvillii*
natalensis nă-ta-*len*-sis. (= *G. psittacinus*).
Of Natal. SE Africa.
primulinus preem-ew-*leen*-us. Primrose
yellow. SE Africa.
psittacinus see *G. natalensis*
recurvus re-*kur*-vus. Curved back (the

Gladiolus (continued)
     perianth lobes). S Africa.
     tristis tris-tis. Sad. Natal.
     undulatus un-dew-*lah*-tus. (= *G.
     cuspidatus*). Wavy margined (the perianth
     lobes). S Africa.

Glaucidium glow-*kid*-ee-um *Ranunculaceae*.
From *Glaucium* q.v. referring to the similar
flowers. Perennial herb.
     palmatum pahl-*mah*-tum. Lobed like a hand
     (the leaves). Japan.

Glaucium glow-kee-um *Papaveraceae*. From
Gk. *glaukos* (grey-green) referring to the
colour of the leaves. Annual, biennial or
perennial herbs.
     corniculatum kor-nik-ew-*lah*-tum. Horned.
     Red Horned Poppy. Europe, W Asia.
     flavum flah-vum. Yellow. Yellow Horned
     Poppy. Europe, N Africa, W Asia.

Glechoma glay-*kō*-ma *Labiatae*. From Gk.
glechon, a kind of mint, referring to the scent
of the leaves. Perennial herb.
     hederacea he-de-*rah*-kee-a. Like *Hedera*.
     Ground Ivy. Europe, Asia.

Gleditsia gle-*dits*-ee-a *Leguminosae*. After
Gottlieb Gleditsch (died 1786), German
botanist. Deciduous trees.
     caspica kās-pi-ka. Of the region of the
     Caspian Sea. Caspian Locust.
     triacanthos tree-a-*kǎnth*-os. Three-spined.
     Honey Locust. N America.

Globe Amaranth see *Gomphrena globosa*
Globe Daisy see *Globularia*
Globe Thistle see *Echinops*
Globeflower see *Trollius*

Globularia glob-ew-*lah*-ree-a *Globulariaceae*.
From L. *globulus* (a small ball) referring to
the flower heads. Perennial herbs and sub-
shrubs. Globe Daisy.
     cordifolia kor-di-*fo*-lee-a. With heart-
     shaped leaves. Alps.
     incanescens in-kah-*nes*-enz. Greyish (the
     leaves). Italy.
     meridionalis me-ree-dee-o-*nah*-lis. (= *G.
     bellidifolia*). Flowering at mid-day. Alps to
     SE Europe.
     nudicaulis new-di-*kaw*-lis. Bare-stemmed.
     Alps, Pyrenees.
     punctata punk-*tah*-ta. (= *G. elongata*).
     Spotted. Europe.

Gloriosa glō-ree-*ō*-sa *Liliaceae*. From L.
gloriosus (glorious). Tender, climbing lilies.
Glory Lily.
     rothschildiana roths-chield-ee-*ah*-na. After
     Lionel Walter, 2nd Baron Rothschild
     (1868–1937). Tropical Africa.
     superba soo-*per*-ba. Superb. Tropical
     Africa, Asia.

Glory Bower see *Clerodendrum speciosissimum*
Glory Bush see *Tibouchina urvilleana*
Glory of Texas see *Thelocactus bicolor*
Glory Lily see *Gloriosa*
Glory Pea see *Clianthus puniceus*

Gloxinia gloks-*in*-ee-a *Gesneriaceae*. After
Benjamin Peter Gloxin. Tender perennial
herb. For the common Gloxinia of cultivation
see *Sinningia speciosa*.
     perennis pe-*ren*-is. Perennial. Colombia to
     Peru.

Glyceria gli-*se*-ree-a classically gloo-*ke*-ree-a
Gramineae. From Gk. *glykis* (sweet) referring
to the edible seeds of one species. Perennial
Grass.
     maxima mahk-si-ma. Larger. Reed Grass.
     Europe, Asia.

Glyptostrobus glip-*to*-stro-bus *Taxodiaceae*.
From Gk. *glypto* (to carve) and *strobilus* (a
cone) referring to the pitted cone scales.
Deciduous conifer.
     lineatus lin-ee-*ah*-tus. Linear (the leaves).
     SE China.

Goat's Beard see *Aruncus dioicus*
Goat's Rue see *Galega officinalis*
Godetia see *Clarkia*
Gold-rayed Lily see *Lilium auratum*
Golden Alyssum see *Aurinia saxatilis*
Golden Bell see *Forsythia suspensa*
Golden Club see *Orontium aquaticum*
Golden Column see *Trichocereus spachianus*
Golden Drop see *Onosma tauricum*
Golden Rod see *Solidago*
Golden Tom Thumb see *Parodia aureispina*
Golden Trumpet see *Allamanda cathartica*

Gomphrena gom-*free*-na *Amaranthaceae*. L.
name for this or a similar plant. Annual
herb.
     globosa glo-*bō*-sa. Spherical (the flower
     heads). Globe Amaranth. Old World
     tropics.

Goniolimon gŏn-ee-ō-*lee*-mon
Plumbaginaceae. From Gk. *gonio* (angled)
and *Limonium* a related genus. Perennial
herbs.
   callicomum ka-*li*-ko-mum. (= *Limonium
   incanum*). With beautiful hair, referring to
   the flower heads. Siberia.
   tataricum ta-*tah*-ri-kum. (= *Limonium
   tataricum*). Of Tatary, C Asia. S Europe,
   N Africa, Caucasus, Russia.

Good King Henry see *Chenopodium bonus-
henricus*
Good Luck Plant see *Cordyline terminalis*

Goodyera gud-*ye*-ra *Orchidaceae.* After John
Goodyer (1592–1664). Hardy orchids.
   pubescens pew-*bes*-enz. Hairy. E N
   America.
   repens ree-penz. Creeping. Himalaya,
   E Asia, N America.
   tesselata tes-e-*lah*-ta. Chequered (the
   leaves). E N America.

Gooseberry see *Ribes uva-crispa*

Gordonia gor-*don*-ee-a *Theaceae.* After
James Gordon (died 1781), a nurseryman
who introduced *Ginkgo biloba.* Evergreen
shrub or small tree.
   alatamaha see *Franklinia alatamaha*
   axillaris aks-il-*lah*-ris. In the leaf axils (the
   flowers). China, Taiwan.

Gorse see *Ulex*
   Common see *U. europaeus*
   Dwarf see *U. minor*
   Spanish see *Genista hispanica*

Gossypium go-*sip*-ee-um *Malvaceae.* From
gossypion the L. name. Tender shrubs and
herbs.
   arboreum ar-*bo*-ree-um. Tree-like. Tree
   Cotton. Tropical Asia.
   herbaceum her-*bah*-kee-um. Herbaceous.
   Levant Cotton. Africa, W Asia, India.

Granadilla see *Passiflora edulis*
   Giant see *P. quadrangularis*
   Red see *P. coccinea*
   Yellow see *P. laurifolia*
Grape, Northern Fox see *Vitis labrusca*
Grape Hyacinth see *Muscari*
   Oxford and Cambridge see *M. tubergeniana*
Grape Ivy see *Cissus rhombifolia*
   Miniature see *Cissus striata*
Grape Vine, Common see *Vitis vinifera*
Grapefruit see *Citrus* × *paradisi*

Graptopetalum grăp-tō-*pe*-ta-lum
Crassulaceae. From Gk. *graptos* (written
upon) and *petalon* (a petal) referring to the
markings on the petals. Tender succulents.
   amethystinum ă-me-*this*-ti-num. (=
   Pachyphytum amethystinum). Violet (the
   leaves). Mexico.
   pachyphyllum pă-kee-*fil*-lum. Thick-leaved.
   Mexico.
   paraguayense pă-ra-gwie-*en*-see. Of
   Paraguay. Ghost Plant, Mother of Pearl
   Plant. Cult.

Grass of Parnassus see *Parnassia palustris*
Greater Spearwort see *Ranunculus lingua*

Grevillea gre-*vil*-ee-a *Proteaceae.* After
Charles Francis Greville (1749–1809). Semi-
hardy and tender, evergreen shrubs.
   robusta rō-*bus*-ta. Robust. Silky Oak.
   E Australia.
   rosmarinifolia rōs-ma-reen-i-*fo*-lee-a.
   Rosmarinus-leaved. New South Wales.
   sulphurea sul-*fu*-ree-a. Sulphur-yellow (the
   flowers). New South Wales.

Grey Sage Brush see *Atriplex canescens*

Grindelia grin-*del*-ee-a *Compositae.* After
David H. Grindel (1776–1836), German
botanist. Evergreen shrub.
   chiloensis ki-lō-*en*-sis. Of Chiloe, where it
   doesn't grow. Argentina, Chile.

Griselinia gri-se-*leen*-ee-a *Cornaceae.* After
Francesco Griselini (1717–83), Venetian
naturalist. Semi-hardy, evergreen shrub or
tree.
   littoralis li-to-*rah*-lis. Growing near the sea
   shore (where it was originally collected).
   New Zealand.

Ground Ivy see *Glechoma hederacea*
Groundsel Tree see *Baccharis halimifolia*
Guava see *Psidium guajava*
Guelder Rose see *Viburnum opulus*
Guernsey Lily see *Nerine sarniensis*
Guinea Goldvine see *Hibbertia scandens*

Gunnera gun-*e*-ra *Gunneraceae.* After Ernst
Gunnerus (1718–83), Norwegian bishop and
botanist. Perennial herbs.
   chilensis chi-*len*-sis. (= *G. tinctoria*). Of
   Chile. Patagonia.
   manicata măn-i-*kah*-ta. With long sleeves.
   Colombia.

Guzmania guz-*mahn*-ee-a *Bromeliaceae.*
After Anastasio Guzman, 18th-century
Spanish naturalist. Tender, evergreen
perennials.
  lingulata ling-gew-*lah*-ta. Tongue-like (the
  bracts). Tropical America.
  monostachia mon-ō-*stăk*-ee-a. With one
  spike. Tropical America.
  musaica mu-*sah*-i-ka. Like a mosaic (the
  leaves). Colombia.
  zahnii zahn-ee-ee. After Gottlieb Zahn who
  collected for Veitch in C America.
  Panama.

Gymnocalycium gim-nō-ka-*li*-kee-um
Cactaceae. From Gk. *gymnos* (naked) and
calyx (a bud) referring to the naked flower
buds.
  baldianum băl-dee-*ah*-num. (= *G.
  venturianum*). After Bald. Argentina.
  damsii dămz-ee-ee. After Dams. Paraguay.
  denudatum day-new-*dah*-tum. Naked, it is
  barely tubercled as most species. Spider
  Cactus. Brazil, Argentina.
  gibbosum gi-*bō*-sum. Swollen on one side.
  Argentina.
  mihanovichii mi-hahn-no-*vich*-ee-ee. After
  Mihanovich, a plant collector. Plain
  Cactus. Paraguay.
  multiflorum mul-tee-*flō*-rum. Many-
  flowered. Brazil, Argentina.
  platense pla-*ten*-see. From near the River
  Plate, Argentina.
  quehlianum kwel-ee-*ah*-num. After Leopold
  Quehl (1849–1923). Argentina.
  saglione săg-lee-ō-nee. After the Frenchman
  who was the first to grow it in Europe.
  N Argentina.

Gymnocarpium gim-nō-*kar*-pee-um
Aspidiaceae. From Gk. *gymnos* (naked) and
karpos (a fruit), the sori are not covered with
an indusium. Ferns.
  dryopteris dree-*op*-te-ris. Oak fern. Oak
  Fern. N America, Europe, Asia.
  robertianum ro-bert-ee-*ah*-num. Resembling
  Geranium robertianum. Limestone
  Polypody. N America, Europe.

Gymnocladus gim-no-kla-dus *Leguminosae.*
From Gk. *gymnos* (naked) and *klados* (a
branch) referring to its deciduous nature.
Deciduous tree.
  dioica dee-ō-*ee*-ka. Dioecious. Kentucky
  Coffee Tree. E and C United States.

Gynura gin-*ew*-ra *Compositae.* From Gk.
gyne (female) and *oura* (a tail) referring to
the long stigma. Tender, scandent herb.
  aurantiaca ow-răn-tee-*ah*-ka. (= *G.
  sarmentosa* hort.). Orange (the flowers).
  Velvet Plant, Purple Passion Vine. Java.

Gypsophila gip-*sof*-i-la *Caryophyllaceae.*
From Gk. *gypsos* (gypsum) and *philos* (loving),
some species grow on lime. Annual and
perennial herbs.
  aretioides a-ray-tee-*oi*-deez. Like *Aretia*
  (now included in *Androsace*). N Iran.
  cerastioides ke-răs-tee-*oi*-deez. Like
  Cerastium. Himalaya.
  elegans ay-le-gahnz. Elegant. Caucasus,
  W Asia.
  paniculata pa-nik-ew-*lah*-ta. With flowers
  in panicles. Europe to C Asia.
  repens ree-penz. Creeping. Europe.

H

Haageocereus hahg-ee-ō-*kay*-ree-us
Cactaceae. After J. N. Haage (1826–78) and
Cereus q.v. Peru.
  decumbens day-*kum*-benz. Prostrate.
  versicolor ver-si-ko-lor. Variably coloured
  (the spines).

Haberlea ha-*ber*-lee-a *Gesneriaceae.* After
Carl Constantin Haberle (1764–1832).
Perennial herbs.
  ferdinandii-coburgii fer-di-*nahn*-dee-ee-kō-
  burg-ee-ee. After King Ferdinand of
  Bulgaria. C Bulgaria.
  rhodopensis ro-do-*pen*-sis. Of the Rhodope
  Mountains, Bulgaria. Bulgaria, N Greece.

Habranthus ha-*brănth*-us *Amaryllidaceae.*
From Gk. *habros* (graceful) and *anthos* (a
flower). Tender, bulbous herbs.
  andersonii ăn-der-*son*-ee-ee. After
  Anderson. S America.
  robustus rō-*bus*-tus. (= *Zephyranthes
  robusta*). Robust. Argentina.

Hackberry see *Celtis occidentalis*

Hacquetia hă-*kay*-tee-a *Umbelliferae.* After
Balthasar Hacquet (1740–1815), Austrian
botanical writer. Perennial herb.
  epipactis e-pi-*păk*-tis. Gk. name for a plant.
  E Europe.

Haemanthus hiem-*ănth*-us *Amaryllidaceae*. From Gk. *haima* (blood) and *anthos* (a flower) referring to the colour of the flowers. Tender bulbous herbs. Blood Lily, Red Cape Tulip.
albiflos ăl-bi-flōs. With white flowers. White Paint Brush. S Africa.
coccineus kok-*kin*-ee-us. Scarlet (the flowers). Ox-tongue Lily. S Africa.
katharinae kath-a-*rin*-ie. After Mrs Katherine Saunders who collected in Natal. Blood Flower. S Africa.
magnificus mahg-*ni*-fi-kus. Magnificent. Royal Paint Brush. S Africa.
multiflorus mul-tee-*flō*-rus. Many-flowered. Salmon Blood Lily. Tropical Africa.
puniceus pew-*ni*-kee-us. Reddish-purple (the flowers). S Africa.

Hair Grass see *Eleocharis acicularis*

Hakea hăk-ee-a *Proteaceae*. After Baron Christian Ludwig von Hake (1745–1818) a German patron of botany. Evergreen hardy and tender shrubs.
laurina low-*reen*-a. Like *Laurus*. Pincushion Flower, Sea Urchin. W Australia.
lissosperma lis-ō-*sperm*-a. (= *H. sericea* hort.). With smooth seeds. Tasmania, SE Australia.
microcarpa mik-rō-*kar*-pa. With small fruits. E Australia, Tasmania.

Hakonechloa ho-kōn-ee-*klō*-a *Gramineae*. From *Hakone*, a region of Japan, and Gk. *chloa* (a grass). Perennial grass.
macra măk-ra. Large. Japan.

Halesia haylz-ee-a *Styracaceae*. After Dr Stephen Hales (1677–1761), English scientist and inventor. Deciduous trees and shrubs. Silver Bell, Snowdrop Tree. SE United States.
carolina kă-ro-*leen*-a. Of Carolina.
diptera dip-te-ra. Two-winged (the fruit).
  magniflora mahg-ni-*flō*-ra. Large-flowered.
monticola mon-*ti*-ko-la. Mountain-loving. Mountain Snowdrop Tree.

× **Halimiocistus** ha-lim-ee-ō-*kis*-tus *Cistaceae*. Intergeneric hybrids, from the names of the parents. *Cistus* × *Halimium*. Semi-hardy, evergreen shrubs.
'Ingwersenii' ing-gwa-*sen*-ee-ee. *Cistus salviifolius* × *Halimium alyssoides*. After W. E. Th. Ingwersen (1883–1960),

× **Halimiocistus** (continued)
nurseryman and plant collector specialising in alpines who discovered it in Portugal about 1929.
sahucii sa-*hook*-ee-ee. *Cistus salviifolius* × *Halimium umbellatum*. After M. Sahuc, a member of the party that discovered it.
wintonensis win-ton-*en*-sis. Of Winchester, where it was raised by Hillier's.

Halimium ha-*lim*-ee-um *Cistaceae*. From *Atriplex halimus*, some species resemble it in leaf. Semi-hardy, evergreen shrubs.
alyssoides ă-lis-*oi*-deez. Like *Alyssum*. SW Europe.
halimifolium ha-lim-i-*fo*-lee-um. With leaves like *Atriplex halimus*. Mediterranean region, N Africa.
lasianthum lă-see-*ănth*-um. With woolly flowers. S Portugal, S Spain.
  formosum for-*mō*-sum. (= *H. lasianthum* hort.). Beautiful. S Portugal.
ocymoides ō-kim-*oi*-deez. Like *Ocimum*. Spain, Portugal.
umbellatum um-bel-*ah*-tum. With flowers in umbels.

Halimodendron ha-lim-ō-*den*-dron *Leguminosae*. From Gk. *halimum* (maritime) and *dendron* (a tree), it grows in salt-rich soils. Deciduous shrub.
halodendron hă-lō-*den*-dron. Salt tree. Salt Tree. C Asia to Mongolia.

Hamamelis hăm-a-*may*-lis *Hamamelidaceae*. Gk. name for another plant. Deciduous shrubs. Witch Hazel.
× *intermedia* in-ter-*med*-ee-a. *H. japonica* × *H. mollis*. Intermediate (between the parents).
japonica ja-*pon*-i-ka. Of Japan. Japanese Witch Hazel.
mollis mol-lis. Softly hairy (the young shoots and leaves). Chinese Witch Hazel. W China.
vernalis ver-*nah*-lis. Of spring (flowering). SE United States.

Hamatocactus hamatacanthus see *Ferocactus hamatacanthus*
  setispinus see *Ferocactus setispinus*
Handkerchief Tree see *Davidia involucrata*

Hardenbergia hard-an-*berg*-ee-a *Leguminosae*. After Franziska, Countess von Hardenberg. Tender evergreen climber.
comptoniana komp-ton-ee-*ah*-na. Compton, the family name of Lady Northampton

Hardenbergia (continued)
who grew it c. 1810. W Australia.
violacea vee-o-*lah*-kee-a. Violet (the
flowers). E Australia, Tasmania.

Harebell see *Campanula rotundifolia*
Hare's-tail Grass see *Lagurus ovatus*
Harlequin Flower see *Sparaxis tricolor*
Harry Lauder's Walking Stick see *Corylus
avellana* 'Contorta'

Hatiora hă-tee-*o*-ra *Cactaceae*. After Thomas
Hariot (1560–1621), mathematician and
cartographer. Originally called *Hariota*, a
name that had already been used. Brazil.
bambusoides băm-bew-*soi*-deez. Like
Bambusa (a bamboo).
salicornioides să-li-korn-ee-oi-deez. Like
Salicornia, a salt-marsh plant. Drunkard's
Dream.

Haworthia hay-*werth*-ee-a *Liliaceae*. After
Adrian Hardy Haworth (1768–1833),
entomologist and botanist who grew
succulents at Chelsea. Tender succulents.
S Africa.
attenuata a-ten-ew-*ah*-ta. Drawn out (the
leaves).
cuspidata kus-pi-*dah*-ta. With a sharp, short
point (the leaves).
limifolia leem-i-*fo*-lee-a. With file-like
leaves.
margaritifera mar-ga-ri-*ti*-fe-ra. Pearl-
bearing (the leaves). Pearl Plant.
maughanii mawn-ee-ee. After Dr R.
Maughan Brown.
papillosa pă-pi-*lō*-sa. With papillae, the
white tubercles on the leaves.
reinwardtii rien-*vart*-ee-ee. After
Reinwardt, see *Reinwardtia*.
tesselata te-se-*lah*-ta. Chequered. Wart
Plant.
truncata trun-*kah*-ta. Abruptly cut off (the
leaves).

Hawthorn see *Crataegus*
 Common see *C. monogyna*
 Midland see *C. laevigata*
Hazel see *Corylus avellana*
 Turkish see *C. colurna*
Heart of Jesus see *Caladium*
Heartsease see *Viola tricolor*
Hearts on a String see *Ceropegia woodii*
Heath see *Erica*
 Cornish see *E. vagans*
 Cross-leaved see *E. tetralix*
 Spanish see *E. australis*
 Tree see *E. arborea*

Heather see *Calluna vulgaris*
 Bell see *Erica cineria*

Hebe hay-bay *Scrophulariaceae*. After Hebe,
goddess of youth. Evergreen shrubs. New
Zealand, apart from hybrids and cultivars.
albicans ăl-bi-kănz. Whitish (the leaves).
'Andersonii' ăn-der-*son*-ee-ee. After Isaac
Anderson-Henry (1800–84) who raised it.
anomala a-*nom*-a-la. Abnormal, the type
specimen has the calyx three lobed (it is
normally four-lobed).
armstrongii hort. see *H. ochracea*
brachysiphon bră-kee-*see*-fon. (= *H. travers*
hort.). With a short tube (the corolla).
'Carnea' *kar*-nee-a. Flesh-coloured (the
flowers).
cattaractae see *Parahebe cattaractae*
chathamica cha-*tăm*-i-ka. Of the Chatham
Islands.
cupressoides kew-pres-*oi*-deez. Like
Cupressus.
darwiniana see *H. glaucophylla*
'Edinensis' ed-in-*en*-sis. Of Edinburgh,
where it was raised.
× *franciscana* frăn-sis-*kah*-na. *H. elliptica*
× *H. speciosa*. (= *H. elliptica* hort.). Of
San Francisco, it was described from plant
in the Golden Gate Park.
glaucophylla glow-kō-*fil*-la. (= *H.
darwiniana*). With glaucous leaves.
hulkeana hŭlk-ee-*ah*-na. After Mr T. H.
Hulke, who is said to have discovered it.
New Zealand Lilac.
'Lindsayi' *lind*-zay-ee. After Mr R. Lindsay
who raised it.
macrantha ma-*krănth*-a. Large-flowered.
ochracea ok-*rah*-kee-a. (= *H. armstrongii*
hort.). Ochre-coloured (the foliage).
pimeleoides pi-me-lay-*oi*-deez. Like
Pimelea.
pinguifolia ping-gwi-*fo*-lee-a. With fat
leaves.
rakaiensis ră-kie-*en*-sis. (= *H. subalpina*
hort.). Of the Rakai Valley, Canterbury.
recurva re-*kur*-va. Curved back (the leaves).
salicifolia să-li-ki-*fo*-lee-a. *Salix*-leaved.
Also in Chile.
subalpina hort. see *H. rakaiensis*
traversii hort. see *H. brachysiphon*

Hedera he-de-ra *Araliaceae*. The L. name.
Evergreen climbers. Ivy.
algeriensis ăl-ge-ree-*en*-sis. (= *H. canariensis*
hort.). Of Algeria.
'Margino-maculata' *mar*-gi-nō-măk-ew-
lah-ta. With spotted margins (the

Hedera (continued)
    leaves). Often seen labelled *H. helix*
    'Marmorata'.
  azorica a-zo-ri-ka. (= *H. canariensis*
  azorica). Of the Azores.
  colchica kol-ki-ka. Of Colchis, the E coast
  of the Black Sea. Caspian region,
  Caucasus, Turkey.
    dentata den-*tah*-ta. Toothed (the leaves).
  helix he-liks. Winding around, perhaps
  referring to it being wound around a staff
  carried by Bacchus or his attendants.
  Common Ivy. Europe, Caucasus.
    'Deltoidea' see *H. hibernica* 'Deltoidea'
    'Digitata' see *H. hibernica* 'Digitata'
    'Hibernica' see *H. hibernica*
    'Luzii' *luts*-ee-ee. After the German
    nursery of Ernst Luz. Often sold as
    'Marmorata'.
    'Pedata' pe-*dah*-ta. Like a bird's foot (the
    leaves).
    poetica pō-*et*-i-ka. Of poets, who used it
    for wreaths. Poet's Ivy.
  hibernica hi-*bern*-i-ka. (= *H. helix*
  'Hibernica'). Irish. Irish Ivy. W Europe.
    'Deltoidea' del-*toi*-dee-a. (= *H. helix*
    'Deltoidea'). Deltoid i.e. shaped like the
    Gk. letter delta (Δ), the leaves.
    'Digitata' di-gi-*tah*-ta. (= *H. helix*
    'Digitata'). With finger-like lobes (the
    leaves).

Hedychium hay-*di*-kee-um *Zingiberaceae*.
From Gk. *hedys* (sweet) and *chion* (snow)
referring to the fragrant, white flowers of *H.
coronarium*. Semi-hardy to tender, herbaceous
perennials. Ginger Lily.
  coccineum kok-*kin*-ee-um. Scarlet. Scarlet
  Ginger Lily. NE India, Indo-China.
  coronarium ko-rō-*nah*-ree-um. Used in
  garlands. Butterfly Lily, Garland Flower.
  Tropical Himalaya, SE Asia.
  densiflorum dens-i-*flō*-rum. Densely
  flowered. Temperate. Himalaya, Assam.
  flavescens flah-*ves*-enz. (= *H. flavum*).
  Yellow (the flowers). E Himalaya.
  gardnerianum gard-na-ree-*ah*-num. After
  Edward Gardner (born 1784), political
  resident in Nepal. Kahili Ginger.
  Himalaya, Assam.
  greenei green-ee-ee. After Mr H. F. Green
  in whose garden the originally discovered
  plants were grown. Bhutan, Assam.

Hedyotis hay-*dee*-o-tis *Rubiaceae*. From Gk.
hedys (sweet) and *otos* (an ear). Perennial
herbs.
  caerulea kie-*ru*-lee-a. (= *Houstonia*

Hedyotis (continued)
  caerulea). Blue (the flowers). Bluets. E N
  America.
  michauxii mee-*shō*-ee-ee. (= *Houstonia
  serpyllifolia*). After André Michaux, see
  Michauxia. SE United States.
  purpurea pur-*pewr*-ree-a. (= *Houstonia
  purpurea*). Purple (the flowers). E N
  America.

Hedysarum hay-*dis*-a-rum *Leguminosae*.
From Gk. *hedys* (sweet) referring to the
fragrant flowers of *H. coronarium*. Perennial
herb and shrub.
  coronarium ko-rō-*nah*-ree-um. Used in
  garlands. French Honeysuckle. Europe.
  multijugum mul-tee-*yoo*-gum. With many
  joined together, referring to the many
  leaflets. W Mongolia, Gansu, E Tibet
    apiculatum a-pik-ew-*lah*-tum. With an
    abrupt, short point.

Helenium he-*le*-nee-um *Compositae*. From
helenion the Gk. name for another plant after
Helen of Troy. Perennial herb.
  autumnale ow-tum-*nah*-lee. Of autumn
  (flowering). Sneezeweed. N America.

Helianthemum hay-lee-*ănth*-e-mum
Cistaceae. From Gk. *helios* (the sun) and
anthemon (a flower). Dwarf, evergreen
shrubs. Sun Rose.
  apenninum ă-pe-*neen*-um. Of the
  Apennines. SW Europe, N Africa.
  lunulatum loon-ew-*lah*-tum. Crescent-
  shaped (the blotch at the base of the
  petals). N Italy.
  nummularium num-ew-*lah*-ree-um. (= *H.
  chamaecistus. H. vulgare*). With coin-
  shaped leaves (in the type specimen).
  Europe, W Asia, Caucasus.
  oelandicum ur-länd-i-kum. Of Öland, SE
  Sweden.
    alpestre ăl-*pes*-tree. (= *H. alpestre*). Of
    the lower mountains. C and S Europe.

Helianthus hay-lee-*ănth*-us *Compositae*.
From Gk. *helios* (the sun) and *anthos* (a
flower). Annual and perennial herbs.
  annuus *ăn*-ew-us. Annual. Sunflower.
  N America, N Mexico.
  decapetalus dek-a-*pe*-ta-lus. Ten-petalled.
  E N America.
  × *multiflorus* mul-tee-*flō*-rus. *H. annuus* ×
  H. decapetalus. Many-flowered.
  tuberosus tew-be-*rō*-sus. Tuberous (the
  rhizome). Jerusalem Artichoke. E N
  America.

× **Heliaporus** hay-lee-a-*po*-rus *Cactaceae*.
Intergeneric hybrid, from the names of the
parents. *Aporocactus* × *Heliocereus*.
    smithii smith-ee-ee. *Aporocactus flagelliformis*
    × *Heliocereus speciosus*. (= *Aporocactus* ×
    mallisonii). After Smith.

Helichrysum hay-li-*kris*-um *Compositae*.
From Gk. *helios* (the sun) and *chryson* (golden).
Annual and perennial herbs and shrubs.
    angustifolium see *H. italicum*
    bellidioides bel-i-dee-*oi*-deez. Like *Bellis*.
    New Zealand.
    bracteatum bräk-tee-*ah*-tum. With
    conspicuous bracts. Everlasting.
    Australia.
    coralloides see *Ozothamnus coralloides*
    frigidum fri-gi-dum. Of cold regions.
    Corsica, Sardinia.
    italicum ee-*tä*-li-kum. (= *H. angustifolium*).
    Of Italy, S Europe.
        serotinum se-*ro*-ti-num. (= *H. serotinum*).
        Late flowering. Curry Plant. SW
        Europe.
    ledifolium see *Ozothamnus ledifolius*
    microphyllum hort. see *Plecostachys*
    serpyllifolia
    milfordiae mil-*ford*-ee-ie. After Mrs Helen
    A. Milford (died 1940) who collected in
    S Africa. S Africa.
    orientale o-ree-en-*tah*-lee. Eastern. SE
    Europe, W Asia.
    petiolare pe-tee-o-*lah*-ree. (= *H. petiolatum*
    hort.). With conspicuous petioles.
    S Africa.
    plicatum pli-*kah*-tum. Pleated (the leaves).
    SE Europe.
    rosmarinifolium see *Ozothamnus*
    rosmarinifolius
    selago see *Ozothamnus selago*
    sibthorpii sib-*thorp*-ee-ee. (= *H. virgineum*).
    After John Sibthorp (1758–96). Greece.
    splendidum splen-di-dum. Splendid.
    S Africa.

Helicodiceros muscivorum see *Dracunculus*
muscivorus

Helictotrichon he-lik-tō-*tri*-kon *Gramineae*.
From Gk. *helix* (spiral) and *trichos* (a hair).
Perennial grass.
    sempervirens sem-per-*vi*-rens. (= *Avena
    candida. Avena sempervirens*). Evergreen.
    SW Alps.

Heliocereus hay-lee-ō-*kay*-ree-us *Cactaceae*.
From Gk. *helios* (the sun) and *Cereus* q.v.

Heliocereus (continued)
    cinnabarinus kin-a-ba-*reen*-us. Cinnabar red
    (the flowers). Guatemala.
    speciosus spek-ee-ō-sus. Showy. Sun
    Cactus. Mexico.
        amecamensis a-me-ka-*men*-sis. Of Ameca.
        C Mexico.

Heliophila hay-lee-o-fi-la *Cruciferae*. From
Gk. *helios* (the sun) and *philos* (loving).
Annual herbs. S Africa.
    leptophylla lep-tō-*fil*-la. Narrow-leaved.
    linearifolia lin-ee-ah-ri-*fo*-lee-a. With linear
    leaves.
    longifolia long-gi-*fo*-lee-a. Long-leaved.

Heliopsis hay-lee-*op*-sis *Compositae*. From
Gk. *helios* (the sun) and -*opsis* indicating
resemblance, referring to the flower heads.
Perennial herbs.
    helianthoides hay-lee-änth-*oi*-deez. Like
    Helianthus. E. United States.
        scabra skäb-ra. (= *H. scabra*). Rough (the
        leaves).
        'Patula' *pät*-ew-la. (= *H. patula*).
        Spreading.
        'Zinniiflora' zin-ee-i-*flō*-ra. Zinnia-
        flowered.

Heliotropium hay-lee-o-trō-pee-um
Boraginaceae. From Gk. *helios* (the sun) and
trope (to turn) referring to the old belief that
the flowerheads turned with the sun. Annual
herb.
    arborescens ar-bo-*res*-enz. (= *H. hybridum.
    H. peruvianum*). Becoming tree-like.
    Heliotrope, Cherry Pie. Peru.

Helipterum hay-*lip*-te-rum *Compositae*.
From Gk. *helios* (the sun) and *pteron* (a wing)
referring to their sun-loving nature and the
feathery pappus bristles. Annual herbs.
Australia.
    albicans äl-bi-känz. Whitish, the leaves and
    flower heads.
    humboldtianum hum-bolt-ee-*ah*-num. After
    W. H. Alexander von Humboldt
    (1769–1859).
    manglesii mang-*galz*-ee-ee. (= *Rhodanthe
    manglesii*). After Mangles, see *Anigozanthus
    manglesii*.
    roseum ro-see-um. Rose-coloured (the
    flower bracts).

Helleborus he-*le*-bo-rus *Ranunculaceae*.
From *helleboros* the Gk. name for *H.*
orientale. Perennial herbs. Hellebore.

Helleborus (continued)
  abchasicus ăb-*kăs*-i-kus. Of Abchasia.
  Caucasus.
  argutifolius ar-gew-ti-*fo*-lee-us. (= *H.*
  corsicus. *H. lividus corsicus*). With sharply-
  toothed leaves. Corsica, Sardinia.
  atrorubens aht-rō-*ru*-benz. Deep red (the
  flowers). Yugoslavia.
  corsicus see *H. argutifolius*
  foetidus foy-ti-dus. Stinking. Stinking
  Hellebore. SW Europe.
  lividus lee-vi-dus. Lead-coloured.
  W Mediterranean islands.
    corsicus see *H. argutifolius*
  niger ni-ger. Black (the roots). Christmas
  Rose. E Alps, N Italy, Yugoslavia.
  × *nigercors* ni-ger-kors. *H. argutifolius* × *H.*
  niger. From the names of the parents.
  orientalis o-ree-en-*tah*-lis. Eastern. Lenten
  Rose. SE Europe, W Asia.
  purpurascens pur-pew-*răs*-enz. Purplish.
  E Europe.
  × *sternii stern*-ee-ee. *H. argutifolius* × *H.*
  lividus. After Sir Frederick Stern
  (1884–1967) of Highdown.
  viridis vi-ri-dis. Green (the sepals). Green
  Hellebore. C and W Europe.

Helxine soleirolii see *Soleirolia soleirolii*

Hemerocallis hay-me-rō-*kă*-lis *Liliaceae*.
From Gk. *hemera* (day) and *kallos* (beauty),
the flowers only last for one day. Perennial
herbs. Day Lily.
  citrina ki-*tree*-na. Lemon-yellow. China.
  dumortieri dew-mor-tee-*e*-ree. After B. C.
  Dumortier (1797–1828). Japan.
  fulva ful-va. Tawny. S Europe to China.
  middendorfii mid-an-*dorf*-ee-ee. After
  Alexander Theodor von Middendorf
  (1815–94), Russian traveller and plant
  collector. E Asia.
  minor mi-nor. Smaller. NE Asia.
  multiflora mul-tee-*flō*-ra. Many-flowered.
  China.
  thunbergii thun-*berg*-ee-ee. After Thunberg,
  see *Thunbergia*. Japan.

Hemigraphis hay-mi-*grăf*-is *Acanthaceae*.
From Gk. *hemi* (half) and *graphis* (a brush)
referring to the hairy filaments of the outer
stamens. Tender herb.
  alternata ăl-ter-*nah*-ta. Alternate. Red Ivy.
  SE Asia.

Hemlock see *Tsuga*
  Eastern see *T. canadensis*
  Mountain see *T. mertensiana*

Western see *T. heterophylla*
Hemp see *Cannabis sativa*
Hemp Agrimony see *Eupatorium cannabinum*
Hen and Chickens see *Sempervivum tectorum*
Henbane see *Hyoscyamus niger*

Hepatica he-*pă*-ti-ka *Ranunculaceae*. From
Gk. *hepar* (the liver) referring to the shape
and colour of the leaves. Perennial herbs.
  americana a-me-ri-*kah*-na. American. E N
  America.
  × *media* me-dee-a. *H. nobilis* × *H.*
  transsilvanica. Intermediate (between the
  parents).
  nobilis nō-bi-lis. (= *H. triloba*). Notable.
  Europe, W Asia.
  transsilvanica trahns-sil-*vah*-ni-ka. (= *H.*
  angulosa). Of Transylvania, Romania.

Heptapleurum arboricolum see *Schefflera*
arboricola

Heracleum hay-ra-*klee*-um *Umbelliferae*.
After Hercules (Herakles). Biennial herb.
  mantegazzianum măn-tee-găts-ee-*ah*-num.
  After Paolo Mantegazzi (1831–1910).
  Giant Hogweed, Cartwheel Flower.
  Caucasus.

Herald's Trumpet see *Beaumontia grandiflora*
Herb Christopher see *Actaea spicata*

Hermodactylus her-mō-*dăk*-ti-lus *Iridaceae*.
From Hermes (Gk. name for Mercury) and
daktylos (a finger) referring to the finger-like
tubers. Perennial herb.
  tuberosus tew-be-*rō*-sus. (= *Iris tuberosa*).
  Tuberous. Snake's-head Iris. S Europe,
  W Asia.

Herniaria her-nee-*ah*-ree-a *Caryophyllaceae*.
From *hernia* (a rupture) referring to
supposed medicinal properties. Perennial
herb.
  glabra glă-bra. Glabrous. Rupturewort.
  Europe, N Africa, W and C Asia.

Hertia her-tee-a *Compositae*. Derivation not
traced. Evergreen sub-shrub.
  cheirifolia kay-ri-*fo*-lee-a. (= *Othonnopsis*
  cheirifolia). With leaves like *Cheiranthus*
  cheiri. N Africa.

Hesperis hes-pe-ris *Cruciferae*. From Gk.
hespera (the evening), the flowers are fragrant
in the evening. Biennial or perennial herb.
  matronalis mah-trō-*nah*-lis. Of matrons,

Hesperis (continued)
from an old name, Mother of the Evening.
Sweet Rocket, Dame's Violet. Europe,
W and C Asia.

Heterocentron he-te-rō-*ken*-tron
Melastomataceae. From Gk. *heteros*
(different) and *kentron* (a spur) referring to
the different spurs of the anthers. Tender
herbs and sub-shrubs.
elegans ay-le-gahnz. (= *Schizocentron
elegans*). Spanish Shawl. Mexico.
C America.
macrostachyum măk-ro-*stăk*-ee-um. (= *H.
roseum*). With large spikes. Mexico.
subtriplinervium sub-trip-li-*nerv*-ee-um.
Somewhat three-nerved. Mexico.

Heuchera hoy-ka-ra *Saxifragaceae*. After
Johann Heinrich von Heucher (1677–1747).
Perennial herb. Alum Root.
sanguinea sang-*gwin*-ee-a. Blood-red (the
flowers). SW United States. Mexico.

× **Heucherella** hoy-ke-*rel*-la *Saxifragaceae*.
Intergeneric hybrid, from the names of the
parents. *Heuchera* × *Tiarella*. Perennial herb.
tiarelloides tee-a-rel-*loi*-deez. *Heuchera*
hybrid × *Tiarella cordifolia*. Like *Tiarella*.

Hibbertia hi-*bert*-ee-a *Dilleniaceae*. After
George Hibbert (1757–1837), merchant who
had a botanic garden at Clapham. Tender,
evergreen shrubs.
dentata den-*tah*-ta. Toothed (the leaves). SE
Australia.
scandens skăn-denz. Climbing. Guinea Gold
Vine. Snake Vine. Australia.

Hibiscus hi-*bis*-kus *Malvaceae*. Gk. name for
mallow. Tender annuals and perennials,
hardy and tender shrubs.
× *archeri* ar-cha-ree. *H. rosa-sinensis* × *H.
schizopetalus*. After the raiser, Mr A. S.
Archer of Antigua.
moscheutos mos-*kew*-tas. Musk-scented. SE
United States.
mutabilis mew-*tah*-bi-lis. Changeable (the
flower colour). Cotton Rose. S China,
Taiwan, Japan.
rosa-sinensis ro-sa-si-*nen*-sis. As the common
name, Rose of China. Origin uncertain.
schizopetalus ski-zo-*pe*-ta-lus. With split
petals. Japanese Hibiscus. E Africa.
sino-syriacus see-nō-si-ree-*ah*-kus. The
Chinese *H. syriacus* q.v. C China.
syriacus si-ree-*ah*-kus. Of Syria where it was

Hibiscus (continued)
originally thought to be native. China,
Taiwan.
trionum tree-ō-num. Three-coloured.
Flower of an Hour. Old World tropics.

Hickory see *Carya*
  Bitternut see *C. cordiformis*
  Mockernut see *C. tomentosa*
  Pignut see *C. glabra*
  Shagbark see *C. ovata*

Hieracium hee-e-*rah*-kee-um *Compositae*.
Gk. and L. name for these or similar plants.
Perennial herbs.
aurantiacum ow-răn-tee-*ah*-kum. Orange
(the flowers). Devil's Paintbrush. Europe.
maculatum măk-ew-*lah*-tum. Spotted (the
leaves). Europe.
villosum vi-*lō*-sum. Softly hairy. Europe.
waldsteinii văld-*stien*-ee-ee. After
Waldstein, see *Waldsteinia*. Yugoslavia.

Hippeastrum hi-pee-*ăs*-trum *Amaryllidaceae*.
From Gk. *hippos* (a horse), the inflorescence
of *H. puniceum* was likened to a horse's head.
Tender bulbous herbs. Amaryllis.
argentinum ar-jen-*teen*-um. (= *H.
candidum*). Of Argentina.
aulicum ow-li-kum. Of the court. Lily of
the Palace. Brazil to Paraguay.
× *johnsonii* jon-*son*-ee-ee. *H. reginae* × *H.
vittatum*. After Mr Johnson, the raiser. St
Joseph's Lily.
puniceum pew-*ni*-kee-um. (= *H. equestre*).
Reddish-purple (the flowers). Tropical
America.
reticulatum ray-tik-ew-*lah*-tum. Net-veined
(the perianth lobes). Brazil.
striatum stree-*ah*-tum. (= *H. rutilum*).
Striped (the flowers). Brazil.
vittatum vi-tah-tum. Banded (the flowers).
Peru.

Hippocrepis hi-pō-*kre*-pis *Leguminosae*.
From Gk. *hippos* (a horse) and *krepis* (a shoe)
referring to the horse-shoe shaped segments
of the pods.
comosa ko-*mō*-sa. Tufted. Horseshoe
Vetch. Europe.

Hippophae hi-*po*-fa-ee *Elaeagnaceae*. The
Gk. name for another plant. Deciduous
shrub or small tree.
rhamnoides răm-*noi*-deez. Like *Rhamnus*.
Sea Buckthorn. Europe, Asia.

Hippuris hi-*pewr*-ris *Hippuridaceae*. From Gk. *hippos* (a horse) and *oura* (a tail). Aquatic perennial.
 vulgaris vul-*gah*-ris. Common. Mare's Tail. Widely distributed.

Hoheria hō-*he*-ree-a *Malvaceae*. From *houhere* the Maori name for *H. populnea*. Semi-hardy, deciduous and evergreen, trees and shrubs. New Zealand.
 angustifolia ang-gust-i-*fo*-lee-a. Narrow-leaved.
 glabrata glåb-*rah*-ta. Rather glabrous.
 lyallii lie-*ål*-ee-ee. After David Lyall (1817–95), naval surgeon and naturalist, who collected the type specimen.
 populnea pō-*pul*-nee-a. *Populus*-like (the leaves).
 sexstylosa seks-sti-*lō*-sa. With six styles.

Holboellia hol-*burl*-ee-a *Lardizabalaceae*. After F. L. Holboell (1765–1829), a superintendent of the Copenhagen Botanic Garden. Evergreen climber.
 coriacea ko-ree-*ah*-kee-a. Leathery (the leaflets). China.

Holcus hol-kus *Gramineae*. The Gk. name for Sorghum. Perennial grass.
 mollis mol-lis. Softly hairy. Europe.

Holly see *Ilex*
 Blue see *I.* × *meserveae*
 Common see *I. aquifolium*
 Horned see *I. cornuta*
Hollyhock see *Alcea rosea*
 Fig-leaved see *Alcea ficifolia*

Holmskioldia holm-*shol*-dee-a *Verbenaceae*. After Theodor Holmskiold (1733–94), Danish botanist. Tender, evergreen shrub.
 sanguinea sang-*gwin*-ee-a. Blood red (the flowers). Chinese-hat Plant. Himalaya.

Holodiscus ho-lō-*dis*-kus *Rosaceae*. From Gk. *holos* (entire) and *diskos* (a disc) referring to the unlobed disc. Deciduous shrub.
 discolor dis-ko-lor. Two-coloured, the leaves are grey-hairy beneath. Ocean Spray. W N America.

Honeysuckle see *Lonicera*
 Common see *L. periclymenum*
 Perfoliate see *L. caprifolium*
 Trumpet see *L. sempervirens*
Hop see *Humulus lupulus*
Hop Hornbeam see *Ostrya carpinifolia*
Hop Tree see *Ptelea trifoliata*

Hordeum hor-dee-um *Gramineae*. The L. name for *H. vulgare* (barley). Perennial grass.
 jubatum yoo-*bah*-tum. Maned (the inflorescence). Squirrel-tail Grass. N temperate regions.

Horminum hor-*meen*-um *Labiatae*. Gk. name for sage. Perennial herb.
 pyrenaicum pi-ray-*nah*-i-kum. Of the Pyrenees. Alps, Pyrenees.

Hornbeam see *Carpinus*
 American see *C. caroliniana*
 Common see *C. betulus*
Hornwort see *Ceratophyllum*
Horse Briar see *Smilax rotundifolia*
Horse-chestnut see *Aesculus*
 Common see *A. hippocastanum*
 Indian see *A. indica*
 Japanese see *A. turbinata*
 Red see *A.* × *carnea*
Horse-radish see *Armoracia rusticana*
Horseshoe Vetch see *Hippocrepis comosa*

Hosta host-a *Liliaceae*. After Nicolas Tomas Host (1761–1834). Herbaceous perennials. Plantain Lily.
 albomarginata see *H. sieboldii*
 crispula krisp-ew-la. Wavy-margined (the leaves). Cult.
 decorata de-ko-*rah*-ta. Decorative. Origin uncertain.
 elata ay-*lah*-ta. Tall. Japan.
 fortunei for-*tewn*-ee-ee. After Fortune, see *Fortunella*. Japan.
 rectifolia rek-ti-*fo*-lee-a. With erect leaves. Japan.
 sieboldiana see-bōld-ee-*ah*-na. After Siebold, see *Acanthopanax sieboldianus*. Japan.
 sieboldii see-*bōld*-ee-ee. (= *H. albomarginata*). As above. Japan.
 tardiflora tar-di-*flō*-ra. Late flowering. Japan.
 undulata un-dew-*lah*-ta. Wavy-margined (the leaves). Cult.
  erromena e-*rō*-me-na. Vigorous.
 ventricosa ven-tri-*kō*-sa. Swollen on one side (the corolla). E Asia.

Hottentot Fig see *Carpobrotus edulis*

Hottonia ho-*ton*-ee-a *Primulaceae*. After Peter Hotton (1648–1709), Dutch physician and botanist. Floating aquatic perennial herb.

Hottonia (continued)
palustris pa-*lus*-tris. Growing in marshes. Water Violet. Europe, Asia.

Hot Water Plant see *Achimenes*
Hound's Tongue see *Cynoglossum officinale*
Houseleek see *Sempervivum*
   Cobweb see *S. arachnoideum*
   Common see *S. tectorum*
   Hen and Chickens see *Jovibarba sobolifera*
Houstonia see *Hedyotis*
   serpyllifolia see *Hedyotis michauxii*

Houttuynia hoo-*tie*-nee-a *Saururaceae*. After Martin Houttuyn (1720–94), Dutch naturalist. Herbaceous perennial.
cordata kor-*dah*-ta. Heart-shaped (the leaves). Himalaya, SE Asia.

Howea *how*-ee-a *Palmae*. After Lord Howe, as was the island where they grow. Tender palms. Sentry Palm. Lord Howe Island (Australia).
belmoreana bel-mor-ree-*ah*-na. (= *Kentia belmoreana*). After De Belmore, a Governor of New South Wales. Curly Sentry Palm.
forsteriana for-sta-ree-*ah*-na. (= *Kentia forsteriana*). After William Forster, a senator of New South Wales. Paradise Palm.

Hoya *hoy*-a *Asclepiadaceae*. After Thomas Hoy (c. 1750–1822), gardener at Syon House. Tender, evergreen climbers.
australis ow-*strah*-lis. Southern. Australia.
bella be-la. Pretty. Miniature Wax Plant. India.
carnosa kar-nō-sa. Fleshy. Honey Plant. Wax Plant. SE Asia to Australia.
imperialis im-pe-ree-*ah*-lis. Showy. Borneo.

Huernia hoo-*ern*-ee-a *Asclepiadaceae*. After Justin Heurnius (1587–1652), Dutch missionary who collected in S Africa. Tender succulent.
keniensis ken-ee-*en*-sis. Of Kenya.

Humble Plant see *Mimosa pudica*
Humea elegans see *Calomeria amaranthoides*

Humulus *hum*-ew-lus *Cannabaceae*. L. version of *humela* the old German name. Perennial climbers.
japonicus ja-*pon*-i-kus. Of Japan. E Asia.
lupulus lup-ew-lus. A small wolf, from an old name, willow-wolf, referring to its habit of climbing over willows. N temperate regions.

Hunnemannia hūn-ee-*măn*-ee-a *Papaveraceae*. After John Hunneman (c. 1760–1839), London bookseller and introducer of plants. Semi-hardy perennial or annual herb.
fumariifolia few-mah-ree-i-*fo*-lee-a. With leaves like *Fumaria*. Mexican Tulip Poppy. Mexico.

Huon Pine see *Dacrydium franklinii*

Hutchinsia hūch-*inz*-ee-a *Cruciferae*. After Ellen Hutchins (1785–1815). Perennial herbs.
alpina ăl-peen-a. Alpine. C and S Europe.
auerswaldii ow-erz-*văld*-ee-ee. (= *H. auerswaldii*). After Auerswald.

Hyacinthoides hee-a-kinth-*oi*-deez *Liliaceae*. From *Hyacinthus* q.v. and Gk. -*oides* indicating resemblance. Bulbous perennial herbs.
hispanica his-*pah*-ni-ka. (= *Endymion hispanicus*. *Scilla hispanica*). Spanish. Spanish Bluebell. SW Europe, N Africa.
non-scripta non-*skrip*-ta. (= *Endymion non-scriptus*. *Scilla non-scripta*). Unmarked. Bluebell. W Europe.

Hyacinthus hee-a-*kinth*-us *Liliaceae*. The Gk. name. Bulbous herb.
amethystinus see *Brimeura amethystina*
azureus see *Muscari azureum*
candicans see *Galtonia candicans*
orientalis o-ree-en-*tah*-lis. Eastern. Dutch Hyacinth. W Asia.

Hydrangea hi-*drang*-gee-a *Hydrangeaceae*. From Gk. *hydor* (water) and *aggos* (a jar) referring to the cup-shaped fruits. Deciduous shrubs and climbers.
arborescens ar-bo-*res*-enz. Becoming tree-like. E United States.
   discolor dis-ko-lor. (= *H. cinerea*). Two coloured, the leaves are grey-white beneath.
aspera a-*spe*-ra. (= *H. villosa*). Rough (the leaves). Himalaya, E Asia.
cinerea see *H. arborescens discolor*
heteromalla het-e-*ro*-mal-la. Variably hairy. Himalaya, China.
   'Bretschneideri' bret-*shnie*-da-ree. After Dr Bretschneider who introduced it.
integerrima see *H. serratifolia*

Hydrangea (continued)
involucrata in-vo-loo-*krah*-ta. With an
involucre. Japan, Taiwan.
'Hortensis' hor-*ten*-sis. Of gardens.
macrophylla măk-rō-*fil*-la. Large-leaved.
Japan (Cult.).
Hortensia hor-*ten*-see-a. Possibly after
Hortense, the daughter of the Prince of
Nassau-Siegen.
'Mariesii' ma-*reez*-ee-ee. After Maries
who introduced it, see *Davallia mariesii*.
paniculata pa-nik-ew-*lah*-ta. With flowers
in panicles. Japan, China.
'Praecox' *prie*-koks. Early flowering.
'Tardiva' *tar*-di-va. Late flowering.
petiolaris pe-tee-o-*lah*-ris. With conspicuous
petioles. Japan, S Korea, Taiwan.
quercifolia kwer-ki-*fo*-lee-a. *Quercus*-leaved.
Oak-leaved Hydrangea. SE United States.
sargentiana sar-jen-tee-*ah*-na. After
Sargent, see *Prunus sargentii*. China.
serrata se-*rah*-ta. Saw-toothed (the leaves).
Japan, S Korea.
serratifolia se-rah-ti-*fo*-lee-a. (= *H.
integerrima*). With saw-toothed leaves.
Chile, Argentina.
villosa see *H. aspera*.

Hydrilla hid-ril-la *Hydrocharitaceae*. From
Gk. *hydor* (water) referring to its habitat.
Aquarium plant.
verticillata ver-ti-ki-*lah*-ta. Whorled (the
leaves). Old World.

Hydrocharis hi-*dro*-ka-ris *Hydrocharitaceae*.
From Gk. *hydor* (water) and *charis* (grace).
Floating aquatic herb.
morsus-ranae mor-sus-*rah*-nie. The common
name. Frog's bit. Europe, Asia.

Hydrocleys hi-*dro*-klee-is *Butomaceae*. From
Gk. *hydor* (water) and *kleis* (a key). Tender
aquatic herb.
nymphoides nimf-*oi*-deez. Like *Nymphaea*.
Water Poppy. Tropical S America.

Hygrophila hi-*gro*-fi-la *Acanthaceae*. From
Gk. *hygros* (moist) and *philos* (loving), they
grow in wet places. Aquarium plants.
difformis di-*for*-mis. (= *Synnema triflorum*).
Of differing shapes (the leaves). SE Asia.
polysperma po-lee-*sperm*-a. With many
seeds. India, Bhutan.

Hylocereus hi-lō-*kay*-ree-us *Cactaceae*.
From Gk. *hyle* (a wood) and *Cereus* q.v.
referring to their epiphytic habit. Climbing
cacti.

Hylocereus (continued)
purpusii pur-*pus*-ee-ee. After one of the
Purpus brothers who collected in
C America. Mexico.
trigonus tri-*gō*-nus. Three angled (the
stems). W Indies.
undatus un-*dah*-tus. Wavy (the stem wings).
Tropical America.

Hymenanthera hi-mayn-an-*the*-ra *Violaceae*.
From Gk. *hymen* (a membrane) and *anthera*
(an anther), the anthers are joined by a
membrane. Semi-hardy evergreen shrubs.
angustifolia ang-gust-i-*fo*-lee-a. Narrow-
leaved. New Zealand.
crassifolia krăs-i-*fo*-lee-a. Thick-leaved. SE
Australia, Tasmania, New Zealand.

Hymenocallis hi-mayn-ō-*kăl*-is
Amaryllidaceae. From Gk. *hymen* (a
membrane) and *kallos* (beauty), the stamens
are united by a membrane. Tender, bulbous
herbs. Spider Lily.
amancaes a-*măn*-kies. The Peruvian name.
Peru.
caribaea kă-ri-*bie*-a. Of the Caribbean.
W Indies.
× *festalis* fay-*stah*-lis. *H. longipetala* × *H.
narcissiflora*. Festive, gay.
harrisiana hă-ris-ee-*ah*-na. After Mr T.
Harris who imported it c. 1840. Mexico.
littoralis li-to-*rah*-lis. Of the shore. Tropical
America.
× *macrostephana* măk-rō-ste-*fah*-na. *H.
narcissiflora* × *H. speciosa*. Large-crowned.
narcissiflora nar-kis-i-*flō*-ra. (= *H.
calathina*). *Narcissus*-flowered. Peru,
Bolivia.

Hymenophyllum hi-mayn-ō-*fil*-lum
Hymenophyllaceae. From Gk. *hymen* (a
membrane) and *phyllon* (a leaf) referring to
the very thin fronds. Filmy Ferns. Widely
distributed.
tunbrigense tūn-brij-*en*-see. Of Tunbridge
Wells. Tunbridge Filmy Fern.
wilsonii wil-*son*-ee-ee. After Wilson.
Wilson's Filmy Fern.

Hyoscyamus hee-ō-*skee*-a-mus *Solanaceae*.
The Gk. name. Poisonous annual or biennial
herb.
niger ni-ger. Black, referring to the
poisonous properties, it was once believed
that any part of the body touched with it
would turn black and rot. Henbane.
Europe, W Asia, N Africa.

Hypericum hi-pe-*ree*-kum *Hypericaceae*.
From Gk. *hyper* (above) and *eikon* (a
picture), it was hung above pictures to ward
off evil spirits. Perennial herbs and shrubs.
   androsaemum ăn-dros-*ie*-mum. Gk. name
   for a plant with red sap from *andros* (man)
   and *haima* (blood). Tutsan. Europe,
   N Africa.
   beanii been-ee-ee. After Bean, see *Cytisus* ×
   beanii. China.
   calycinum kăl-i-*kee*-num. With a
   conspicuous calyx. Rose of Sharon,
   Aaron's Beard. SE Europe, Turkey.
   cerastioides ke-răs-tee-*oi*-deez. Like
   Cerastium. SE Europe, NW Turkey.
   coris *ko*-ris. From the resemblance of the
   leaves to those of *Coris*. Alps, N Italy.
   empetrifolium em-pet-ri-*fo*-lee-um.
   Empetrum-leaved. Greece.
      oliganthum o-lig-*ănth*-um. Few-flowered.
      Crete.
   forrestii fo-*rest*-ee-ee. After Forrest who
   introduced it, see *Abies delavayi forrestii*.
   Himalaya, China.
   fragile hort. see *H. olympicum*
   × *inodorum* in-o-*dō*-rum. *H. androsaemum*
   × *H. hircinum*. Not scented. *H. hircinum*
   smells of goats as the name suggests.
   × *moserianum* mō-za-ree-*ah*-num. *H.*
   calycinum × *H. patulum*. After Moser's
   nursery, Versailles where it was raised c.
   1887.
   olympicum o-*lim*-pi-kum. (= *H. fragile* hort.
   H. polyphyllum hort.). Of Mt Olympus,
   applied to several mountains in SE Europe
   and W Asia. SE Europe, Turkey.
   polyphyllum hort. see *H. olympicum*
   reptans rep-tănz. Creeping. Himalaya.
   xylosteifolium zi-los-tee-i-*fo*-lee-um. With
   leaves like *Lonicera xylosteum*. NW
   Turkey, SW Georgia.

Hypoestes hi-pō-*es*-teez *Acanthaceae*. From
Gk. *hypo* (below) and *estia* (a house), the
bracts cover the calyx. Tender herbs.
Madagascar.
   phyllostachya fil-lō-*stăk*-ee-a. (= *H.*
   sanguinolenta hort.). With leafy spikes.
   Polka-dot Plant, Baby's Tears, Freckle
   Face.
   taeniata tie-nee-*ah*-ta. From *Taenia* a genus
   of tapeworms, referring to the worm-like
   flowers.

Hypoxis hi-*poks*-is *Hypoxidaceae*. Gk. name
for a plant. Perennial herbs.

Hypoxis (continued)
   hirsuta hir-*soo*-ta. Hairy (the leaves).
   N America.
   hygrometrica hi-gro-*met*-ri-ka. Measuring
   moisture, the flowers close in cloudy
   weather. Australia.

Hypsela hip-*say*-la *Lobeliaceae*. From Gk.
hypselos (high), the following grows at high
altitudes. Perennial herb.
   reniformis ray-ni-*form*-is. Kidney-shaped
   (the leaves). S America.

Hyssopus hi-*sōp*-us *Labiatae*. Old name for
another plant. Evergreen shrub.
   officinalis o-fi-ki-*nah*-lis. Sold as a herb.
   Hyssop. S Europe, W Asia.

I

Iberis i-*be*-ris *Cruciferae*. Gk. *iberis*, from
Iberia. Annual herbs and sub-shrubs.
Candytuft.
   amara a-*mah*-ra. Bitter. Rocket Candytuft.
   Europe.
   'Correifolia' ko-ree-i-*fo*-lee-a. With leaves
   like *Correa*.
   gibraltarica ji-brawl-*tah*-ri-ca. Of Gibraltar.
   Gibraltar Candytuft. S Spain, Morocco.
   saxatilis săks-*ah*-ti-lis. Growing on rocks.
   S Europe.
   sempervirens sem-per-*vi*-renz. Evergreen.
   S Europe, W Asia.
   umbellata um-bel-*ah*-ta. With flowers in
   umbels. Common Candytuft.
   Mediterranean region.

Ice Plant see *Mesembryanthemum crystallinum*

Idesia i-*deez*-ee-a *Flacourtiaceae*. After E. I.
Ides a Dutch traveller who visited China in
the 18th century. Deciduous tree.
   polycarpa po-lee-*kar*-pa. With many fruits.
   Japan, China.

Ilex *ee*-leks *Aquifoliaceae*. The L. name for
Quercus ilex. Evergreen trees and shrubs.
Holly.
   × *altaclerensis* ăl-ta-kle-*ren*-sis. *I. aquifolium*
   × *I. perado*. Of Highclere (Alta Clera),
   Berks. where it is thought to have been first
   raised.
      'Camelliifolia' ka-mel-ee-i-*fo*-lee-a. With
      leaves like *Camellia*.
      'Hendersonii' hen-der-*son*-ee-ee. After
      Mr Henderson, a friend of Mr Shepherd

Ilex (continued)

who supplied material to the
Handsworth nursery of Fisher and
Holmes.
'Hodginsii' ho-*jinz*-ee-ee. After Edward
Hodgins, the raiser.
'Lawsoniana' law-son-ee-*ah*-na. After the
Lawson nursery of Edinburgh who
distributed it.
aquifolium ă-kwi-*fo*-lee-um. The L. name.
Common Holly. Europe, W Asia.
cornuta kor-*new*-ta. Horned, referring to
the leaf spines. Horned Holly. China,
Korea.
crenata kray-*nah*-ta. With shallow, rounded
teeth (the leaves). Japan, Korea.
'Mariesii' ma-*reez*-ee-ee. After Maries
who introduced it in about 1879, see
Davallia mariesii.
× *meserveae* me-*serv*-ee-ie. *I. aquifolium* ×
I. rugosa. After Mrs Kathleen Meserve
who raised it. Blue Holly.
pernyi per-nee-ee. After the Abbé Paul
Hubert Perny (1818–1907), French
missionary in China who discovered it in
1858. C and W China.

Illicium il-*lik*-ee-um *Illiciaceae*. From L.
illicio (to attract) referring to the fragrance.
Evergreen trees and shrubs.
anisatum ăn-i-*sah*-tum. Anise-scented (the
leaves). China, Japan.
floridanum flo-ri-*dah*-num. Of Florida.
S United States.
henryi hen-ree-ee. After its discoverer,
Augustine Henry (1857–1930), an Irish
doctor who collected in China. W China.

Immortelle see *Xeranthemum annuum*

Impatiens im-*păt*-ee-enz *Balsaminaceae*. L.
for impatient, referring to the explosive
release of the seed when a ripe capsule is
touched. Hardy and tender, annual and
perennial herbs.
balfourii băl-*for*-ree-ee. After Sir Isaac
Bayley Balfour (1853–1922), Regius
Keeper of the Edinburgh Botanic Garden.
W Himalaya.
balsamina băl-sa-*meen*-a. Bearing balsam.
SE Asia.
capensis ka-*pen*-sis. (= *I. biflora*). Originally
thought to be from the Cape of Good
Hope. Jewel Weed. N America.
glandulifera glănd-ew-*li*-fe-ra. (= *I. roylei*).
Glandular. Himalayan Balsam,
Policeman's Helmet. W Himalaya.
hawkeri hawk-a-ree. (= *I. petersiana* hort.).

Impatiens (continued)

After Lieutenant Hawker who sent plants
from the South Sea Islands. New Guinea.
noli-tangere nŏ-lee-tang-*ge*-ree. Do not
touch (the pods). Touch-me-not. Europe,
Asia.
walleriana wo-la-ree-*ah*-na. (= *I. holstii*. *I.
sultanii*). After the Rev. Horace Waller
(1833–96), a missionary in C Africa. Busy
Lizzie. E Africa.

Incarvillea in-kar-*vil*-ee-a *Bignoniaceae*.
After Pierre d'Incarville (1706–57), French
missionary and plant collector in China.
Perennial herbs.
delavayi del-a-*vay*-ee. After Delavay, see
Abies delavayi. China.
mairei mair-ree-ee. After Edouard Ernest
Maire (born 1848), French missionary in
China. China.
grandiflora grănd-i-*flŏ*-ra. (= *I.
grandiflora*). Large-flowered. Himalaya,
W China.
olgae ol-gie. After Olga Fedtschenko
(1845–1921). C Asia.

Incense Cedar see *Calocedrus decurrens*
Inch Plant see *Callisia*, *Tradescantia albiflora*
  Flowering see *Tradescantia blossfeldiana*
  Striped see *Callisia elegans*
Indian Bean Tree see *Catalpa bignonioides*
Indian Currant see *Symphoricarpus rivularis*
Indian Hawthorn see *Rhaphiolepis indica*
Indian Physic see *Gillenia trifoliata*
Indian Shot see *Canna indica*
Indian Turnip see *Arisaema triphyllum*
Indigo, False see *Amorpha fruticosa*, *Baptisia
australis*
  Wild see *Baptisia tinctoria*

Indigofera in-di-go-fe-ra *Leguminosae*. From
indigo and L. *fero* (to bear), indigo is
obtained from *I. tinctoria*. Deciduous shrubs.
decora de-*kŏ*-ra. Beautiful. China, Japan.
heterantha he-te-*rănth*-a. (= *I. gerardiana*).
With different flowers. NW Himalaya.

Inula *in*-ew-la *Compositae*. The L. name for
I. helenium. Perennial herbs.
acaulis a-*kaw*-lis. Stemless. W Asia.
ensifolia ayns-i-*fo*-lee-a. With sword-shaped
leaves. Europe.
helenium he-*len*-ee-um. From the
resemblance to *Helenium*. Elecampane.
C Asia.
hookeri huk-a-ree. After Sir Joseph Dalton
Hooker (1817–1911). Himalaya, Burma,
W China.

Inula (continued)
  magnifica mahg-*ni*-fi-ka. Splendid.
  Caucasus.
  oculus-christi ok-ew-lus-*kris*-tee. Literally,
  the eye of Christ. E Europe.
  orientalis o-ree-en-*tah*-lis. Eastern.
  Caucasus.
  royleana royl-ee-*ah*-na. After John Forbes
  Royle (1798–1858) who collected in India
  and the Himalaya. Himalaya.

Iochroma ee-ō-krō-ma *Solanaceae*. From Gk.
ion (violet) and *chroma* (colour) referring to
the colour of the flowers. Tender shrubs.
  coccineum kok-*kin*-ee-um. Scarlet (the
  flowers). C America.
  cyaneum see-*án*-ee-um. (= *I. tubulosum*).
  Blue (the flowers). S America.
  grandiflorum gránd-i-*flō*-rum. Large-
  flowered. Ecuador, Peru.

Ionopsidium ee-on-op-*sid*-ee-um *Cruciferae*.
From Gk. *ion* (violet) and *-opsis* indicating
resemblance, from the resemblance to a
violet. Annual herb.
  acaule a-*kaw*-lee. Stemless. Violet Cress.
  Portugal.

Ipheion i-*fay*-on *Amaryllidaceae*. Derivation
obscure. Bulbous herb.
  uniflorum ew-ni-*flō*-rum. (= *Brodiaea
  uniflora. Triteleia uniflora*). One-flowered.
  Argentina, Uruguay.

Ipomoea i-pom-*oy*-a *Polemoniaceae*. From
Gk. *ips* (a worm) and *homoios* (resembling).
Tender annual and perennial climbers.
Morning Glory.
  acuminata a-kew-min-*ah*-ta. (= *I. learii*).
  Acuminate (the leaves). Tropical America.
  alba ál-ba. White (the flowers). Tropical
  America.
  batatas ba-*tah*-tas. The Haitian name.
  Sweet Potato Vine. S America.
  bonariensis bo-nah-ree-*en*-sis. Of Buenos
  Aires. S America.
  coccinea kok-*kin*-ee-a. (= *Quamoclit
  coccinea*). Scarlet (the flowers). E United
  States.
  hederacea he-de-*rah*-kee-a. Like *Hedera* (the
  leaves). Tropical America.
  horsfalliae hors-*fál*-ee-ie. After Mrs Charles
  Horsfall who painted for the Botanical
  Magazine. W Indies.
  learii see *I. acuminata*
  lobata see *Mina lobata*
  nil nil. Violet (the flowers). Tropics.
  pes-caprae pays-*káp*-rie. Like a goat's-foot

Ipomoea (continued)
  (the leaves). Tropics.
  purpurea pur-*pewr*-ree-a. Purple (the
  flowers). Tropical America.
  quamoclit kwah-mo-klit. (= *Quamoclit
  pennata*). The Mexican name. Cypress
  Vine. Tropical America.
  rubrocaerulea see *I. tricolor*
  tricolor tri-ko-lor. (= *I. rubrocaerulea. I.
  violacea* hort.). Three-coloured (the
  flowers). Tropical America.
  violacea hort. see *I tricolor*

Ipomopsis i-pom-*op*-sis *Polemoniaceae*.
From Gk. *ips* (a worm) and *-opsis* indicating
resemblance. Annual herb.
  rubra rub-ra. (= *Gilia rubra*). Red (the
  flowers). Standing Cypress. S United
  States.

Iresine ee-res-*ee*-nay *Amaranthaceae*. From
Gk. *eiresione*, a branch wound round with
wool. Tender perennials.
  herbstii hairbst-ee-ee. After Hermann Carl
  Gottlieb Herbst (c. 1830–1904), Director
  of the Rio de Janeiro Botanic Gardens, later
  a nurseryman of Richmond, Surrey.
  S America.
  lindenii lin-*den*-ee-ee. After Linden,
  Belgian nurseryman. Ecuador.

Iris ee-ris *Iridaceae*. After the Gk. goddess of
the rainbow. Rhizomatous or bulbous
perennials.
  bakeriana see *I. reticulata bakeriana*
  bucharica bew-*kah*-ri-ka. Of Bokhara,
  Turkestan. C Asia.
  bulleyana bul-ee-*ah*-na. After Arthur Kilpin
  Bulley (1861–1942), founder of Bees
  nursery, who employed Forrest and
  Kingdon-Ward as plant collectors.
  W China.
  chamaeiris see *I. lutescens*
  chrysographes kris-*o*-gra-feez. With gold
  lines (on the falls). W China.
  clarkei klark-ee-ee. After Charles Baron
  Clarke (1832–1906) who collected in
  India. Himalaya.
  cristata kris-*tah*-ta. Crested (the falls).
  E United States.
  danfordiae dán-*ford*-ee-ie. After Mrs C. G.
  Danford who collected bulbs in W Asia,
  1876–9. C Turkey.
  douglasiana dŭg-las-ee-*ah*-na. After David
  Douglas (1798–1834), who introduced
  many plants from WN America. W N
  America.
  ensata ayn-*sah*-ta. Sword-like (the leaves).

Iris (continued)
C to E Asia.

florentina see *I. germanica florentina*

foetidissima foy-ti-*dis*-i-ma. Very foetid.
Stinking Iris, Gladdon. S and W Europe,
N Africa.

forrestii fo-*rest*-ee-ee. After Forrest, see
Abies delavayi forrestii. SW China.

fulva ful-va. Tawny (the flowers). Red Iris.
E N America.

gatesii gayts-ee-ee. After the Rev. F. S.
Gates of the American Mission in Armenia,
c 1888. W Asia.

germanica ger-*mahn*-i-ka. Of Germany.
Cult.

    florentina flō-ren-*teen*-a. (= *I. florentina*).
    Of Florence. Orris. Cult.

gracilipes gra-*kil*-i-pays. Slender-stalked.
Japan.

graebneriana grayb-na-ree-*ah*-na. After
Karl Graebner (1871–1933). Turkestan.

graminea grah-*min*-ee-a. Grass-like (the
leaves). C and S Europe, Caucasus.

histrio his-tree-ō. L. *histrio* (an actor) from
the colouring of the flowers. W Asia.

    aintabensis ien-ta-*ben*-sis. Of Gaziantep
    (Aintab). W Asia.

histrioides his-tree-*oi*-deez. Like *I. histrio*.
C Turkey.

    'Major' *mah*-yor. Larger.

hoogiana hoog-ee-*ah*-na. After John Hoog
of the van Tubergen nursery. Turkestan.

innominata in-nom-i-*nah*-ta. Nameless.
W United States.

japonica ja-*pon*-i-ka. Of Japan. Japan,
China.

kaempferi kemp-fa-ree. After Engelbert
Kaempfer (1651–1716). Japan.

laevigata lie-vi-*gah*-ta. Smooth (the leaves).
E Asia.

latifolia lah-tee-*fo*-lee-a. (= *I. xiphioides*).
Broad-leaved. English Iris. Pyrenees,
N Spain.

lutescens loo-*tes*-enz. (= *I. chamaeiris*).
Yellowish (the flowers of some forms). SW
Europe.

missouriensis mi-sur-ree-*en*-sis. Of the
source of the Missouri River. C and W N
America.

ochroleuca see *I. spuria ochroleuca*

orchioides or-kee-*oi*-deez. Orchid-like.
C Asia.

pallida pă-li-da. Pale (the flowers).
Yugoslavia.

pseudocorus sood-*ă*-ko-rus. False *Acorus*.
Yellow Flag. S and W Europe, N Africa.

pumila pew-mi-la. Dwarf. E Europe,
W Asia.

Iris (continued)

reticulata ray-tik-ew-*lah*-ta. Netted (the
bulb). W Asia.

    bakeriana bay-ka-ree-*ah*-na. (= *I.
    bakeriana*). After J. G. Baker
    (1834–1920), Kew botanist and authority
    on bulbs. Iran.

ruthenica roo-*then*-i-ka. Of Ruthenia, SW
Russia. E Europe to China.

setosa say-*tō*-sa. Bristly (the apex of the
standards). NE Asia, Alaska.

sibirica si-*bi*-ri-ka. Siberian. C Europe to
Russia.

spuria spewr-ree-a. False. Europe,
N Africa, W Asia.

    ochroleuca ok-rō-*loo*-ka. (= *I. ochroleuca*).
    Yellowish-white. Greece, W Asia.

stolonifera sto-lōn-*i*-fe-ra. Bearing stolons.
C Asia.

stylosa see *I. unguicularis*.

susiana soo-see-*ah*-na. Of Shush (Susa),
Iran. W Asia.

tectorum tek-*to*-rum. Growing on roofs.
Roof Iris, Wall Iris. China.

tenax ten-ahks. Tough (the leaves).
W United States.

tingitana ting-gi-*tah*-na. Of Tangier (Tingi).
Morocco.

tuberosa see *Hermodactylus tuberosus*

unguicularis un-gwik-ew-*lah*-ris. (= *I.
stylosa*). Clawed. SE Europe, N Africa,
W Asia.

verna ver-na. Of Spring. E United States.

versicolor ver-*si*-ko-lor. Variously coloured.
EN America.

xiphioides see *I. latifolia*

xiphium zi-fee-um. Gk. name for *Gladiolus*
from *xiphos* (a sword) referring to the
leaves. SW Europe, N Africa.

Ironweed see *Vernonia*
Iron Wood see *Ostrya virginiana*

Isatis ee-sa-tis *Cruciferae*. The Gk. name.
Biennial herb.

    tinctoria tink-*to*-ree-a. Used in dyeing.
    Woad. C and S Europe.

Itea ee-tee-a *Escalloniaceae*. The Gk. name
for willow. Deciduous and evergreen shrubs.

    ilicifolia ee-lik-i-*fo*-lee-a. *Ilex*-leaved.
    W China.

    virginica vir-*jin*-i-ka. Of Virginia. E United
    States.

Ivy see *Hedera*
    Common see *H. helix*
    Irish see *H. hibernica*
    Poet's see *H. helix poetica*

Ivy-leaved Toadflax see *Cymbalaria muralis*

Ixia *iks*-ee-a *Iridaceae*. From Gk. *ixia* (bird lime) referring to the sticky sap. Semi-hardy, cormous perennial. Corn Lily.
viridiflora vi-ri-di-*flō*-ra. With green flowers. S Africa.

Ixora iks-*o*-ra *Rubiaceae*. Portuguese version of *Iswara* a Malabar deity to whom the flowers were offered. Tender, evergreen shrubs.
chinensis chin-*en*-sis. Of China. SE Asia.
coccinea kok-*kin*-ee-a. Scarlet (the flowers). Flame of the Woods. India.

J

Jacaranda jăk-a-*rănd*-a *Bignoniaceae*. The Portuguese name, derived from the Brazilian Indian name. Tender, semi-evergreen tree.
mimosifolia mee-mo-si-*fo*-lee-a. (= *J. ovalifolia*. *J. acutifolia* hort.). Mimosa-leaved. NW Argentina.

Jack-in-the-Pulpit see *Arisaema triphyllum*
Jacobinia carnea see *Justicea carnea*
   coccinea see *Pachystachys coccinea*
   ghiesbreghtiana see *Justicea ghiesbreghtiana*
   pauciflora see *Justicea rizzinii*
   pohliana see *Justicea carnea*
Jacob's Coat see *Acalypha wilkesiana*
Jacob's Ladder see *Pedilanthes tithymaloides smallii, Polemonium caeruleum*
Jade Plant see *Crassula argentea*
Japanese Foam Flower see *Tanakaea radicans*
Japan Pepper see *Zanthoxylum piperitum*

Jasione yă-see-*ō*-nay *Campanulaceae*. Gk. name for another plant. Perennial herbs. Sheep's Bit.
crispa kris-pa. (= *J. humilis*). Crisped (the stem hairs). Pyrenees.
laevis lie-vis. (= *J. perennis*). Smooth. W Europe.
montana mon-*tah*-na. (= *J. jankae*). Of mountains. Europe.

Jasminum yăs-*meen*-um *Oleaceae*. From *yasmin* the Persian name. Deciduous and evergreen shrubs and climbers. Jasmine, Jessamine.
beesianum beez-ee-*ah*-num. After Bees

Jasminum (continued)
   nursery for whom Forrest introduced it in 1906. China.
humile hum-i-lee. Low growing. Himalaya, China.
   'Revolutum' re-vo-*loo*-tum. Rolled back (the leaf margins).
mesnyi mez-nee-ee. After William Mesny (1842–1919), a Major-General in the Chinese National Army who collected in China. Primrose Jasmine. W China.
nudiflorum new-di-*flō*-rum. Flowering naked, it flowers in winter when it is leafless. Winter Jasmine. China.
officinale o-fik-i-*nah*-lee. Sold as a herb. Common Jasmine. Caucasus, Himalaya, China.
parkeri park-a-ree. After R. N. Parker who discovered it in 1919 and later introduced it. Indian Himalaya.
polyanthum po-lee-*ănth*-um. Many-flowered. China.
× *stephanense* stef-an-*en*-see. *J. beesianum* × *J. officinale*. Of Saint Etienne, France where it was raised.

Jeffersonia jef-er-*son*-ee-a *Berberidaceae*. After Thomas Jefferson (1743–1826), 3rd President of the United States and botanical patron.
diphylla di-*fil*-la. Two-leaved. E N America.
dubia dub-ee-a. Doubtful. Vladivostok, Korea.

Jelly Beans see *Sedum pachyphyllum*
Jerusalem Cherry see *Solanum pseudocapsicum*
   False see *S. capsicastrum*
Jerusalem Cross see *Lychnis chalcedonica*
Jerusalem Sage see *Phlomis fruticosa*
Jerusalem Thorn see *Parkinsonia aculeata*
Jessamine see *Jasminum*
Jewel Weed see *Impatiens capensis*
Job's Tears see *Coix lacryma-jobi*
Joe-pye Weed see *Eupatorium maculatum, E. purpureum*
Joseph's Coat see *Amaranthus tricolor*

Jovellana ho-vel-*lah*-na *Scrophulariaceae*. After Gaspar Melchor de Jovellanos, 18th-century patron of botany in Peru. Semi-hardy shrub.
violacea vee-o-*lah*-kee-a. Violet (the corolla). Chile.

Jovibarba yov-i-*bar*-ba *Crassulaceae*. From L. *Jovis* (Jupiter) and *barba* (a beard) presumably referring to the fringed petals which distinguishes the genus from *Sempervivum*. Succulent herbs.

sobolifera so-bo-*li*-fe-ra. (= *Sempervivum soboliferum*). Bearing offspring. Hen and Chickens Houseleek. Alps.

Joy Weed see *Alternanthera*
Judas Tree see *Cercis siliquastrum*

Juglans *yoo*-glahnz *Juglandaceae*. The L. name for *J. regia* from *jovis* (of Jupiter) and *glans* (an acorn or nut). Deciduous trees.

ailantifolia ie-län-ti-*fo*-lee-a. *Ailanthus*-leaved. Japanese Walnut. Japan.
cinerea kin-*e*-ree-a. Grey (the bark). Butter-nut. E N America.
microcarpa mik-rō-*kar*-pa. With small fruit. Texan Walnut. SW United States, N Mexico.
nigra *nig*-ra. Black (the bark). Black Walnut. E and C United States.
regia ray-gee-a. Royal. Common Walnut. S Europe to China.

Juncus yung-kus *Juncaceae*. L. name for rush. Perennial herb.

effusus e-*few*-sus. Loose. Widely distributed.
'Spiralis' spee-*rah*-lis. Spiral (the stems). Corkscrew Rush.

Juniperus yoo-*ni*-pe-rus *Cupressaceae*. The L. name. Evergreen conifers. Juniper.

chinensis chin-*en*-sis. Of China. China, Mongolia, Japan.
sargentii sar-*jent*-ee-ee. (= *J. sargentii*). After Sargent who discovered and introduced it, see *Prunus sargentii*.
communis kom-*ew*-nis. Common. Common Juniper. N temperate regions.
conferta kōn-*fer*-ta. Crowded (the leaves). Japan.
drupacea droo-*pah*-kee-a. With fleshy fruit. SE Europe, W Asia.
horizontalis ho-ri-zon-*tah*-lis. Horizontal, the prostrate habit. N America.
'Douglasii' dūg-*läs*-ee-ee. After the Douglas nursery, Waukegon, Illinois. Waukegon Juniper.
'Glauca' *glow*-ka. Glaucous (the foliage).
'Plumosa' ploo-*mō*-sa. Feathery (the foliage).
'Wiltonii' wil-*ton*-ee-ee. After South

Juniperus (continued)
Wilton Nurseries, Wilton, Connecticut.
× *media* me-dee-a. *J. chinensis* × *J. sabina*. Intermediate (between the parents).
'Pfitzeriana' fits-a-ree-*ah*-na. After W. Pfitzer, a Stuttgart nurseryman.
procumbens prō-*kum*-benz. Prostrate. Japan.
recurva re-*kur*-va. Curved downwards (the shoots). Himalayan Juniper. Himalaya, W China.
coxii *koks*-ee-ee. After E. H. M. Cox who, with Farrer, introduced it in 1920.
rigida *ri*-gi-da. Rigid (the leaves). Japan.
sabina sa-*been*-a. The L. name. Savin. C and S Europe, W Russia.
tamariscifolia tàm-a-risk-i-*fo*-lee-a. *Tamarix*-leaved.
sargentii see *J. chinensis sargentii*
scopulorum skop-ew-*lo*-rum. Growing, on cliffs. Rocky Mountain Juniper. W United States.
squamata skwah-*mah*-ta. Scaly (the bark). Himalaya, China.
'Meyeri' *may*-a-ree. After F. N. Meyer who introduced it to the United States in 1910.
virginiana vir-jin-ee-*ah*-na. Of Virginia. Eastern Red Cedar. E and CN America.

Jupiter's Beard see *Anthyllis barba-jovis*

Justicea jūs-*tis*-ee-a *Acanthaceae*. After James Justice 18th-century Scottish botanist. Tender, evergreen shrubs.

brandegeana bränd-ee-gee-*ah*-na. (= *Beloperone guttata*). After Brandegee. Shrimp Plant. Mexico.
carnea *kar*-nee-a. (= *Jacobinia carnea*. *Jacobinia pohliana*). Flesh-coloured (the flowers). Brazilian Plume. S America.
ghiesbreghtiana geez-brekt-ee-*ah*-na. (*Jacobinia ghiesbreghtiana*). After Augustine Boniface Ghiesbreght (1810–93). Mexico.
rizzinii riz-*in*-ee-ee. (= *Jacobinia pauciflora*). After Carlos Toledo Rizzini. Brazil.

K

Kaffir Lily see *Clivia miniata*, *Schizostylis coccinea*
Kahili Ginger see *Hedychium gardnerianum*
Kaki see *Diospyros kaki*

Kalanchoe ka-*lăn*-kō-ee *Crassulaceae*. From the Chinese name of one species. Tender succulents.

beharensis bee-hah-*ren*-sis. Of Behara, Madagascar. Felt Bush. Velvet Leaf.

bentii bent-ee-ee. (= *K. teretifolia*). After Theodore Bent who collected in S Arabia about 1894. Arabia.

blossfeldiana blos-feld-ee-*ah*-na. After Robert Blossfeld, German nurseryman. Madagascar.

crenata kray-*nah*-ta. (= *Bryophyllum crenatum*). With shallow, rounded teeth (the leaves). Malaysia.

daigremontiana day-gre-mont-ee-*ah*-na. (= *Bryophyllum daigremontianum*). After Mme. and M. Daigremont. Devil's Backbone, Mother of Thousands. Madagascar.

fedtschenkoi fet-*shenk*-ō-ee. After Boris Fedtschenko (1872–1933), Russian traveller. Madagascar.

flammea flăm-ee-a. Flame-coloured (the flowers). Somalia.

× *kewensis* kew-*en*-sis. *K. bentii* × *K. flammea*. Of Kew.

marmorata mar-mo-*rah*-ta. Marbled (the leaves). Pen Wiper. NE Africa.

millotii mi-*lot*-ee-ee. After Prof. Millot of the Natural History Museum, Paris. Madagascar.

pinnata pin-*ah*-ta. (= *Bryophyllum pinnatum*). Pinnate (the leaves). Air Plant. Origin uncertain.

pumila pew-mi-la. Dwarf. Madagascar.

teretifolia see *K. bentii*

tomentosa tō-men-*tō*-sa. Hairy. Panda Plant, Pussy Ears. Madagascar.

tubiflora tew-bi-*flō*-ra. (= *Bryophyllum tubiflorum*). With tubular flowers. Chandelier Plant. S Africa, Madagascar.

uniflora ew-ni-*flō*-ra. (= *Bryophyllum uniflorum*). One-flowered. Madagascar.

Kale, Ornamental see *Brassica oleracea* Acephala

Kalmia kăl-mee-a *Ericaceae*. After Pehr Kalm (1715–79), Finnish student of Linnaeus. Evergreen shrubs.

angustifolia ang-gust-i-*fo*-lee-a. Narrow-leaved. Sheep Laurel. EN America.

latifolia lah-tee-*fo*-lee-a. Broad-leaved. Calico Bush, Mountain Laurel. E N America.

polifolia pol-i-*fo*-lee-a. With leaves like *Teucrium polium*. N America.

Kalmiopsis kăl-mee-*op*-sis *Ericaceae*. From *Kalmia* q.v. and Gk. -*opsis* indicating resemblance. Evergreen shrub.

leachiana leech-ee-*ah*-na. After Mr and Mrs Leach who discovered it in 1930. Oregon.

Kalopanax kăl-ō-*păn*-ăks *Araliaceae*. From Gk. *kalos* (beautiful) and *Panax* a related genus, see *Acanthopanax*. Deciduous tree.

septemlobus sep-*tem*-lo-bus. (= *K. pictus*). Seven-lobed (the leaves). NE Asia.

Kangaroo Apple see *Solanum aviculare*
Kangaroo Vine see *Cissus antarctica*
Kangaroo's Paw see *Anigozanthus*
Katsura Tree see *Cercidiphyllum japonicum*
Kauri Pine see *Agathis australis*
Kentia see *Howea*
Kentucky Coffee Tree see *Gymnocladus dioica*

Kerria ke-ree-a *Rosaceae*. After William Kerr (died 1814), Kew gardener who introduced *K. japonica* 'Pleniflora'. Deciduous shrub.

japonica ja-*pon*-i-ka. Of Japan, where it is grown. China.

'Pleniflora' play-ni-*flō*-ra. With double flowers.

King Cup see *Caltha palustris*
King of Bromeliads see *Vriesia hieroglyphica*
King of the Alps see *Eritrichium nanum*
King Plant see *Anoectochilus regalis*
King William Pine see *Athrotaxis selaginoides*
Kingfisher Daisy see *Felicia bergeriana*
King's Spear see *Asphodeline lutea*

Kirengeshoma ki-reng-ge-*shō*-ma *Saxifragaceae*. The Japanese name. Herbaceous perennial.

palmata pahl-*mah*-ta. Lobed like a hand (the leaves). Japan.

Kiwi Fruit see *Actinidia chinensis*
Kleinia see *Senecio*
  neriifolia see *S. kleinia*
  repens see *S. serpens*
  tomentosa see *S. haworthii*

Kniphofia nee-*fof*-ee-a *Liliaceae*. After Johann Hieronymus Kniphof (1704–63). Perennial herbs. Red Hot Poker, Torch Lily.

caulescens kaw-*les*-enz. With a stem. S Africa.

'Erecta' e-*rek*-ta. Erect (the flowers).

foliosa fo-lee-ō-sa. Leafy. Ethiopia.

Kniphofia (continued)
galpinii găl-*pin*-ee-ee. After Ernst E. Galpin, a plant collector of Barberton, SE Transvaal. S Africa.
nelsonii nel-*son*-ee-ee. After Mr William Nelson (1852–1922), who collected in S Africa. S Africa.
northiae north-ee-ie. After Miss Marianne North (1830–90), botanical artist. S Africa.
uvaria oo-*vah*-ree-a. Like a bunch of grapes. S Africa.

Kochia kok-ee-a *Chenopodiaceae*. After Wilhelm Daniel Josef Koch (1771–1849). Annual herb.
scoparia skō-*pah*-ree-a. Broom-like. Summer Cypress, Burning Bush. S Europe to Japan.
trichophylla tri-ko-*fil*-la. With hair-like leaves.

Koelreuteria kurl-roy-*te*-ree-a *Sapindaceae*. After Joseph Gottlieb Koelreuter (1733–1806), German professor of botany. Deciduous tree.
paniculata pa-nik-ew-*lah*-ta. With flowers in panicles. China.

Kohleria kō-*le*-ree-a *Gesneriaceae*. After Michael Kohler, 19th-century natural history teacher in Zurich. Tender perennial herbs and shrubs.
amabilis a-*mah*-bi-lis. Beautiful. Colombia.
bogotensis bo-go-*ten*-sis. Of Bogota. Colombia.
digitaliflora di-gi-*tah*-li-flō-ra. With flowers like *Digitalis*. Colombia.
eriantha e-ree-*ănth*-a. With woolly flowers. Colombia.
lindeniana lin-den-ee-*ah*-na. After Linden, a Belgian nurseryman. Ecuador.
tubiflora tew-bi-*flō*-ra. With tubular flowers. Colombia.

Kohl Rabi see *Brassica oleracea* Gongylodes

Kolkwitzia kol-*kwitz*-ee-a *Caprifoliaceae*. After Richard Kolkwitz (born 1873), German professor of botany. Deciduous shrub.
amabilis a-*mah*-bi-lis. Beautiful. Beauty Bush. China.

Kowhai see *Sophora tetraptera*
Kris Plant see *Alocasia lindeniana*
Kumquat see *Fortunella japonica*

L

Labrador Tea see *Ledum groenlandicum*

+ Laburnocytisus la-burn-ō-*sit*-i-sus *Leguminosae*. Graft hybrid, from the names of the parents. *Cytisus* × *Laburnum*. Deciduous tree.
adamii a-*dăm*-ee-ee. *Cytisus purpureus* + *Laburnum anagyroides*. After its raiser, Jean Louis Adam.

Laburnum la-*burn*-um *Leguminosae*. The L. name. Deciduous trees.
alpinum ăl-*peen*-um. Alpine. Scotch Laburnum. Alps, Italy, Czechoslavakia, Yugoslavia.
anagyroides ăn-a-gi-*roi*-deez. Like *Anagyris*. Common Laburnum. C and S Europe.
× *watereri waw*-ta-ra-ree. *L. alpinum* × *L. anagyroides*. After the Waterer nursery where one form was raised.

Lace Flower Vine see *Episcia dianthiflora*
Lace Trumpets see *Sarracenia leucophylla*

Lachenalia lah-shen-*ahl*-ee-a *Liliaceae*. After Werner de La Chenal (1736–1800), professor of botany at Basle. Tender, bulbous herbs. Cape Cowslip. S Africa.
aloides ă-lō-*ee*-deez. Like *Aloe*.
'Nelsonii' nel-*son*-ee-ee. After the Rev. John Gudgeon Nelson (1818–82) who raised it.
bulbifera bul-*bi*-fe-ra. Bulb-bearing.
glaucina glow-*keen*-a. Glaucous. Opal Lachenalia.
mutabilis mew-*tah*-bi-lis. Changeable (the flower colour). Fairy Lachenalia.
orchioides or-kee-*oi*-deez. Orchid-like.

Lactuca lăk-*too*-ka *Compositae*. From L. *lac* (milk) referring to the white sap. Annual herb.
alpina see *Cicerbita alpina*
plumieri see *Cicerbita plumieri*
sativa sa-*tee*-va. Cultivated. Lettuce. Cult.

Ladies' Fingers see *Abelmoschus esculentus*
Lady of the Night see *Brassavola nodosa*, *Brunfelsia americana*
Lady's Mantle see *Alchemilla*
Alpine see *A. alpina*
Lady's Smock see *Cardamine pratensis*

Laelia *lie*-lee-a *Orchidaceae*. After Laelia, one of the vestal virgins. Greenhouse orchids.
  anceps ăn-keps. Two-edged (the pseudobulbs). Mexico.
  autumnalis ow-tum-*nah*-lis. Of autumn (flowering). Mexico.
  cinnabarina kin-a-ba-*reen*-a. Cinnabar-red. Brazil.
  grandis gránd-is. Large (the flowers). Brazil.
    tenebrosa ten-e-*brŏ*-sa. Growing in shady places.
  × *harpophylla* harp-ŏ-*fil*-la. With sickle-shaped leaves.
  pumila pew-mi-la. Dwarf. Brazil.
  purpurata pur-pew-rah-ta. Purple (the flowers). Brazil.
  speciosa spek-ee-ŏ-sa. Showy. Mexico.
  xanthina zănth-*ee*-na. Yellow (the flowers). Brazil.

× **Laeliocattleya** lie-lee-ŏ-*kăt*-lee-a *Orchidaceae*. Intergeneric hybrids, from the names of the parents. *Cattleya* × *Laelia*. Greenhouse orchids.

Lagarosiphon lă-ga-rŏ-*see*-fon *Hydrocharitaceae*. From Gk. *lagaros* (narrow) and *siphon* (a tube) referring to the narrow perianth tube. Aquatic herb.
  major mah-yor. (= *Elodea crispa*). Larger. S Africa.

Lagenaria lă-gen-*ah*-ree-a *Cucurbitaceae*. From Gk. *lagenos* (a flask) referring to the shape and use of the fruits. Annual herb.
  siceraria see-ke-*rah*-ree-a. From L. *sicera* (an intoxicating drink). Bottle Gourd. Old World tropics.

Lagerstroemia lah-ger-*sturm*-ee-a *Lythraceae*. After Magnus von Lagerström (1696–1759), Swedish merchant and friend of Linnaeus. Semi-hardy, deciduous tree.
  indica in-di-ka. Indian. Crape Myrtle. China, Korea.

Lagurus la-*gew*-rus *Gramineae*. From Gk. *lagos* (a hare) and *oura* (a tail) referring to the inflorescence. Annual grass.
  ovatus ŏ-*vah*-tus. Ovate (the inflorescence). Hare's Tail Grass. Mediterranean region.

Lamb's Lettuce see *Valerianella locusta*
Lamb's Tongue see *Stachys byzantina*

Lamiastrum galeobdolon see *Galeobdolon luteum*

Lamium lă-mee-um *Labiatae*. The L. name. Perennial herbs. Dead Nettle.
  galeobdolon see *Galeobdolon luteum*
  garganicum gar-*gah*-ni-kum. Of Monte Gargano, S Italy. S Italy, SE Europe.
  maculatum măk-ew-*lah*-tum. Spotted (the leaves). Spotted Dead Nettle. Europe, Asia.
  orvala or-*vah*-la. From French *orvale* (= L. *aureis galli*) a kind of sage. S Europe.

Lampranthus lăm-*pránth*-us *Aizoaceae*. From Gk. *lampros* (shining) and *anthos* (a flower). Tender, succulent sub-shrubs. S Africa.
  amoenus a-*moy*-nus. Pleasant.
  aurantiacus ow-răn-tee-*ah*-kus. Orange.
  aureus ow-ree-us. Golden.
  blandus blănd-us. Charming.
  brownii brown-ee-ee. After Robert Brown, naturalist and first Keeper of Botany at the British Museum.
  coccineus kok-*kin*-ee-us. Scarlet.
  conspicuus kon-*spik*-ew-us. Conspicuous.
  multiradiatus mul-tee-ră-dee-*ah*-tus. (= L. *roseus*). With many rays.
  spectabilis spek-*tah*-bi-lis. Spectacular.
  zeyheri zay-ha-ree. After Carl L. P. Zeyher who collected in S Africa.

Lantana lăn-*tah*-na *Verbenaceae*. L. name for *Viburnum*, from the similar inflorescence. Tender shrubs.
  camara ka-*mah*-ra. The S American name. Yellow Sage. Tropical America.
  montevidensis mon-tee-vid-*en*-sis. Of Montevideo. S America.

Lapageria lah-pa-*zhe*-ree-a *Philesiaceae*. After Josephine de la Pagerie, wife of Napoleon Bonaparte. Evergreen climber.
  rosea ros-ee-a. Rose-coloured (the flowers). Chile, Argentina.

Lapeirousia lah-pay-*rooz*-ee-a *Iridaceae*. After Baron Phillipe de la Peyrouse (1744–1818), French botanist. Semi-hardy, cormous herb.
  laxa lăks-a. (= L. *cruenta*). Loose (the inflorescence). S Africa.

Larch see *Larix*
  Common, European see *L. decidua*
  Dunkeld see *L.* × *eurolepis*

Larch (continued)
  Golden see *Pseudolarix amabilis*
  Japanese see *L. kaempferi*

Larix lă-riks *Pinaceae*. The L. name.
Deciduous conifers. Larch.
  decidua de-*kid*-ew-a. (= *L. europaea*).
  Deciduous. Common Larch, European
  Larch. Europe.
  × *eurolepis* oy-rō-*lep*-is. *L. decidua* × *L.*
  kaempferi. From the names of the parents
  (see synonyms). Dunkeld Larch.
  kaempferi kemp-fa-ree. (= *L. leptolepis*).
  After Engelbert Kaempfer (1651–1716),
  German physician and botanist. Japanese
  Larch. Japan.

Larkspur see *Consolida*

Lathyrus lă-thi-rus *Leguminosae*. Gk. name
for the pea. Annual and perennial herbs.
  cirrhosus ki-*rō*-sus. Climbing by tendrils.
  Pyrenees, Cevennes.
  grandiflorus gránd-i-*flō*-rus. Large-flowered.
  S Europe.
  latifolius lah-tee-*fo*-lee-us. Broad-leaved.
  Europe.
  magellanicus mă-gel-*lăn*-i-kus. Of the region
  of the Magellan Straits.
  odoratus o-dō-*rah*-tus. Scented. Sweet Pea.
  S Italy, Sicily.
  rotundifolius ro-tund-i-*fo*-lee-us. With
  round leaves (leaflets). E Europe, W Asia.
  splendens splen-denz. Splendid.
  S California, N Mexico.
  tingitanus ting-gi-*tahn*-us. Of Tangier
  (Tingi). S Europe, N Africa.
  tuberosus tew-be-*rō*-sus. Tuberous. Europe,
  W Asia.
  vernus ver-nus. Of spring (flowering).
  Europe.

Laurus *low*-rus *Lauraceae*. L. name for *L.*
nobilis. Evergreen trees or shrubs.
  azorica a-*zo*-ri-ka. Of the Azores, Canary
  Islands, Azores.
  nobilis nō-bi-lis. Notable. Bay Laurel.
  Mediterranean region.

Laurustinus see *Viburnum tinus*

Lavandula la-văn-dew-la *Labiatae*. From L.
lavo (to wash) from its use in soaps.
Evergreen shrubs. Lavender.
  angustifolia ıng-gust-i-*fo*-lee-a. (= *L. spica*).
  Narrow-leaved. Common Lavender.
  W Mediterranean region.
  dentata den-*tah*-ta. Toothed (the leaves).

Lavandula (continued)
  Mediterranean region, Spain.
  × *intermedia* in-ter-*med*-ee-a. *L. angustifolia*
  × *L. latifolia*. Intermediate (between the
  parents). Many cultivated lavenders belong
  here.
  lanata lah-*nah*-ta. Woolly (the leaves and
  shoots). Spain.
  stoechas stoy-kas. Of the Stoechades (now
  the Isles d'Hyères) off the S coast of France
  near Toulon. SW Europe.

Lavatera lah-va-*te*-ra *Malvaceae*. After the
Lavater brothers, 16th-century Zurich
naturalists. Annual herbs and shrubs.
  arborea ar-*bo*-ree-a. Tree-like. Tree
  Mallow. S and W Europe.
  maritima ma-*ri*-ti-ma. Growing near the sea.
  W Mediterranean region, NW Africa.
  olbia ol-bee-a. *Olbia*, the L. name of several
  towns including one in S France and one
  in Sardinia. W Mediterranean coasts.
  trimestris tri-*mays*-tris. Of three months (the
  flowering period). Mediterranean region.

Lavender see *Lavandula*
  Common see *L. angustifolia*
  French see *L. stoechas*
Lavender Cotton see *Santolina
chamaecyparissus*

Layia lay-ee-a *Compositae*. After G.
Tradescant Lay (died 1845). Annual herb.
  platyglossa plăt-ee-*glos*-a. (= *L. elegans*).
  Broad-tongued. Tidy Tips. California.

Ledebouria lay-da-*bour*-ree-a *Liliaceae*.
After Carl Friedrich von Ledebour
(1785–1851). Tender, bulbous herbs.
S Africa.
  ovalifolia ō-vah-li-*fo*-lee-a. (= *Scilla
  ovalifolia*). With oval leaves.
  socialis so-kee-*ah*-lis. (= *Scilla violacea*).
  Growing in colonies. Silver Squill.

Ledum *lay*-dum *Ericaceae*. From *ledon* the
Gk. name for *Cistus*. Evergreen shrubs.
  groenlandicum grurn-*lănd*-i-kum. Of
  Greenland. Labrador Tea. N America,
  Greenland.
  palustre pa-*lus*-tree. Growing in marshes.
  Marsh Ledum. Arctic and alpine N
  hemisphere.

Leek see *Allium porrum*

Legousia le-*goos*-ee-a *Campanulaceae*.
Derivation not traced. Annual herb.

Legousia (continued)
speculum-veneris spek-ew-lum-*ven*-e-ris. (= *Specularia speculum-veneris*). As the common name. Venus's Looking Glass. C and S Europe.

Leiophyllum lay-ō-*fil*-lum *Ericaceae*. From Gk. *leios* (smooth) and *phyllon* (a leaf) referring to the glossy leaves. Evergreen shrub.
buxifolium buks-i-*fo*-lee-um. *Buxus*-leaved. Sand Myrtle. E United States.

Lemaireocereus la-mair-ree-ō-*kay*-ree-us *Cactaceae*. After Charles Lemaire (1801–71), French cactus specialist, and *Cereus* q.v.
marginatus mar-gi-*nah*-tus. Margined, the ribs are margined with white wool. C Mexico.
pruinosus proo-in-ō-sus. Bloomed (the shoots). S Mexico.
thurberi thur-ba-ree. After George Thurber (1821–1890) who collected in the SW United States and Mexico. S Arizona, W Mexico.
treleasei tra-*lees*-ee-ee. After William Trelease (1857–1945), American botanist. S Mexico.

Lemon see *Citrus limon*
Lemon Balm see *Melissa officinalis*
Lemon Mint see *Monarda citriodora*
Lemon-scented Gum see *Eucalyptus citriodora*
Lemon verbena see *Aloysia triphylla*
Lenten Rose see *Helleborus orientalis*

Leonotis lee-ō-*nō*-tis *Labiatae*. From Gk. *leon* (a lion) and *otis* (an ear), the corolla has been likened to a lion's ear. Tender shrub.
leonurus lee-ō-*new*-rus. Like a lion's tail (the flowering shoots). Lion's Ear. S Africa.

Leontopodium lee-on-to-*pod*-ee-um *Compositae*. Classical name for another plant from Gk. *leon* (a lion) and *podion* (a foot). Perennial herb.
alpinum ăl-*peen*-um. Alpine. Edelweiss. Europe (mountains).

Leopard Flower see *Belamcanda chinensis*
Leopard's Bane see *Arnica montana*, *Doronicum*

Leptospermum lep-to-*sperm*-um *Myrtaceae*. From Gk. *leptos* (slender) and *sperma* (a seed) referring to the narrow seeds. Semi-hardy, evergreen shrubs.

Leptospermum (continued)
humifusum hum-i-*few*-sum. (= *L. scoparium prostratum* hort.). Prostrate. Tasmania.
lanigerum lah-*ni*-ge-rum. (= *L. cunninghamii*). Woolly (the leaves and shoots). Australia, Tasmania.
scoparium skō-*pah*-ree-um. Broom-like. Manuka, Tea Tree. Australia, New Zealand.
'Chapmanii' chăp-*măn*-ee-ee. After Sir Frederick Chapman who found it in 1889.
'Keatleyi' *keet*-lee-ee. After Captain Keatley who found it.
'Nicholsii' ni-*kolz*-ee-ee. After Mr William Nichols who supplied a nursery with the seed from which it was raised.

Lespedeza les-pe-*dee*-za *Leguminosae*. After Vincente Manuel de Céspedes, governor of Florida in about 1790. Deciduous shrubs. Bush Clover.
bicolor bi-ko-lor. Two-coloured. N China, Japan.
thunbergii thun-*berg*-ee-ee. After Thunberg, see *Thunbergia*. N China, Japan.

Lettuce see *Lactuca sativa*

Leucadendron loo-ka-*den*-dron *Proteaceae*. From Gk. *leukos* (white) and *dendron* (a tree) referring to the silvery foliage. Tender tree.
argenteum ar-*gen*-tee-um. Silvery. Silver Tree. S Africa.

Leuchtenbergia loyk-tan-*berg*-ee-a *Cactaceae*. After Prince Maximilian E. Y. N. von Beauharnais (1817–52), Duke of Leuchtenberg, Germany.
principis pring-ki-pis. Distinguished. Mexico.

Leucocoryne loo-kō-*ko*-ri-nay *Amaryllidaceae*. From Gk. *leukos* (white) and *coryne* (a club). Bulbous herb.
ixioides iks-ee-*oi*-deez. Like *Ixia*. Chile.

Leucojum loo-*kō*-yum *Amaryllidaceae*. Gk. name for a plant from *leukon* (white) and *ion* (violet). Bulbous herbs. Snowflake.
aestivum ies-ti-vum. Of summer. Summer Snowflake. Europe, W Asia.
autumnale ow-tum-*nah*-lee. Of autumn. W Mediterranean region, N Africa.
roseum ros-ee-um. Rose-coloured (the flowers). Corsica, Sardinia.
vernum ver-num. Of spring. Spring Snowflake. Europe.

Leucothoe loo-*ko*-thō-ee *Ericaceae*. After Leucothoe of Gk. mythology, daughter of Orchamus, king of Babylon and Eurynome, who is said to have been changed into a shrub by her lover, Apollo, after being buried alive by his rival, Clytia. Evergreen and deciduous shrubs.

davisiae day-*vis*-ee-ie. After Miss N. J. Davis. California, Oregon.

fontanesiana font-a-neez-ee-*ah*-na. After Desfontaines, see *Desfontainea*. SE United States.

grayana gray-*ah*-na. After Asa Gray (1810–88). Japan.

keiskei kies-kee-ee. After Keisuke Ito (1803–1901). Japanese physician and botanist. Japan.

Levisticum le-*vis*-ti-kum *Umbelliferae*. Probably derived from *Ligusticum*, a related genus. Perennial herb.

officinale o-fi-ki-*nah*-lee. Sold as a herb. Lovage. S Europe.

Lewisia loo-*is*-ee-a *Portulacaceae*. After Captain Meriwether Lewis (1774–1809) of the Lewis and Clark expedition 1806–7. Perennial herbs.

columbiana ko-lūm-bee-*ah*-na. Of British Columbia. W N America.

cotyledon ko-ti-*lay*-don. From the resemblance to *Cotyledon*. California, Oregon.

howellii how-*el*-ee-ee. After Thomas Jefferson Howell (1842–1912).

nevadensis ne-va-*den*-sis. Of the Sierra Nevada, California. W United States.

pygmaea pig-*mie*-a. Dwarf. W United States.

rediviva re-di-*vee*-va. Brought back to life, the original herbarium specimen continued to grow after pressing. W United States.

tweedyi twee-dee-ee. After Tweedy. NW United States.

Leycesteria lest-*e*-ree-a *Caprifoliaceae*. After William Leycester. Chief Justice in Bengal during the early 19th century. Deciduous shrub.

formosa for-*mō*-sa. Beautiful. Himalaya, W China.

Liatris lee-*aht*-ris *Compositae*. Derivation obscure. Perennial herbs. Gay Feather.

graminifolia grah-min-i-*fo*-lee-a. Grass-leaved. SE United States.

Liatris (continued)

pycnostachya pik-nō-*stāk*-ee-a. With dense spikes. Button Snake Root. SE United States.

spicata spee-*kah*-ta. (= *L. callilepis* hort.). With flowers in a spike. E United States.

Libertia lee-*bert*-ee-a *Iridaceae*. After Marie A. Libert (1782–1863), Belgian botanist. Evergreen perennial herbs.

formosa for-*mō*-sa. Beautiful. Chile.

grandiflora grănd-i-*flō*-ra. Large-flowered. New Zealand.

ixioides iks-ee-*oi*-deez. Like *Ixia*. New Zealand.

pulchella pul-*kel*-la. Pretty. New Zealand, Australia, Tasmania.

Libocedrus chilensis see *Austrocedrus chilensis*
decurrens see *Calocedrus decurrens*

Ligularia lig-ew-*lah*-ree-a *Compositae*. From L. *ligula* (a strap) referring to the strap-like ray florets. Perennial herbs.

dentata den-*tah*-ta. (= *L. clivorum*. *Senecio clivorum*). Toothed (the leaves). China, Japan.

× *hessei* hes-ee-ee. *L. dentata* × *L. wilsoniana*. After Hesse.

hodgsonii hoj-son-ee-ee. After C. P. Hodgson, Consul in Japan c. 1840, who discovered it. Japan.

przewalskii sha-*văl*-skee-ee. (= *Senecio przewalskii*). After its discoverer, the Russian explorer Nicolai M. Przewalski. N China.

stenocephala sten-ō-*kef*-a-la. Narrow-headed. E Asia.

tangutica see *Senecio tanguticus*

tussilaginea tus-i-lah-*gin*-ee-a. Like *Tussilago*. E Asia.

veitchiana veech-ee-*ah*-na. After the Veitch nursery. China.

wilsoniana wil-son-ee-*ah*-na. After Ernest Wilson, see *Magnolia wilsonii*. China.

Ligustrum li-*gus*-trum *Oleaceae*. The L. name. Evergreen and semi-evergreen shrubs and trees. Privet.

delavayanum del-a-vay-*ah*-num. After Delavay who introduced it, see *Abies delavayi*. China.

henryi hen-ree-ee. After Henry who discovered it, see *Illicium henryi*. China.

japonicum ja-*pon*-i-kum. Of Japan. N China, Korea, Japan.

Ligustrum (continued)
 lucidum loo-ki-dum. Glossy (the leaves).
 China.
 obtusifolium ob-tew-si-*fo*-lee-um. Blunt-
 leaved. Japan.
 ovalifolium ō-vah-li-*fo*-lee-um. Oval-leaved.
 Japan.
 quihoui kee-*hoo*-ee. After M. Quihou.
 China.
 sinense si-*nen*-see. Of China.
 vulgare vul-*gah*-ree. Common. Common
 Privet. Europe, N Africa, SW Asia.

Lilac see *Syringa*
 Common see *S. vulgaris*
 Persian see *S.* × *persica*
 Rouen see *S.* × *chinensis*

Lilium *lee*-lee-um Liliaceae. The L. name.
Bulbous perennial herbs. Lily.
 auratum ow-*rah*-tum. Marked with gold
 (the flowers). Gold-rayed Lily. Japan.
 brownii brown-ee-ee. After Mr F. E. Brown
 of Slough in whose nursery it first flowered,
 in 1837. S China.
 bulbiferum bul-*bi*-fe-rum. Bearing bulbs (in
 the upper leaf axils). C Europe.
 canadense kan-a-*den*-see. Of Canada or NE
 North America. Canada Lily. E N
 America.
 candidum *kăn*-di-dum. White (the flowers).
 Madonna Lily. SE Europe to Lebanon.
 cernuum *kern*-ew-um. Nodding (the
 flowers). NE Asia.
 chalcedonicum kăl-kay-dō-ni-kum. Of
 Chalcedon (now Kadikoy) nr. Istanbul.
 Greece, Albania.
 davidii dă-*vid*-ee-ee. After David, see
 Davidia. W China.
 davuricum see *L. pensylvanicum*
 formosanum for-mō-*sah*-num. Of Taiwan
 (Formosa).
 hansonii hăn-*son*-ee-ee. After Peter Hanson
 of New York, artist and grower of lilies.
 henryi hen-*ree*-ee. After Henry, see *Illicium
 henryi*. C China.
 japonicum ja-*pon*-i-kum. Of Japan.
 lancifolium lăn-ki-*fo*-lee-um. (= *L.
 tigrinum*). With lance-shaped leaves. Tiger
 Lily. Cult., probably a hybrid.
 leichtlinii liekt-*lin*-ee-ee. After Max
 Leichtlin. Japan.
 longiflorum long-gi-*flō*-rum. With long
 flowers. Easter Lily. Japan.
 mackliniae ma-*klin*-ee-ie. Discovered by
 Kingdon Ward and named after his
 second wife, Jean Macklin. Manipur.
 martagon mar-ta-gon. A Turkish word used

Lilium (continued)
 for a sort of turban and this plant. Turk's-
 cap Lily. S Europe to Siberia.
 monadelphum mon-a-*delf*-um. (= *L.
 szovitsianum*). From Gk. *monos* (one) and
 adelphos (a brother), referring to the united
 stamens. Crimea, Caucasus.
 nepalense ne-pa-*len*-see. Of Nepal.
 Himalaya.
 pardalinum par-da-*leen*-um. Spotted like a
 leopard. Panther Lily. W N America.
 pensylvanicum pen-sil-*vahn*-i-kum. (= *L.
 davuricum*). Of Pennsylvania. N America.
 philippinense fil-i-peen-*en*-see. Of the
 Philippines.
 pumilum *pew*-mi-lum. Dwarf. Siberia,
 E China.
 pyrenaicum pi-ray-*nah*-i-kum. Of the
 Pyrenees.
 regale ray-*gah*-lee. Royal. W China.
 rubellum rub-*el*-lum. Reddish (the flowers).
 Japan.
 sargentiae sar-*jent*-ee-ie. After Charles
 Sargent's wife. Japan.
 speciosum spek-ee-ō-sum. Showy. S Japan.
 superbum soo-*per*-bum. Superb. E United
 States.
 szovitsianum see *L. monadelphum*
 × *testaceum* test-*ah*-kee-um. *L. candidum* ×
 L. chalcedonicum. Brick-coloured.
 tigrinum see *L. lancifolium*
 tsingtauense tsing-tow-*en*-see. Of Qingdao
 (Tsingtao), E China. Korea, NE China.
 wardii ward-ee-ee. After Kingdon Ward,
 see *Cassiope wardii*. Tibet.

Lily of China see *Rohdea japonica*
Lily of the Palace see *Hippeastrum aulicum*
Lily of the Valley see *Convallaria majalis*
Lily of the Valley Tree see *Clethra arborea*
Lily Tree see *Magnolia denudata*
Lime see *Citrus aurantiifolia, Tilia*
 Broad-leaved see *Tilia platyphyllos*
 Common see *T.* × *europaea*
 European White see *T. tomentosa*
 Red-twigged see *T. platyphyllos* 'Rubra'
 Small-leaved see *T. cordata*
 Weeping Silver see *T.* 'Petiolaris'

Limnanthes lim-*nănth*-eez Limnanthaceae.
From Gk. *limne* (a marsh) and *anthos* (a
flower). Annual herb.
 douglasii dūg-*lăs*-ee-ee. After David
 Douglas, see *Iris douglasii*. Poached Egg
 Flower. California, Oregon.

Limnophila lim-*no*-fil-a Scrophulariaceae.
From Gk. *limne* (a marsh) and *phileo* (to

Limnophila (continued)
love). Aquarium plants. SE Asia.
   heterophylla he-te-rō-*fil*-la. With variable
leaves, the leaves above and below the
water are different.
   sessiliflora se-si-li-*flō*-ra. With un-stalked
flowers.

Limonium lee-*mō*-nee-um *Plumbaginaceae*.
From Gk. *leimon* (a meadow), referring to
their habitat. Annual, biennial and perennial
herbs.
   bellidifolium bel-i-di-*fo*-lee-um. With leaves
like *Bellis*. The true species is rarely
grown. Europe, Asia (coasts).
   bonduellei bon-*dwel*-ee-ee. After M.
Bonduelle, French army surgeon (c
1813–70). Algerian Statice. Algeria.
   cosyrense ko-si-*ren*-see. Of Pantellaria
(Cossyra) an island between Sicily and
Tunisia. S Europe.
   incanum see *Goniolimon callicomum*
   latifolium lah-tee-*fo*-lee-um. Broad-leaved.
E Europe.
   sinuatum sin-ew-*ah*-tum. Wavy-edged (the
leaves). Mediterranean region, W Asia.
   spicatum see *Psylliostachys spicata*
   suworowii see *Psylliostachys suworowii*
   tataricum see *Goniolimon tataricum*

Linanthus leen-*ănth*-us *Polemoniaceae*. From
Gk. *linon* (flax) and *anthos* (a flower). Annual
herb.
   androsaceus ăn-dros-*ah*-kee-us. Like
Androsace. California.
      luteus loo-tee-us. (= *Gilia lutea*). Yellow
(the flowers).

Linaria leen-*ah*-ree-a *Scrophulariaceae*. From
Gk. *linon* (flax) referring to the similar
leaves. Annual and perennial herbs. Toadflax.
   alpina ăl-*peen*-a. Alpine. Alpine Toadflax.
Alps.
   cymbalaria see *Cymbalaria muralis*
   genistifolia gen-i-sti-*fo*-lee-a. *Genista*-leaved.
E Europe.
      dalmatica dăl-*mă*-ti-ka. (= *L. dalmatica*.
L. madedonica). Of Dalmatia. SE Europe.
   maroccana mă-ro-*kah*-na. Of Morocco.
Bunny Rabbits.
   origanifolia see *Chaenorhinum origanifolium*
   purpurea pur-*pewr*-ree-a. Purple (the
flowers). Italy, Sicily.
   reticulata ray-tik-ew-*lah*-ta. Net-veined (the
corolla). N Africa.
   triornithophora tree-or-ni-*tho*-fo-ra. From
Gk. *tri-* (three), *ornis* (a bird) and *phorea*
(to bear). The flowers are in threes and

Linaria (continued)
resemble three birds with long tails. Three
Birds Flying. Spain, Portugal.
   vulgaris vul-*gah*-ris. Common. Common
Toadflax. Europe.

Lindera lin-*de*-ra *Lauraceae*. After Johann
Linder (1676–1723), Swedish botanist.
Deciduous shrubs or trees.
   benzoin ben-zō-in. From the Arabic name
for another plant. Spice Bush. E N
America.
   obtusiloba ob-tew-si-*lō*-ba. Bluntly-lobed
(the leaves). E Asia.

Ling see *Calluna vulgaris*

Linnaea lin-*ie*-a *Caprifoliaceae*. After
Linnaeus (1707–78), who popularised the
binomial system of naming plants, i.e. using
one name for the genus and one for the
species. Evergreen sub-shrub.
   borealis bo-ree-*ah*-lis. Northern. Twin
Flower. Northerly latitudes of the
N hemisphere.

Linum leen-um *Linaceae*. The L. name for
flax. Annual and perennial herbs and shrubs.
   alpinum see *L. perenne alpinum*
   arboreum ar-*bo*-ree-um. Tree-like. Tree
Flax. Greece, Turkey.
   austriacum ow-stree-*ah*-kum. Austrian.
C and S Europe.
   campanulatum kăm-păn-ew-*lah*-tum. Bell-
shaped (the flowers). W Mediterranean
region.
   flavum flah-vum. Yellow (the flowers).
Golden Flax. C and SE Europe.
   grandiflorum gränd-i-*flō*-rum. Large-
flowered. N Africa.
   monogynum mon-*o*-gi-num. With one pistil.
New Zealand.
   narbonense nar-bon-*en*-see. Of Narbonne.
SW Europe.
   perenne pe-*ren*-ee. Perennial. Europe.
      alpinum ăl-*peen*-um. (= *L. alpinum*).
Alpine.
   salsoloides see *L. suffruticosum salsoloides*
   suffruticosum su-froo-ti-*kō*-sum. Sub-
shrubby. Spain.
      salsoloides sal-so-*loi*-deez. (= *L.
salsoloides*). Like *Salsola*. SW Europe.
   usitatissimum ew-see-tah-*tis*-i-mum. Most
useful. Flax. Cult.

Lion's Ear see *Leonotis leonurus*
Lippia citriodora see *Aloysia triphylla*
   nodiflora see *Phyla nodiflora*

Lipstick Vine see *Aeschynanthus radicans*

Liquidambar li-kwid-*ăm*-bar
Hamamelidaceae. From. L. *liquidus* (liquid)
and *ambar* (amber), referring to the resin
obtained from the bark. Deciduous trees.
  formosana for-mō-*sah*-na. Of Taiwan
  (Formosa). SE Asia.
    Monticola mon-*ti*-ko-la. Growing on
    mountains. China.
  styraciflua sti-ra-*ki*-floo-a. Flowing with
  styrax (storax, an aromatic balsam used in
  medicine and perfumery). Sweet Gum. E
  United States, Mexico, Guatemala.

Liriodendron li-ree-ō-*den*-dron
Magnoliaceae. From Gk. *leiron* (a lily) and
dendron (a tree). Deciduous trees.
  chinense chin-*en*-see. Of China. Chinese
  Tulip Tree.
  tulipifera tew-lip-*i*-fe-ra. Tulip-bearing,
  referring to the tulip-like flowers. Tulip
  Tree. E N America.

Liriope lee-*ree*-o-pay *Liliaceae*. After Liriope
a fountain nymph and the mother of
Narcissus. Evergreen, perennial herbs.
  graminifolia grah-min-i-*fo*-lee-a. Grass-
  leaved. China, Vietnam.
  muscari mus-*kah*-ree. From the
  resemblance to *Muscari*. Japan, China.
  spicata spee-*kah*-ta. With flowers in spikes.
  China, Vietnam.

Lithocarpus lith-ō-*kar*-pus *Fagaceae*. From
Gk. *lithos* (stone) and *karpos* (a fruit)
referring to the hard acorns. Evergreen trees.
  densiflorus dens-i-*flō*-rus. Densely flowered.
  Tanbark Oak. California, Oregon.
  edulis ed-*ew*-lis. Edible. Japan.
  henryi hen-*ree*-ee. After Henry, see *Illicium
  henryi*. Japan.

Lithodora lith-ō-*do*-ra *Boraginaceae*. From
Gk. *lithos* (a stone) and *dorea* (a gift).
Evergreen shrubs.
  diffusa di-*few*-sa. (= *Lithospermum
  diffusum*). Spreading. SW Europe.
  oleifolia o-lee-i-*fo*-lee-a. (= *Lithospermum
  oleifolium*). With leaves like *Olea*. Spain.
  rosmarinifolia rōs-ma-reen-i-*fo*-lee-a. (=
  Lithospermum rosmarinifolium).
  Rosmarinus-leaved. S Italy, Sicily.

Lithops *lith*-ops *Aizoaceae*. From Gk. *lithos*
(stone) and *ops* (appearance). Tender
succulents that resemble stones. Living

Lithops (continued)
Stones. S and SW Africa.
  bella bel-la. Pretty.
  bromfieldii brom-*feeld*-ee-ee. After Mr H.
  Bromfield.
  comptonii komp-*ton*-ee-ee. After Dr R. H.
  Compton, Director of Kirstenbosch
  Botanic Garden, S Africa.
  erniana ern-ee-*ah*-na. After F. Erni.
  fulleri ful-a-ree. After Mr E. R. Fuller,
  Postmaster at Prieska, Cape Province.
  karasmontana kă-ras-mon-*tah*-na. Of the
  Little Karasberg Mountains, Namibia.
    mickbergensis mik-berg-*en*-sis. (= *L.
    mickbergensis*). Of Mickberg, Namibia.
  lesliei lez-lee-ee. After its discoverer, Mr T.
  N. Leslie who collected in the Transvaal.
  marmorata mar-mo-*rah*-ta. Marbled.
  marthae mar-thie. After the discoverer's
  wife, Martha Erni.
  mickbergensis see *L. karasmontana
  mickbergensis*
  olivacea o-li-*vah*-kee-a. Olive green.
  optica op-ti-ka. Seeing, referring to the
  translucent, eye-like window in the leaves.
  pseudotruncatella soo-dō-trunk-a-*tel*-la.
  False *L. truncatella*.
    mundtii munt-ee-ee. After Mundt of
    Windhuk.
  salicola sa-*li*-ko-la. Growing in salty places.
  turbiniformis tur-bin-i-*form*-is. Top-shaped.

Lithospermum diffusum see *Lithodora diffusa*
  doerfleri see *Moltkia doerfleri*
  oleifolium see *Lithodora oleifolium*
  purpurocaeruleum see *Buglossoides
  purpurocaerulea*
  rosmarinifolium see *Lithodora
  rosmarinifolia*
Little Candles see *Mammillaria prolifera*

Littonia li-*ton*-ee-a *Liliaceae*. After Samuel
Litton (1781–1847), professor of botany at
Dublin. Tender perennial herb.
  modesta mo-*des*-ta. Modest. S Africa.

Living Stones see *Lithops*
Livingstone Daisy see *Dorotheanthus
bellidiformis*

Livistona li-vi-*ston*-a *Palmae*. After Patrick
Murray, Baron of Livingston (died 1671)
whose plant collection helped to found
Edinburgh Botanic Garden in 1670. Tender
palms.
  australis ow-*strah*-lis. Southern. Australian
  Fan Palm. E Australia.
  chinensis chin-*en*-sis. Of China. Chinese
  Fan Palm. S Japan, China.

Lobelia lō-*bel*-ee-a *Lobeliaceae*. After Mathias de l'Obel (1538–1616), Flemish botanist. Annual and perennial herbs.
cardinalis kar-di-*nah*-lis. Scarlet (the flowers). Cardinal Flower. E United States.
erinus e-ri-nus. Gk. name of a plant. Edging Lobelia. S Africa.
siphilitica si-fi-*li*-ti-ka. A reference to supposed medicinal properties. E United States.
splendens splen-denz. (= *L. fulgens*). Splendid. Mexico.
tenuior ten-*ew*-ee-or. Slender. W Australia.

Lobivia lo-*biv*-ee-a *Cactaceae*. An anagram of Bolivia where some species grow. Cob Cactus.
allegraiana a-leg-ray-ah-na. Perhaps a reference to Puerto Alegre, Bolivia. S Peru.
bruchii bruk-ee-ee. (= *Soehrensia bruchii*). After Bruch. N Argentina.
famatimensis fā-ma-tee-*men*-sis. Of Famatina
  densispina dens-i-*speen*-a. (= *L. densispina*). Densely spined.
ferox see *Echinopsis ferox*
hertichiana her-trik-ee-*ah*-na. After William Hertrich. Peru.
jajoiana ha-hō-ee-*ah*-na. Probably a reference to Jujuy. N Argentina.
pygmaea pig-*mie*-a. (= *Rebutia pygmaea*). Dwarf. Bolivia, N Argentina.

Lobularia lob-ew-*lah*-ree-a *Cruciferae*. From L. *lobulus* (a small pod) referring to the fruit. Annual or perennial herb.
maritima ma-*ri*-ti-ma. (= *Alyssum maritimum*). Growing near the sea. Sweet Alyssum. S Europe.

Locust see *Robinia pseudacacia*
  Caspian see *R. caspica*
  Honey see *Gleditsia triacanthos*
Loganberry see *Rubus loganobaccus*

Loiseleuria lwū-ze-*lur*-ree-a *Ericaceae*. After Jean Louis Auguste Loiseleur-Deslongchamps (1774–1849). Prostrate, evergreen shrub.
procumbens prō-*kum*-benz. Prostrate. Alpine Azalea. Arctic-alpine N hemisphere.

Lollipop Plant see *Pachystachys lutea*

Lomatia lō-*mah*-tee-a *Proteaceae*. From Gk. *loma* (a border), the seeds are edged with a wing. Semi-hardy, evergreen shrubs.
ferruginea fe-roo-*gin*-ee-a. Rusty (the hairs of the shoots). Chile, Argentina.
myricoides mi-ri-*koi*-deez. Like *Myrica*. SE Australia.
tinctoria tink-*to*-ree-a. Used in dyeing. Tasmania.

London Pride see *Saxifraga* × *urbium*

Lonicera lon-i-*se*-ra *Caprifoliaceae*. After Adam Lonitzer (1528–86), German naturalist. Deciduous and evergreen shrubs and climbers. Honeysuckle.
  × *americana* a-me-ri-*kah*-na *L. caprifolium* × *L. etrusca*. Of America where it was once thought to be native.
  × *brownii brown*-ee-ee. *L. hirsuta* × *L. sempervirens*. After Brown.
    'Fuchsioides' fuks-ee-*oi*-deez. *Fuchsia*-like.
    'Plantierensis' plon-tee-e-*ren*-sis. From the Simon-Louis nursery at Plantières near Metz, France.
  caprifolium kăp-ri-*fo*-lee-um. Old generic name from L. *capra* (a goat) and *folium* (a leaf). Perfoliate Honeysuckle. Europe, Caucasus, Turkey.
  etrusca e-*troos*-ka. Of Tuscany. Mediterranean region.
  fragrantissima frah-gran-*tis*-i-ma. Very fragrant. China.
  × *heckrottii* hek-*rot*-ee-ee. *L.*×*americana* × *L. sempervirens*. After Heckrott.
  henryi hen-ree-ee. After Henry, see *Illicium henryi*. W China.
  hildebrandiana hil-da-brănd-ee-*ah*-na. After Mr A. H. Hildebrand (1852–1918) who collected in India. Burma, Thailand, China.
  involucrata in-vo-loo-*krah*-ta. With an involucre, the conspicuous bracts around the flowers which enlarge in fruit. W N America, Mexico, S Canada.
  japonica ja-*pon*-i-ka. Of Japan. Japan, China, Korea.
    'Aureoreticulata' ow-ree-ō-ray-tik-ew-*lah*-ta. Veined with gold (the leaves).
    'Halliana' hawl-ee-*ah*-na. After Dr George Hall who introduced it to the United States in 1862.
    repens ree-penz. Creeping.
  nitida ni-ti-da. Glossy (the leaves). China.
    'Fertilis' *fer*-ti-lis. Fertile, i.e. bearing fruit.

Lonicera (continued)
periclymenum pe-ree-*klim*-en-um. From
periklymenon the Gk. name for
honeysuckle. Common Honeysuckle,
Woodbine. Europe, N Africa.
  'Belgica' *bel*-gi-ka. Belgian.
  'Serotina' se-*ro*-ti-na. Late flowering.
pileata pi-lee-*ah*-ta. Capped (the fruit).
China.
sempervirens sem-per-*vi*-renz. Evergreen.
Trumpet Honeysuckle. S and E United
States.
standishii stăn-*dish*-ee-ee. After John
Standish (1814–75), of the Standish and
Noble nursery, for whom Fortune
introduced it. China.
syringantha si-ring-*gănth*-a. With flowers
like *Syringa*. China, Tibet.
tatarica ta-*tah*-ri-ka. Of Tatary, C Asia.
× *tellmanniana* tel-măn-ee-*ah*-na. *L.*
tragophylla × *L. sempervirens* 'Superba'.
After Tellmann.
tragophylla tră-go-*fil*-la. From Gk. *tragos* (a
goat) and *phyllon* (a leaf) it is related to *L.*
caprifolium q.v. China.

Loofah Gourd see *Luffa aegyptiaca*
Loosestrife see *Lysimachia*
  Purple see *Lythrum salicaria*
  Yellow see *Lysimachia vulgaris*

Lophophora lo-fo-*fo*-ra *Cactaceae*. From Gk.
lophos (a crest) and *phoreo* (to bear) referring
to the tufts of woolly hairs on the areoles.
  williamsii wil-*yămz*-ee-ee. After Williams.
  Dumpling Cactus, Mescal Button, Peyote.
  Texas, N Mexico.

Loquat see *Eriobotrya japonica*
Lords and Ladies see *Arum maculatum*

Lotus lō-tus *Leguminosae*. A Gk. name used
for several plants. Perennial herb.
  corniculatus kor-nik-ew-*lah*-tus. With small
  horns. Bird's-foot Trefoil. Europe.

Lotus see *Nelumbo*
  American see *N. lutea*
  Blue Egyptian see *N. caerulea*
  Indian or Sacred see *N. nucifera*
  White Egyptian see *N. lotus*
Lovage see *Levisticum officinale*
Love Grass see *Eragrostis elegans*
  Japanese see *E. amabilis*
Love-in-a-Mist see *Nigella damascena*
Love-in-a-Puff see *Cardiospermum
halicacabum*
Love-lies-Bleeding see *Amaranthus caudatus*

Luculia lu-*kew*-lee-a *Rubiaceae*. From *lukuli
swa*, the native name of *L. gratissima*.
Tender, evergreen shrubs.
  grandifolia grănd-i-*fo*-lee-a. Large-leaved.
  Bhutan.
  gratissima grah-*tis*-i-ma. Very pleasing.
  Himalaya, China.
  pinceana pin-see-*ah*-na. After Robert
  Taylor Pince (c 1804–71), Exeter
  nurseryman and partner of Lucombe.
  Nepal, Assam.

Luffa *luf*-a *Cucurbitaceae*. From the Arabic
name. Annual climber.
  aegyptiaca ie-gip-tee-*ah*-ka. (= *L.
  cylindrica*). Egyptian. Loofah Gourd. Old
  World tropics.

Lunaria loon-*ah*-ree-a *Cruciferae*. From L.
luna (the moon) referring to the pods.
Annual and perennial herbs.
  annua ăn-ew-a. Annual. Honesty. SE
  Europe.
  rediviva re-di-*veev*-a. Coming back to life,
  i.e. perennial. Europe.

Lungwort see *Pulmonaria*

Lupinus lu-*peen*-us *Leguminosae*. The L.
name. Annual and perennial herbs. Lupin.
  arboreus ar-*bo*-ree-us. Tree-like. California.
  chamissonis sha-mi-*sō*-nis. After Ludolf
  Adalbert von Chamisso (1781–1838).
  California.
  densiflorus dens-i-*flō*-rus. Densely-flowered.
  California.
  excubitus eks-*kew*-bi-tus. Presumably
  likening its habit to that of a sentinel
  (*excubitor*). S California.
  hartwegii hart-*weg*-ee-ee. After Karl
  Theodore Hartweg (1812–71). Mexico.
  luteus loo-tee-us. Yellow. Mediterranean
  region.
  mutabilis mew-*tah*-bi-lis. Changeable (the
  flower colour). Andes.
  polyphyllus po-lee-*fil*-lus. With many leaves
  (leaflets). W N America.
  pubescens pew-*bes*-enz. Hairy. Mexico,
  C America.
  subcarnosus sub-kar-*nō*-sus. Somewhat
  fleshy. Texas.
  texensis teks-*en*-sis. Of Texas. Texas
  Bluebonnet.

Luronium loo-ron-ee-um *Alismataceae*. From
Gk. *luron*, another name for *Alisma*. Aquatic
perennial herb.

Luronium (continued)
natans nă-tănz. (= *Alisma natans*). Floating. Europe.

Luzula luz-ew-la *Juncaceae*. From Italian *lucciola* (a firefly). Perennial herbs.
nivea ni-vee-a. Snow-white (the flowers). W and C Europe (mountains).
sylvatica sil-vă-ti-ka. (= *L. maxima*). Of woods. Europe, Caucasus, W Asia.

Lycaste li-kăs-tay *Orchidaceae*. After Lycaste, daughter of Priam, the last King of Troy. Greenhouse orchids.
aromatica ă-rō-mă-ti-ka. Fragrant. Mexico, C America.
brevispatha brev-ee-spăth-a. With a short spathe. C America.
cruenta kroo-en-ta. Blood red (a blotch on the lip). Mexico, C America.
deppei dep-ee-ee. After Ferdinand Deppe (died 1861). Mexico, C America.
fimbriata fim-bree-ah-ta. Fringed (the petal margins). S America.
gigantea gi-găn-tee-a. Very large. S America.
macrophylla măk-rō-fil-la. Large-leaved. C and S America.
schilleriana shil-a-ree-ah-na. After Herr Schiller of Hamburg who owned a large orchid collection. Colombia.
virginalis vir-gin-ah-lis. Virginal (the white flowers). Mexico, C America.

Lychnis lik-nis *Caryophyllaceae*. The classical name, from Gk. *lychnos* (a lamp). Perennial herbs.
alpina ăl-peen-a. (= *Viscaria alpina*). Alpine. Alpine Campion. Europe, Siberia, N America.
× *arkwrightii* ark-riet-ee-ee. *L. chalcedonica* × *L.* × *haageana*. After Arkwright.
chalcedonica kăl-kay-dōn-i-ka. Of Kadikoy (Chalcedon) near Istanbul. Maltese Cross, Jerusalem Cross. Russia.
coeli-rosea see *Silene coeli-rosea*
coronaria ko-rō-nah-ree-a. Used in garlands. Rose Campion. SE Europe, N Africa to C Asia.
coronata ko-rō-nah-ta. Crowned. China.
flos-jovis flōs-yov-is. Flower of Jupiter. Alps.
× *haageana* hah-gee-ah-na. *L. coronata* × *L. fulgens*. After F. A. Haage.
lagascae see *Petrocoptis glaucifolia*
viscaria vis-kah-ree-a. Sticky (the stems). German Catchfly. Europe to E Asia.

Lycium lik-ee-um *Solanaceae*. From the Gk. name for a thorny tree. Deciduous shrub.
barbarum bar-ba-rum. (= *L. chinense*. *L. halimifolium*). Foreign. Box Thorn. China.

Lycopersicum li-kō-per-si-kum *Solanaceae*. Gk. name for another plant from *lykos* (a wolf) and *persicon* (a peach). Annual herb.
esculentum es-kew-lent-um. Edible. Tomato. Andes.

Lycoris li-ko-ris *Amaryllidaceae*. After Lycoris, a beautiful Roman actress and mistress of Marc Antony. Tender, bulbous herbs.
africana ăf-ri-kahn-a. (= *L. aurea*). It was originally thought to be from Africa. Golden Spider Lily. SE Asia.
incarnata in-kar-nah-ta. Flesh pink (the flowers). C China.
radiata răd-ee-ah-ta. Radiating (the long-exserted stamens). Red Spider Lily. China, Japan.
sanguinea sang-gwin-ee-a. Blood red (the flowers). Japan.
squamigera skwah-mi-ge-ra. Scaly (? the bulb). Resurrection Lily. Japan.

Lyme Grass see *Elymus arenarius*

Lyonia lie-on-ee-a *Ericaceae*. After John Lyon (c 1765–1814) a Scottish gardener who travelled and collected in the SE United States. Deciduous shrubs.
ligustrina li-gus-tree-na. Like *Ligustrum*. E N America.
mariana mă-ree-ah-na. Of Maryland. E and S United States.

Lyonothamnus lie-on-ō-thăm-nus *Rosaceae*. After W. S. Lyon (1851–1916) who discovered it in 1884 and Gk. *thamnos* (a shrub). Semi-hardy, evergreen tree.
floribundus flō-ri-bun-dus. Profusely flowering. Santa Catalina Island (California).
asplenifolius a-splay-ni-fo-lee-us. With leaves like *Asplenium*. S California (islands).

Lysichiton li-si-ki-ton or loo-si-ki-ton. *Araceae*.. From Gk. *lysis* (releasing) and *chiton* (a cloak) referring to the shedding of the large spathe. Perennial herbs.
americanum a-me-ri-kah-num. American. Skunk Cabbage. W N America.

Lysichiton (continued)
  camtschatcense kamt-shăt-*ken*-see. Of
  Kamchatka. Kamchatka to N Japan.

Lysimachia li-si-*măk*-ee-a or loo-si-*măk*-ee-a
Primulaceae. After King Lysimachos of
Thrace (c 360–281 B.C.) who is said to have
pacified a bull with a piece of loosestrife. In
Gk. his name means ending strife. Perennial
herbs.
  ciliata ki-lee-*ah*-ta. Fringed with hairs (the
  petals). N America.
  clethroides kleth-*roi*-deez. Like *Clethra*.
  E Asia.
  ephemerum e-*fem*-e-rum. From *ephemeron*
  the L. name of a plant. SW Europe.
  leschenaultii lay-shen-*olt*-ee-ee. After Louis
  Theodore Leschenault de la Tour
  (1773–1826), French botanist. India.
  nemorum ne-mo-rum. Growing in woods.
  Europe, Caucasus.
  nummularia num-ew-*lah*-ree-a. With coin-
  shaped leaves. Creeping Jenny. Europe,
  Caucasus.
  punctata punk-*tah*-ta. Dotted (the
  undersides of the leaves). Europe, W Asia.
  thyrsiflora thurs-i-*flō*-ra. With flowers in a
  thyrse (L. *thyrsus*, a staff) a type of
  inflorescence. Europe, Asia, N America.
  vulgaris vul-*gah*-ris. Common. Yellow
  Loosestrife. Europe, Asia.

Lythrum *lith*-rum *Lythraceae*. From Gk.
lythron (blood) referring to the colour of the
flowers. Herbaceous perennials.
  salicaria săl-i-*kah*-ree-a. Like *Salix*. Purple
  Loosestrife. Europe, Asia.
  virgatum vir-*gah*-tum. Wand-like.
  E Europe, W Asia.

M

Maackia *mahk*-ee-a *Leguminosae*. After
Richard Maack (1825–86), Russian
naturalist. Deciduous tree.
  amurensis ăm-ew-*ren*-sis. Of the Amur
  region, NE Asia. Manchuria.

Macleaya ma-*klay*-a *Papaveraceae*. After
Alexander Macleay (1767–1848).
Herbaceous perennials.
  cordata kor-*dah*-ta. Heart-shaped (the
  leaves). China, Japan.
  microcarpa mik-rō-*kar*-pa. With small
  fruits. Plume Poppy. C China.

Maclura ma-*kloo*-ra *Moraceae*. After William
Maclure (1763–1840), American geologist.
Deciduous tree.
  pomifera pom-*i*-fe-ra. Apple-bearing,
  referring to the fruit. Osage Orange. S and
  C United States.

Madagascar Jasmine see *Stephanotis
floribunda*
Madagascar Lace Plant see *Aponogeton
madagascariensis*
Madonna Lily see *Lilium candidum*
Madrona see *Arbutus menziesii*

Magnolia măg-*nol*-ee-a *Magnoliaceae*. After
Pierre Magnol (1638–1715), French
professor of botany. Deciduous and
evergreen, shrubs and trees.
  acuminata a-kew-min-*ah*-ta. With a long
  point (the leaves). Cucumber Tree. E N
  America.
  campbellii kăm-*bel*-ee-ee. After Dr
  Archibald Campbell (1805–74), Political
  Resident in Darjeeling. Himalaya.
    mollicomata mol-i-kom-*ah*-ta. Softly
    hairy.
  conspicua see *M. denudata*
  delavayi del-a-*vay*-ee. After Delavay who
  discovered it, see *Abies delavayi*. China.
  denudata day-new-*dah*-ta. (= *M. conspicua*.
  M. heptapeta). Naked, the flowers open
  before the leaves emerge. Yulan, Lily Tree.
  China.
  fraseri *fray*-za-ree. After John Fraser
  (1750–1811), a collector of American
  plants who introduced it to England. Ear-
  leaved Umbrella Tree. SE United States.
  grandiflora grănd-i-*flō*-ra. Large-flowered.
  Bull Bay. S United States.
  heptapeta see *M. denudata*
  hypoleuca hi-pō-*loo*-ka. White beneath (the
  leaves). Japan.
  kobus kō-bus. From *kobushi* the Japanese
  name. Japan.
  liliiflora lee-lee-i-*flō*-ra. (= *M. quinquepeta*).
  With flowers like *Lilium*. China.
    'Nigra' *nig*-ra. Black, the very dark
    purple flowers.
  × *loebneri* *lurb*-na-ree. *M. kobus* × *M.
  stellata*. After Max Löbner, German
  botanist.
  macrophylla măk-rō-*fil*-la. Large-leaved. SE
  United States.
  officinalis o-fi-ki-*nah*-lis. Sold as a herb, the
  bark and flower buds are used medicinally
  in China. China.
    biloba bi-lō-ba. Two-lobed (the leaves).
  quinquepeta see *M. liliiflora*

Magnolia (continued)
salicifolia să-li-ki-*fo*-lee-a. *Salix*-leaved.
Japan.
sieboldii see-*bōld*-ee-ee. After Siebold, see
Acanthopanax sieboldii. S Japan, Korea.
sinensis si-*nen*-sis. Of China.
× *soulangiana* soo-lon-zhee-*ah*-na. *M.*
denudata × *M. liliiflora*. After Etienne
Soulange-Bodin, French cavalry officer and
Director of the Royal Institute of
Horticulture who raised it.
   'Brozzonii' bro-*zon*-ee-ee. After Camillo
   Brozzoni in whose garden it was raised.
   'Lennei' *len*-ee-ee. After Peter Joseph
   Lenné (1789–1866), German botanist.
sprengeri spreng-a-ree. After Carl L.
Sprenger, German nurseryman. China.
stellata ste-*lah*-ta. Star-like (the flowers).
Star Magnolia. Japan.
× *thompsoniana* tomp-son-ee-*ah*-na. *M.*
tripetala × *M. virginiana*. After Archibald
Thompson a nurseryman of Mile End,
London who raised it in 1808.
tripetala tri-*pe*-ta-la. With three petals.
Umbrella Tree. E N America.
× *veitchii* veech-ee-ee. *M. campbellii* × *M.*
denudata. After Peter C. M. Veitch
(1850–1929) who raised it in 1907.
virginiana vir-jin-ee-*ah*-na. Of Virginia.
Sweet Bay. E United States.
wilsonii wil-*son*-ee-ee. After its discoverer
and introducer Ernest Henry Wilson
(1876–1936), one of the most prolific of all
plant collectors. Originally sent to China
to introduce *Davidia*, he made many
expeditions there and introduced
numerous fine plants. China.

Mahonia ma-*hon*-ee-a *Berberidaceae*. After
Bernard McMahon (died 1816), American
horticulturist. Evergreen shrubs.
aquifolium ă-kwi-*fo*-lee-um. The L. name
for holly. Oregon Grape. W N America.
bealei beel-ee-ee. After T. C. Beale who
grew plants collected by Fortune in his
garden in Shanghai. China.
fortunei for-*tewn*-ee-ee. After Fortune who
introduced it in 1846, see *Fortunella*.
China.
japonica ja-*pon*-i-ka. Of Japan where it is
cultivated. ? China.
lomariifolia lō-mah-ree-i-*fo*-lee-a. With
leaves like *Lomaria* (now *Blechnum*).
Burma, W China.
× *media me*-dee-a. *M. japonica* × *M.*
lomariifolia. Intermediate (between the
parents).

Mahonia (continued)
pinnata pin-*nah*-ta. Pinnate (the leaves).
California.
pumila pew-mi-la. Dwarf. California,
Oregon.
trifoliolata tri-fo-lee-ō-*lah*-ta. With three
leaflets. S Texas.
   glauca glow-ka. Glaucous. Texas,
   N Mexico.

Maianthemum mah-*yănth*-e-mum *Liliaceae*.
From Gk. *maios* (May) and *anthemon*
(blossom), referring to its flowering time.
Perennial herbs. May Lily.
bifolium bi-*fo*-lee-um. Two leaved. Europe,
Asia.
canadense kăn-a-*den*-see. Of Canada or NE
North America. N America.

Maidenhair Fern see *Adiantum*
   Australian see *A. formosum*
   Common see *A. capillus-veneris*
   Delta see *A. raddianum*
   Giant see *A. trapeziforme*
   Kashmir see *A. venustum*
   N. American see *A. pedatum*
   Rose-fronded see *A. pedatum japonicum*
   Rough see *A hispidulum*
   Trailing see *A. caudatum*
Maidenhair Tree see *Ginkgo biloba*
Maize see *Zea mays*

Malcolmia măl-*kol*-mee-a *Cruciferae*. After
William Malcolm (died 1798) and his son
William (c. 1768–1835), London nurserymen.
Annual herb.
maritima ma-*ri*-ti-ma. Growing near the sea.
Virginia Stock. Greece, Albania.

Mallow see *Malva*
   Hairy see *Anisodontea scabrosa*
   Marsh see *Althaea officinalis*
   Musk see *Malva moschata*
   Poppy see *Callirhoe*
   Tree see *Lavatera arborea*

Malope ma-*lō*-pee *Malvaceae*. Greek name
for a related plant. Annual herb.
trifida tri-fi-da. Three-lobed (the leaves).
Spain, N Africa.

Maltese Cross see *Lychnis chalcedonica*

Malus *mah*-lus *Rosaceae*. L. name for apple.
Deciduous trees. Apple, Crab.
× *atrosanguinea* aht-rō-sang-*gwin*-ee-a. *M.*
halliana × *M. sieboldii*. Deep red (the
flowers).

Malus (continued)

baccata bah-*kah*-ta. Bearing berries. Siberian Crab. NE Asia.

mandshurica mănd-*shu*-ri-ka. Of Manchuria. Manchurian Crab.

coronaria ko-rō-*nah*-ree-a. Used in garlands. E N America.

'Charlottae' shar-*lot*-ie. After Mrs Charlotte de Wolf who found it in 1902.

× *eleyi* e-lee-ee. M. *niedzwetzkyana* × M. *spectabilis*. After Charles Eley who raised it.

floribunda flō-ri-*bun*-da. Profusely flowering. Cult.

halliana hawl-ee-*ah*-na. After Dr G. R. Hall who introduced this and the following to N America in about 1863. Cult.

'Parkmanii' park-*măn*-ee-ee. After Francis Parkman, a historian and friend of Dr Hall.

'Hillieri' *hil*-ee-a-ree. After Hillier's who raised it about 1928.

hupehensis hew-pee-*hen*-sis. Of Hubei (Hupeh). C and W China.

'Lemoinei' la-*mwūn*-ee-ee. After Lemoine who raised it.

purpurea pur-*pewr*-ree-a. M.× *atrosanguinea* × M. *niedzwetzkyana*. Purple (the foliage).

× *robusta* rō-*bust*-a. M. *baccata* × M. *prunifolia*. Robust.

sargentii sar-*jent*-ee-ee. After Sargent who discovered and introduced it, see *Prunus sargentii*. Japan.

spectabilis spek-*tah*-bi-lis. Spectacular. Cult.

tschonoskii chon-*os*-kee-ee. After Tschonoski, a Japanese plant collector. Japan.

Malva *măl*-va *Malvaceae*. L. name for mallow. Biennial or perennial herbs. Mallow.

alcea ăl-kee-a. From Gk. *alkaia* a sort of mallow. Europe.

'Fastigiata' fa-stig-ee-*ah*-ta. Erect-branched.

moschata mos-*kah*-ta. Musk-scented. Musk Mallow. Europe, N Africa.

sylvestris sil-*ves*-tris. Of woods. Europe, N Africa, Asia.

Malvastrum campanulatum see *Sphaeralcea purpurata*

coccineum see *Sphaeralcea coccinea*

scabrosum see *Anisodontea scabrosa*

Malvaviscus măl-va-*vis*-kus *Malvaceae*.

Malvaviscus (continued)

From *Malva* q.v. and L. *viscus* (glue) referring to the pulp around the seeds. Tender shrub.

arboreus ar-*bo*-ree-us. (= M. *mollis*). Tree-like. S America.

Mammillaria măm-i-*lah*-ree-a *Cactaceae*. From L. *mammilla* (a nipple) referring to the shape of the tubercles. Unless otherwise stated C Mexico.

aurihamata ow-ree-hah-*mah*-ta. With golden hooks, referring to the yellow, hooked spines.

bocasana bō-ka-*sah*-na. Of the Sierra de Bocas. Powder Puff, Snowball Cactus.

camptotricha kămp-tō-*tree*-ka. From Gk. *kamptos* (curved) and *trichos* (hair) referring to the slender, curved spines.

candida kăn-di-da. White (the spines).

celsiana kels-ee-*ah*-na. After Jean Francois Cels (1810–88).

compressa kom-*pres*-a. Compressed.

decipiens day-*kip*-ee-enz. Deceptive. N Mexico.

densispina dens-i-*speen*-a. Densely spiny.

elegans ay-le-gahnz. Elegant.

elongata ay-long-*gah*-ta. Elongated. Gold Lace Cactus.

erythrosperma e-rith-rō-*sperm*-a. With red seeds.

gigantea gi-*găn*-tee-a. Very large.

glochidiata glo-kid-ee-*ah*-ta. With barbed bristles.

haageana hahg-ee-*ah*-na. After F. A. Haage.

hahniana hahn-ee-*ah*-na. After Hahn. Old Woman Cactus.

heyderi hay-da-ree. After Heyder (1804–84), German cactus grower. Coral Cactus. Texas.

kunzeana kunts-ee-*ah*-na. After Gustav Kunze (1793–1851).

magnimamma mahg-ni-*măm*-a. With large tubercles.

microhelia mik-rō-*hay*-lee-a. A small sun, perhaps referring to the yellow spine clusters.

microheliopsis mik-rō-hay-lee-*op*-sis. Like M. *microhelia*.

parkinsonii par-kin-*son*-ee-ee. After John Parkinson (c 1772–1847), Consul-General in Mexico in 1838.

plumosa ploo-*mō*-sa. Feathery (the spines). Feather Cactus. N Mexico.

prolifera prō-*li*-fe-ra. Proliferous. Little Candles, Silver Cluster Cactus. W Indies.

pygmaea pig-*mie*-a. Dwarf.

Mammillaria (continued)
schelhasii shel-*hah*-see-ee. After Schelhas.
schiedeana shee-dee-*ah*-na. After Christian
J. W. Schiede (died 1836) who collected
in Mexico.
spinosissima speen-ō-*sis*-i-ma. Very spiny.
tetracantha tet-ra-*kānth*-a. Four-spined.
uncinata un-kee-*nah*-ta. Hooked (the
spines).
zeilmanniana ziel-mǎn-ee-*ah*-na. After
Zeilmann. Rose Pincushion.

Mandarin Lime see *Citrus* × *limonia*
Mandarin Orange see *Citrus reticulata*

Mandevilla mǎn-da-*vil*-a *Apocynaceae*. After
Henry John Mandeville (1773–1861) a
minister in Buenos Aires who introduced *M.*
suaveolens. Evergreen climbers.
boliviensis bo-liv-ee-*en*-sis. (= *Dipladenia*
boliviensis). Of Bolivia. Bolivia, Ecuador.
sanderi sahn-da-ree. (= *Dipladenia sanderi*).
After the Sander nursery of St Albans and
Bruges. Brazil.
splendens splen-denz. (= *Dipladenia*
splendens). Splendid. Brazil.
suaveolens swah-*vee*-o-lenz. Sweetly
scented. Chilean Jasmine. Argentina.

Mandragora mǎn-*drǎg*-o-ra *Solanaceae*. The
Gk. name. Perennial herb.
officinarum o-fi-ki-*nah*-rum. Sold as a herb.
Mandrake. S Europe.

Manettia ma-*net*-ee-a *Rubiaceae*. After
Saveria Manetti (1723–85). Tender,
evergreen climber.
inflata in-*flah*-ta. (= *M. bicolor*). Swollen
(the base of the corolla). Firecracker Vine.
Paraguay, Uruguay.

Manuka see *Leptospermum scoparium*
Manzanita see *Arctostaphylos manzanita*
Maple see *Acer*
  Amur see *A. ginnala*
  Ash-leaved see *A. negundo*
  Field see *A. campestre*
  Hedge see *A. campestre*
  Hornbeam see *A. carpinifolium*
  Japanese see *A. palmatum*
  Montpelier see *A. monspessulanum*
  Nikko see *A. nikoense*
  Norway see *A. platanoides*
  Oregon see *A. macrophyllum*
  Red see *A. rubrum*
  Silver see *A. saccharinum*
  Sugar see *A. saccharum*
  Vine see *A. circinatum*

Maranta ma-*rǎn*-ta *Marantaceae*. After
Bartolommeo Maranti, 16th-century
Venetian botanist. Tender, evergreen herbs.
arundinacea a-run-di-*nah*-kee-a. Reed-like.
Arrowroot. Tropical America.
bicolor bi-ko-lor. Two-coloured (the
leaves). Brazil, Guiana.
leuconeura loo-kō-*newr*-ra. (= *M. leuconeura*
massangeana). White-veined (the leaves).
Prayer Plant. Brazil.
  erythroneura e-rith-rō-*newr*-ra. Red-
  veined (the leaves).
kerchoveana ker-chov-ee-*ah*-na. After
Kerchove. Rabbit's Foot.

Marble Plant see *Neoregelia marmorata*
Mare's Tail see *Hippuris vulgaris*
Marguerite see *Chrysanthemum leucanthemum*
  White see *C. frutescens*

Margyricarpus mar-gi-ri-*kar*-pus *Rosaceae*.
From Gk. *margarites* (a pearl) and *karpos* (a
fruit) referring to the white fruits. Dwarf,
evergreen shrub.
pinnatus pin-*nah*-tus. (= *M. setosus*).
Pinnate (the leaves). Pearl Fruit.
S America.

Marigold see *Tagetes*
  African see *T. erecta*
  French see *T. patula*
  Pot see *Calendula officinalis*
  Signet see *T. tenuifolia*
Mariposa Lily see *Calochortus*
Marjoram, Pot see *Origanum onites*
  Sweet see *O. marjorana*
  Wild see *O. vulgare*
Marmalade Bush see *Streptosolen jamesonii*
Marrow see *Cucurbita pepo*
Marsdenia erecta see *Cionura erecta*
Marsh Marigold see *Caltha palustris*
Martynia louisianica see *Proboscidea*
louisianica
Marvel of Peru see *Mirabilis jalapa*

Masdevallia mǎs-da-*vah*-lee-a *Orchidaceae*.
After José Masdevall (died 1801), Spanish
botanist. Greenhouse orchids.
amabilis a-*mah*-bi-lis. Beautiful. Peru.
bella bel-la. Pretty. Colombia.
caudata kaw-*dah*-ta. With a slender tail (the
sepals). Venezuela, Colombia.
chimaera ki-*mie*-ra. A monster, perhaps
referring to the very large flowers.
Colombia.
coccinea kok-*kin*-ee-a. Scarlet. Colombia,
Peru.

Masdevallia (continued)
  militaris mee-li-*tah*-ris. (= *M. ignea* hort.).
  Military. Colombia.
  muscosa moo-*skō*-sa. Mossy. Perhaps
  muscipula (fly-catching) was intended as
  the lip of the flower traps small insects.
  Colombia.
  simula *sim*-ew-la. Imitating. Colombia.
  tovarensis tō-vah-*ren*-sis. Of Tovar,
  Venezuela. Venezuela, Colombia.

Mask Flower see *Alonsoa*
Masterwort see *Astrantia*
Matricaria eximia hort. see *Tanacetum parthenium*

Matteuccia ma-*too*-kee-a *Aspidiaceae*. After
Carlo Matteucci, 19th-century Italian
physicist. Fern.
  struthiopteris stroo-thee-*op*-te-ris. (= *M. germanica*). Presumably from Gk.
  struihokamelos (an ostrich) and *pteris* (a
  fern), the fronds resemble ostrich feathers.
  Ostrich-feather Fern. Europe, Asia.

Matthiola mă-tee-ō-la *Cruciferae*. After
Pierandrea Mattioli (1500–77), Italian
botanist. Annual or biennial herbs. Stock.
  longipetala long-gi-*pe*-ta-la. Long-petalled.
  S Ukraine.
    bicornis bi-*kor*-nis. (= *M. bicornis*). Two-
    horned (the fruits). Night-scented
    Stock. Greece, Aegean region.
  incana in-*kah*-na. Grey. Brompton Stock.
  S and W Europe (coasts).
    Annua *ăn*-ew-a. Annual. Ten Week
    Stock.

Maurandya maw-*răn*-dee-a *Scrophulariaceae*.
After Catherine Maurandy, wife of Prof. A.
J. Maurandy. Tender perennials. Mexico.
  barclayana bark-lay-*ah*-na. (= *Asarina barclayana*). After Robert Barclay
  (1751–1830).
  erubescens e-roo-*bes*-enz. (= *Asarina erubescens*). Blushing (the pink flowers).
  scandens skăn-denz. (= *Asarina scandens*).
  Climbing.

Maxillaria măks-i-*lah*-ree-a *Orchidaceae*.
From L. *maxilla* (a jaw), in some species the
column and lip resemble the jaw of an insect.
Greenhouse orchids.
  grandiflora grănd-i-*flō*-ra. Large-flowered.
  N Andes.
  longisepala long-gi-*sep*-a-la. With long
  sepals. Venezuela.

Maxillaria (continued)
  luteoalba loo-tee-ō-*ăl*-ba. Yellow-white (the
  flowers). Costa Rica to Ecuador.
  picta *pik*-ta. Painted (the flowers).
  Colombia, Brazil.
  praestans *prie*-stahnz. Distinguished.
  Mexico, Guatemala.
  rufescens roo-*fes*-enz. Reddish (the flowers).
  C and S America.
  sanderiana sahn-da-ree-*ah*-na. After the
  Sander orchid nursery. Ecuador, Peru.
  tenuifolia ten-ew-i-*fo*-lee-a. Slender-leaved.
  C America.
  variabilis vă-ree-*ah*-bi-lis. Variable.
  C America.
  venusta ven-*us*-ta. Charming. S America.

May Apple see *Passiflora incarnata*
May Lily see *Maianthemum*
Maypop see *Passiflora incarnata*

Maytenus may-*ten*-us *Celastraceae*. From
maiten the native name of *M. boaria*.
Evergreen tree.
  boaria bō-*ah*-ree-a. Of cattle, which eat the
  leaves. Chile, Argentina.

Mazus mă-zus *Scrophulariaceae*. From Gk.
mazos (a teat) referring to the swellings in
the throat of the corolla. Perennial herbs.
  pumilio pew-*mi*-lee-o. Dwarf. New
  Zealand, Australia.
  radicans rah-di-kănz. With rooting stems.
  New Zealand.
  reptans *rep*-tănz. Creeping. Himalaya.

Meadow Rue see *Thalictrum*
Meadowsweet see *Filipendula ulmaria*

Meconopsis may-kō-*nop*-sis *Papaveraceae*.
From Gk. *mekon* (a poppy) and -*opsis*
indicating resemblance. Perennial or
monocarpic herbs.
  baileyi see *M. betonicifolia*
  betonicifolia be-ton-i-ki-*fo*-lee-a. (= *M. baileyi*). Betonica-leaved. Himalayan Blue
  Poppy. China.
  cambrica *kăm*-bri-ka. Welsh. Welsh Poppy.
  W Europe.
  chelidoniifolia ke-li-dŏn-ee-i-*fo*-lee-a.
  Chelidonium-leaved. W China.
  grandis *grănd*-is. Large. Himalaya,
  W China.
  horridula ho-*rid*-ew-la. Somewhat spiny.
  Himalaya, W China.
  integrifolia in-teg-ri-*fo*-lee-a. With entire
  leaves. E Himalaya, W China.
  napaulensis nă-paw-*len*-sis. Of Nepal.

Meconopsis (continued)
Himalaya, W China.
quintuplinervia kwin-tup-li-*ner*-vee-a. Five-veined (the leaves). Harebell Poppy. Tibet.
regia ray-gee-a. Royal. Nepal.
simplicifolia sim-pli-ki-*fo*-lee-a. With simple leaves. Himalaya, Tibet.
superba soo-*per*-ba. Superb. Tibet.
villosa vi-*lō*-sa. Softly hairy. Himalaya.

Medinilla me-di-*ni*-la *Melastomataceae*. After José de Medinilla, Governor of the Marianna Islands in 1820. Tender, evergreen shrub.
magnifica mahg-*ni*-fi-ka. Magnificent. Philippines.

Medlar see *Mespilus germanica*
Medusa's Head see *Euphorbia caput-medusae*

Melaleuca me-la-*loo*-ka *Myrtaceae*. From Gk. *melas* (black) and *leukos* (white) referring to the black trunk and white shoots of many species. Semi-hardy and tender, evergreen shrubs.
armillaris arm-i-*lah*-ris. Encircled, the inflorescence encircles the shoot. SE Australia.
elliptica e-*lip*-ti-ka. Elliptic (the leaves). W Australia.
hypericifolia hi-pe-ree-ki-*fo*-lee-a. *Hypericum*-leaved. New South Wales.
wilsonii wil-*son*-ee-ee. After Mr Charles Wilson who discovered it. SE Australia.

Melia me-lee-a *Meliaceae*. Gk. name for the ash, from the similar leaves. Semi-hardy, deciduous tree.
azedarach a-*zed*-a-răk. From the native name. Bead Tree. Iran, Himalaya, China.

Melianthus me-lee-ănth-us *Melianthaceae*. From Gk. *meli* (honey) and *anthos* (a flower) referring to the abundant nectar. Semi-hardy sub-shrub.
major mah-yor. Larger. Honey Flower. S Africa.

Melissa me-*lis*-a *Labiatae*. Gk. name for honeybees which are attracted to the flowers. Herbaceous perennial.
officinalis o-fi-ki-*nah*-lis. Sold as a herb. Lemon Balm. S Europe.

Melittis me-*li*-tis *Labiatae*. Derivation as for *Melissa*. Perennial herb.
melissophyllum me-lis-ō-*fil*-lum. With leaves like *Melissa*. Europe.

Melocactus may-lō-kăk-tus *Cactaceae*. From L. *melopepo* (an apple-shaped melon) and *Cactus* q.v.
broadwayi brord-way-ee. After W. R. Broadway who discovered it in 1914. W Indies.

Melon see *Cucumis melo*

Mentha men-tha *Labiatae*. The L. name. Perennial herbs. Mint. Most hybrids occur naturally.
aquatica a-*kwah*-ti-ka. Growing in or near water. Water Mint. Europe, N Africa, Asia.
× *gentilis* gen-*tee*-lis. *M. arvensis* × *M.* × *spicata*. Related to. Ginger Mint.
longifolia long-gi-*fo*-lee-a. Long-leaved. Horse Mint. Europe.
× *piperita* pi-pe-*ree*-ta. *M. aquatica* × *M.* × *spicata*. Like pepper. Black Peppermint. In the wild with the parents.
officinalis o-fi-ki-*nah*-lis. Sold as a herb. White Peppermint.
citrata ki-*trah*-ta. Lemon-scented. Eau de Cologne Mint.
'Crispa' *kris*-pa. Crisped (the leaves). Curly Mint.
pulegium poo-*leg*-ee-um. The L. name. Pennyroyal. Europe, W Asia.
requienii rek-wee-*en*-ee-ee. After Esprit Requien (1788–1851) who studied the flora of S France and Corsica. Corsican Mint. Italy, Corsica, Sardinia.
× *spicata* spee-kah-ta. *M. longifolia* × *M. suaveolens*. With flowers in spikes. Spearmint, Garden Mint.
suaveolens swah-*vee*-o-lenz. (= *M.* × *rotundifolia* hort.). Sweetly scented. Round-leaved Mint. S and W Europe.
× *villosa* vi-*lō*-sa. *M.* × *spicata* × *M. suaveolens*. Softly hairy.
alopecuroides ă-lō-pek-ew-*roi*-deez. Like *Alopecurus*. Bowles' Mint.

Mentzelia ment-*zel*-ee-a *Loasaceae*. After Christian Mentzel (1622–1701), German botanist. Annual herb.
lindleyi lind-lee-ee. (= *Bartonia aurea*). After John Lindley (1799–1865), professor of botany at London University. Blazing Star. C California.

Menyanthes may-nee-ănth-eez *Menyanthaceae*. From *menanthos* (moonflower) the Gk. name for *Nymphoides peltata*, a related plant. Aquatic or bog

Menyanthes (continued)
garden perennial herb.
   trifoliata tri-fo-lee-*ah*-ta. With three leaves
(leaflets). Buck Bean, Bog Bean.
N hemisphere.

Menziesia men-*zeez*-ee-a *Ericaceae*. After
Archibald Menzies (1754–1842), naval
surgeon and botanist who collected in W N
America. Deciduous shrubs.
   ciliicalyx ki-lee-i-*kǎ*-liks. With the calyx
fringed with hairs. Japan.
   purpurea pur-*pewr*-ree-a. Purple (the
flowers).

Merendera me-ren-*de*-ra *Liliaceae*. From
quita meriendas, the Spanish name for a
Colchicum. Cormous herbs.
   filifolia fee-li-*fo*-lee-a. With thread-like
leaves. SW Europe.
   pyrenaica pi-ray-*nah*-i-ka. (= *M. montana*).
Of the Pyrenees. Spain, Portugal,
Pyrenees.
   robusta rō-*bus*-ta. Robust. N Afghanistan,
Russia.
   sobolifera so-bo-*li*-fe-ra. Bearing offspring.
Balkans, Romania.
   trigyna *tri*-gi-na. With three pistils.
Caucasus, Turkey, N Iran.

Merrybells see *Uvularia*

Mertensia mer-*tenz*-ee-a *Boraginaceae*. After
Franz Karl Mertens (1764–1831), German
botanist. Perennial herbs.
   ciliata ki-lee-*ah*-ta. Fringed with hairs (the
leaves). W United States.
   echioides e-kee-oi-deez. Like *Echium*.
Himalaya.
   longiflora long-gi-*flō*-ra. Long-flowered.
W United States.
   maritima ma-*ri*-ti-ma. Growing near the sea.
N Europe (coasts).
   primuloides preem-ew-*loi*-deez. Like
Primula. Himalaya.
   sibirica si-*bi*-ri-ka. Of Siberia.
   virginica vir-*jin*-i-ka. Of Virginia. E United
States.

Mescal Button see *Lophophora williamsii*

Mesembryanthemum mes-em-bree-*ǎnth*-e-
mum *Aizoaceae*. Previously spelled
Mesembrianthemum meaning flowering at mid-
day, the current spelling refers to the position
of the ovary. Succulent annual.
   cordifolium see *Aptenia cordifolia*
   criniflorum see *Dorotheanthus bellidiformis*

Mesembryanthemum (continued)
   crystallinum kris-ta-*leen*-um. (= *Cryophytum
crystallinum*). Crystalline, the appearance of
the leaves. Ice Plant. S Africa.
   tricolor see *Dorotheanthus tricolor*

Mespilus *mes*-pi-lus *Rosaceae*. The L. name.
Deciduous tree.
   germanica ger-*mah*-ni-ka. Of Germany.
Medlar. Europe, SW Asia.

Metasequoia me-ta-se-*kwoy*-a *Taxodiaceae*.
From Gk. *meta* (changed) *Sequoia* q.v. to
which it is related. Deciduous conifer only
discovered in 1941.
   glyptostroboides glip-to-stro-*boi*-deez. Like
Glyptostrobus. Dawn Redwood. China.

Metrosideros may-trō-si-*day*-ros *Myrtaceae*.
From Gk. *metra* (heart-wood) and *sideros*
(iron) referring to the very hard wood. Semi-
hardy, evergreen trees and shrubs.
   excelsa eks-*kel*-sa. (= *M. tomentosa*). Tall.
New Zealand.
   kermadecensis kerm-a-dek-*en*-sis. Of the
Kermadec Islands.
   robusta rō-*bus*-ta. Robust. New Zealand.
   umbellata um-bel-*ah*-ta. (= *M. lucida*). The
flowers appear to be in umbels. New
Zealand.

Mexican Orange Blossom see *Choisya ternata*
Mexican Sunflower see *Tithonia rotundifolia*
Mezereon see *Daphne mezereum*
Michaelmas Daisy see *Aster novi-belgii*

Michauxia mee-*shō*-ee-a *Campanulaceae*.
After André Michaux (1746–1803), French
traveller and plant collector. Biennial herbs.
   campanuloides kǎm-pǎn-ew-*loi*-dees. Like
Campanula. E Mediterranean region.
   tchihatcheffii chee-ha-*chef*-ee-ee. After
Count Pierre de Tchihatcheff (1808–90),
Russian traveller and writer. Turkey.

Microbiota mik-rō-bee-*o*-ta *Cupressaceae*.
From Gk. *micros* (small) and *Biota* (*Thuja*).
Dwarf, evergreen conifer.
   decussata day-kus-*ah*-ta. With the leaves in
pairs, one pair at right angles to the next.
SE Siberia.

Microcachrys mik-rō-*kǎk*-ris *Podocarpaceae*.
From Gk. *mikros* (small) and *kachrys* (a cone).
Dwarf, evergreen conifer.
   tetragona tet-ra-*gōn*-a. Four-angled (the
shoots).

Microcoelum mik-rō-*koy*-lum *Palmae*. From Gk. *mikros* (small) and *koilos* (a hollow) referring to a small hollow in the endosperm. Tender Palm.
   weddellianum we-del-ee-*ah*-num. (= *Cocos weddelliana*). After Dr H. A. Weddell who collected in S America in the 19th century. Weddell Palm. Brazil.

Microglossa albescens see *Aster albescens*
Mignonette see *Reseda odorata*

Mila *mee*-la *Cactaceae*. An anagram of Lima, capital of Peru.
   caespitosa kie-spi-*tō*-sa. Tufted. Peru.

Milium *mi*-lee-um *Gramineae*. L. name for millet. Perennial grass.
   effusum e-*few*-sum. Spreading. Europe, Asia.
   'Aureum' *ow*-ree-um. Golden. Bowles' Golden Grass.

Milkweed see *Asclepias*

Miltonia mil-*ton*-ee-a *Orchidaceae*. After Charles Fitzwilliam, Viscount Milton (1786–1857), horticultural patron. Greenhouse orchids. Pansy Orchid.
   candida *kán*-di-da. White (the lip). Brazil.
   clowesii klowz-ee-ee. After Clowes, see *Anguloa clowesii*. Brazil.
   endresii en-*dres*-ee-ee. After Senor Endres who collected in Costa Rica in about 1870. Costa Rica, Panama.
   flavescens flah-*ves*-enz. Yellowish. Paraguay.
   phalaenopsis fă-lie-*nop*-sis. Moth-like (the flowers). Colombia.
   regnellii reg-*nel*-ee-ee. After Mr Regnell who introduced it. Brazil.
   roezlii rurz-lee-ee. After Benedict Roezl (c 1824–85), a plant collector. Colombia.
   spectabilis spek-*tah*-bi-lis. Spectacular. Brazil.
   vexillaria veks-i-*lah*-ree-a. Standard-bearing. Colombia.
   warscewiczii var-sha-*vich*-ee-ee. After Warscewicz, see *Alonsoa warscewiczii*. Colombia, Peru.

Mimosa *mee*-*mos*-a *Leguminosae*. From Gk. *mimos* (a mimic) referring to the sensitive leaves. Tender shrub.
   pudica pu-*dee*-ka. Shy. Humble Plant, Sensitive Plant. Tropical America.

Mimosa see *Acacia dealbata*
   Pink see *Albizia julibrissin*

Mimulus *mee*-mew-lus *Scrophulariaceae*. A diminutive of L. *mimus* (a mimic), the flowers resemble a monkey's face. Perennial herbs and shrubs. Monkey Flower.
   aurantiacus ow-răn-tee-*ah*-kus. (= *M. glutinosus*. *Diplacus aurantiacus*). Orange. Oregon, California.
    puniceus pew-ni-kee-us. (= *M. puniceus*. *Diplacus puniceus*). Reddish-purple. S California, N Mexico.
   × *burnetii* bur-*net*-ee-ee. M. *cupreus* × M. *luteus*. After Dr Burnet of Aberdeen who raised it about 1901.
   cardinalis kar-di-*nah*-lis. Scarlet. Cardinal Monkey Flower. W N America.
   cupreus kew-pree-us. Coppery (the flowers). S Chile.
   glutinosus see M. *aurantiacus*
   guttatus gu-*tah*-tus. Spotted (the flowers). W N America, Mexico.
   × *hybridus* hib-ri-dus. M. *guttatus* × M. *luteus*. (= M. *tigrinus* hort.) Hybrid.
   lewisii loo-is-ee-ee. After Lewis, see *Lewisia*. W N America.
   luteus loo-tee-us. Yellow. Chile.
   moschatus mos-*kah*-tus. Musk-scented. N America.
   primuloides preem-ew-*loi*-deez. Like *Primula*. W United States.
   puniceus see M. *aurantiacus puniceus*
   ringens ring-genz. Gaping (the corolla). N America.
   tigrinus hort. see M. × *hybridus*
   variegatus vă-ree-a-*gah*-tus. Variegated (the flowers). Chile.

Mina *mee*-na *Convolvulaceae*. After Joseph Mina of Mexico. Tender climber.
   lobata lo-*bah*-ta. (= *Ipomoea lobata*. *Quamoclit lobata*). Lobed (the leaves). Mexico to S America.

Mind-your-own-Business see *Soleirolia soleirolii*
Mint see *Mentha*
   Bowles' see M. × *villosa alopecuroides*
   Corsican see M. *requienii*
   Curly see M. × *piperita* 'Crispa'
   Eau de Cologne see M. × *piperita citrata*
   Garden see M. × *spicata*
   Ginger see M. × *gentilis*
   Horse see M. *longifolia*
   Round-leaved see M. *suaveolens*
   Water see M. *aquatica*
Mint Bush see *Prostanthera*

Minuartia min-*wah*-tee-a *Caryophyllaceae*. After Juan Minuart (1693–1768) of Barcelona. Perennial herbs.
  laricifolia lă-ri-ki-*fo*-lee-a. (= *Arenaria laricifolia*). With leaves like *Larix*. Europe.
  verna hort. see *Sagina glabra*

Mirabilis mee-*rah*-bi-lis *Nyctaginaceae*. L. for wonderful. Annual herbs.
  jalapa ha-*lah*-pa. Of Jalapa (Xalapa), Mexico. Four o'Clock Plant, Marvel of Peru. Tropical America.
  longiflora long-gi-*flō*-ra. Long-flowered. SW United States, Mexico.
  multiflora mul-tee-*flō*-ra. Many-flowered. SW United States.

Miscanthus mis-*kănth*-us *Gramineae*. From Gk. *miskos* (a stem) and *anthos* (a flower) referring to the stalked spikelets. Perennial grasses.
  sacchariflorus să-ka-ri-*flō*-rus. With flowers like *Saccharum* (sugar cane). Asia.
  sinensis si-*nen*-sis. Of China. E Asia.

Mistflower see *Eupatorium coelestinum*
Mistletoe see *Viscum album*

Mitchella mi-*chel*-la *Rubiaceae*. After Dr John Mitchell (1711–68), Virginian physician and botanist. Dwarf, evergreen shrub.
  repens ree-penz. Creeping. Partridge Berry. E and C North America.

Mitella mi-*tel*-la *Saxifragaceae*. A diminutive of Gk. *mitra* (a cap) referring to the fruit. Perennial herbs. Bishop's Cap.
  breweri broo-a-ree. After Brewer, see *Picea breweriana*. W N America.
  caulescens kaw-*les*-enz. With a stem. W N America.
  diphylla di-*fil*-la. Two-leaved. E N America.

Mitraria mi-*trah*-ree-a *Gesneriaceae*. From Gk. *mitra* (a cap) referring to the fruit. Semi-hardy, evergreen climber.
  coccinea kok-*kin*-ee-a. Scarlet (the flowers). Chile, Argentina.

Mock Orange see *Philadelphus coronarius*
Mole Plant see *Euphorbia lathyris*

Molinia mo-*leen*-ee-a *Gramineae*. After Juan Ignacio Molina (1740–1829), Chilean

Molinia (continued) botanist. Perennial grass.
  caerulea kie-*ru*-lee-a. Blue. Purple Moor Grass. Europe.

Moltkia molt-kee-a *Boraginaceae*. After Count Joachim Gadske Moltke (1746–1818), Danish statesman and naturalist. Herbs and shrubs.
  doerfleri durf-la-ree. (= *Lithospermum doerfleri*). After J. D. Doerfler. Albania.
  × *intermedia* in-ter-*med*-ee-a. *M. petraea* × *M. suffruticosa*. Intermediate (between the parents).
  petraea pe-*trie*-a. Growing on rocks. SE Europe.
  suffruticosa su-froo-ti-*kō*-sa. Sub-shrubby. Italy.

Moluccella mo-lu-*kel*-la *Labiatae*. Derivation obscure, possibly from Molucca. Annual herb.
  laevis lie-vis. Smooth. Bells of Ireland, Shell Flower. W Asia.

Monarch of the East see *Sauromatum guttatum*
Monarch of the Veldt see *Venidium fastuosum*

Monarda mo-*nar*-da *Labiatae*. After Nicholas Monardes (1493–1588), Spanish botanist and physician. Annual, biennial and perennial herbs.
  citriodora kit-ree-o-*dō*-ra. Lemon-scented. Lemon Mint. S United States, Mexico.
  didyma di-di-ma. In pairs (the stamens or the leaves). Oswego Tea, Sweet Bergamot. E United States.
  fistulosa fist-ew-*lō*-sa. Tubular. E N America.
  media me-dee-a. Intermediate. E N America.

Monkey Flower see *Mimulus*
Monkey Puzzle see *Araucaria araucana*
Monkshood see *Aconitum*

Monstera mon-*ste*-ra *Araceae*. Derivation obscure, possibly from the monstrous appearance of the leaves.
  deliciosa day-li-kee-*ō*-sa. Delicious (the fruit). Swiss Cheese Plant. Mexico, C America.

Montbretia see *Crocosmia* × *crocosmiiflora*

Montia *mon*-tee-a *Portulacaceae*. After Giuseppe Monti (1682–1760), Italian

Montia (continued)
professor of botany. Annual herbs.
perfoliata per-fo-lee-*ah*-ta. With leaves
joined around the stem. W N America.
sibirica si-*bi*-ri-ka. Siberian. W N America.

Monvillea mon-*vil*-ee-a *Cactaceae*. After
Monville, a 19th-century, French cactus
specialist.
cavendishii kăv-an-*dish*-ee-ee. After William
Spencer Cavendish, Duke of Devonshire.
S America.
spegazzinii speg-a-*zeen*-ee-ee. After Prof.
Carlos Spegazzini (1858–1926), Argentine
botanist. Paraguay.

Moonstones see *Pachyphytum oviferum*
Sticky see *P. glutinicaule*
Moosewood see *Acer pensylvanicum*
Mop-headed Acacia see *Robinia pseudacacia*
'Umbraculifera'

Moraea mo-*rie*-a *Iridaceae*. After Robert
More (1703–80), amateur botanist. Cormous
perennials. S Africa.
iridioides ee-ri-dee-*oi*-dees. *Iris*-like.
ramosissima rah-mō-*si*-si-ma. Much
branched.
spathacea spa-*thah*-kee-a. Spathe-like.
tricuspidata tri-kus-pi-*dah*-ta. Three-
pointed.

Morina mo-*reen*-a *Dipsacaceae*. After Louis
Pierre Morin (1635–1715), French botanist.
Perennial herb.
longifolia long-gi-*fo*-lee-a. Long-leaved.
Whorl Flower. Himalaya.

Morisia mo-*ris*-ee-a *Cruciferae*. After
Giuseppe Giacinto Moris (1796–1869),
Italian botanist. Perennial herb.
monanthos mon-*ănth*-os. One-flowered.
Corsica, Sardinia.

Morning Glory see *Ipomoea*

Morus mō-rus *Moraceae*. The L. name for
M. nigra. Deciduous trees. Mulberry.
alba ăl-ba. White (the fruits). White
Mulberry. China.
nigra nig-ra. Black (the ripe fruit). Common
Mulberry, Black Mulberry. W Asia.

Mosaic Plant see *Fittonia verschaffeltii*
argyroneura
Mossfern see *Selaginella pallescens*
Mother-in-law's Tongue see *Sansevieria*
trifasciata

Mother of Pearl Plant see *Graptophyllum*
paraguayense
Mother of Thousands see *Kalanchoe*
daigremontiana, *Saxifraga stolonifera*
Mount Atlas Daisy see *Anacyclus depressus*
Mount Wellington Peppermint see
Eucalyptus coccifera
Mountain Ash see *Sorbus aucuparia*
Mountain Avens see *Dryas octopetala*
Mountain Laurel see *Kalmia latifolia*
Mountain Pepper see *Drimys lanceolata*
Mountain Tobacco see *Arnica montana*
Mourning Widow see *Geranium phaeum*
Mouse-tail Plant see *Arisarum proboscideum*
Moutan see *Paeonia suffruticosa*
Mrs Robb's Bonnet see *Euphorbia robbiae*

Muehlenbeckia moo-lan-*bek*-ee-a
Polygonaceae. After Henri Gustave
Muehlenbeck (1789–1845), French
physician and botanist. Deciduous, twining
shrubs.
axillaris ăks-i-*lah*-ris. Axillary, the flowers
are borne in the leaf axils. New Zealand,
Tasmania, Australia.
complexa kom-*pleks*-a. Embraced, the
perianth swells, enclosing the fruit. New
Zealand.
trilobata tri-lo-*bah*-ta. Three-lobed (the
leaves).

Mulberry see *Morus*
Common or Black see *M. nigra*
Paper see *Broussonetia papyrifera*
White see *M. alba*
Mullein see *Verbascum*
Cretan see *V. creticum*
Dark see *V. nigrum*
Moth see *V. blattaria*
Nettle-leaved see *V. chaixii*
Purple see *V. phoeniceum*

Muscari mus-*kah*-ree *Liliaceae*. The Turkish
name. Bulbous herbs. Grape Hyacinth.
armeniacum ar-men-ee-*ah*-kum. Of
Armenia. SE Europe, Caucasus, Turkey.
aucheri *ow*-ka-ree. After P. M. R. Aucher-
Eloy (1792–1838). Turkey.
azureum a-*zew*-ree-um. Sky-blue (the
flowers). Caucasus, NW Turkey.
botryoides bot-ree-*oi*-deez. Like a bunch of
grapes. Europe.
comosum ko-*mō*-sum. With a tuft (of sterile
flowers). Tassel Hyacinth. Europe,
W Asia.

Muscari (continued)
'Plumosum' ploo-*mō*-sum. Feathery (the inflorescence).
latifolium lah-tee-*fo*-lee-um. Broad-leaved. NW Turkey.
macrocarpum măk-rō-*kar*-pum. With large fruit. Greece, W Turkey.
moschatum mos-*kah*-tum. Musk-scented. Musk Hyacinth. W Asia.
neglectum ne-*glek*-tum. (= *M. racemosum*). Overlooked. Europe, N Africa, W Asia, Caucasus.
paradoxum see *Bellevallia paradoxa*
racemosum see *M. neglectum*
tubergenianum tew-ber-gen-ee-*ah*-num. After van Tubergen who introduced it. Oxford and Cambridge Grape Hyacinth. N Iran.

Mutisia mew-*tis*-ee-a *Compositae*. After José Celestino Mutis (1732–1808), Spanish botanist who studied the S American flora. Semi-hardy evergreen climbers. Climbing Gazania.
clematis *kle*-ma-tis. Climbing. Colombia.
decurrens day-*ku*-renz. With the base of the leaf gradually merging with the stem. Chile, Argentina.
ilicifolia ee-li-ki-*fo*-lee-a. *Ilex*-leaved. Chile.
oligodon o-*li*-go-don. Few-toothed. Chile, Argentina.

Myosotidium mee-os-ō-*tid*-ee-um *Boraginaceae*. From *Myosotis* q.v. to which it is related. Evergreen perennial herb.
hortensia hor-*tens*-ee-a. Of gardens (from where it was originally described). Chatham Island Forget-me-not. Chatham Islands.

Myosotis mee-os-ō-*tis* *Boraginaceae*. The Gk. name for another plant, from *mus* (a mouse) and *otos* (an ear) referring to the leaves. Annual, biennial and perennial herbs. Forget-me-not.
alpestris ăl-*pes*-tris. (= *M. rupicola*). Of lower mountains. Europe.
azorica a-*zo*-ri-ka. Of the Azores.
caespitosa kie-spi-*tō*-sa. Tufted. Europe.
scorpioides skor-pee-*oi*-deez. (= *M. palustris*). Like a scorpion's tail (the inflorescence). Europe, Asia.
sylvatica sil-*vă*-ti-ka. Of woods. Europe, Asia.

Myrica mi-*ree*-ka *Myricaceae*. From *myrike* the Gk. name for *Tamarix*. Deciduous and evergreen shrubs.

Myrica (continued)
californica kăl-i-*forn*-i-ka. Of California. Californian Bayberry.
cerifera kay-*ri*-fe-ra. Wax-bearing (the fruits). Wax Myrtle. SE United States.
gale gah-lee. From old English *gagel*. Sweet Gale, Bog Myrtle. Northerly N hemisphere.
pensylvanica pen-sil-*vahn*-i-ka. Of Pennsylvania. Bayberry. E N America.

Myricaria mi-ree-*kah*-ree-a. From Gk. *myrike* (*Tamarix*) which it resembles. Deciduous shrub.
germanica ger-*mahn*-i-ka. (= *Tamarix germanica*). Of Germany. Europe to the Himalaya.

Myriophyllum mi-ree-ō-*fil*-lum *Haloragaceae*. From Gk. *myrios* (many) and *phyllon* (a leaf) referring to the finely divided leaves. Aquatic herbs. Water Milfoil.
aquaticum a-*kwah*-ti-kum. (= *M. proserpinacoides*). Growing in water. S America.
heterophyllum he-te-rō-*fil*-lum. With variable leaves. E North America.
spicatum spee-*kah*-tum. With flowers in spikes. Temperate N hemisphere.
verticillatum ver-ti-ki-*lah*-tum. Whorled (the leaves). Temperate N hemisphere.

Myrrhis *mi*-ris *Umbelliferae*. The Gk. name for a plant. Perennial herb.
odorata o-dō-*rah*-ta. Scented. Sweet Cicely. Europe.

Myrsine *mur*-si-nay *Myrsinaceae*. Gk. name for myrtle. Evergreen shrub.
africana ăf-ri-*kah*-na. African. African Boxwood. Africa, Himalaya, China.

Myrtillocactus mur-ti-lō-*kăk*-tus *Cactaceae*. From L. *myrtillus* (a small myrtle) and *Cactus* q.v. referring to the myrtle-like fruits. Mexico.
geometrizans gee-ō-*met*-ri-zănz. Regularly marked.
schenkii shenk-ee-ee. After Professor H. Schenk, Director of the Darmstadt Botanic Garden.

Myrtle see *Myrtus*
Common see *M. communis*
Tarentum see *M. communis tarentina*
Myrobalan see *Prunus cerasifera*

Myrtus *mur*-tus *Myrtaceae*. The Gk. and L.

Myrtus (continued)
name. Semi-hardy, evergreen trees and
shrubs. Myrtle.
  bullata bu-*lah*-ta. With puckered leaves.
  New Zealand.
  communis kom-*ew*-nis. Common. Common
  Myrtle. W Asia.
    tarentina tă-ren-*teen*-a. Of Taranto,
    S Italy. Tarentum Myrtle.
  luma loo-ma. (= *M. apiculata*). The native
  name. Chile, Argentina.
  nummularia num-ew-*lah*-ree-a. With coin-
  shaped leaves. S South America, Falkland
  Islands.
  ugni un-yee. The native name. Chile.

N

Namaqualand Daisy see *Venidium fastuosum*

Nandina năn-*deen*-a *Berberidaceae*. From
nanten the Japanese name. Evergreen shrub.
  domestica do-*mes*-ti-ka. Cultivated. China.

Narcissus nar-*kis*-us *Amaryllidaceae*. After
Narcissus of Gk. mythology who, it is said,
was turned into this plant after killing himself
because he couldn't reach the person he saw
reflected in a pool. Bulbous perennials.
  asturiensis a-stu-ree-*en*-sis. Of Asturia,
  Spain. N Spain, N Portugal.
  bulbocodium bul-bō-*kō*-dee-um. From Gk.
  bulbos (a bulb) and *kodion* (wool). Hoop-
  petticoat Daffodil. SW Europe, N Africa.
    citrinus ki-*tree*-nus. Lemon-yellow.
    conspicuus kon-*spik*-ew-us. Conspicuous.
    romieuxii see *N. romieuxii*
    tenuifolius ten-ew-i-*fo*-lee-us. Slender-
    leaved.
  canaliculatus hort. see *N. tazetta italicus*
  cantabricus kăn-*tăb*-ri-kus. Of Cantabria,
  Spain. S Spain, N Africa.
    monophyllus mo-nō-*fil*-lus. One-leaved.
    N Africa.
  cyclamineus sik-la-*min*-ee-us. Like
  Cyclamen. NW Spain, NW Portugal.
  jonquilla yong-*kwil*-la. From *junquillo* the
  Spanish name, from *Juncus*, referring to
  the slender leaves. Spain, Portugal.
  juncifolius hort. see *N. requienii*
  minor mi-nor. (= *N. nanus*). Smaller.
  Pyrenees, N Spain.
  nanus see *N. minor*
  × *odorus* o-*dō*-rus. *N. jonquilla* × *N.*
  pseudonarcissus. Scented. Campernelle.
  papyraceus pă-pi-*rah*-kee-us. Paper-like (the

Narcissus (continued)
white flowers). Mediterranean region, SW
Europe.
  poeticus pō-*e*-ti-kus. Of poets. Poet's
  Narcissus. C and S Europe.
    recurvus re-*kur*-vus. Curved back (the
    perianth segments). Pheasant's-eye
    Daffodil.
  pseudonarcissus soo-dō-nar-*kis*-us. False
  Narcissus. Wild Daffodil. W Europe.
  requienii rek-wee-*en*-ee-ee. (= *N. juncifolius*
  hort.). After Requien, see *Mentha*
  requienii. S France, Spain.
  romieuxii rom-*ew*-ee-ee. (= *N. bulbocodium*
  romieuxii). After Romieux of Geneva who
  grew it. Morocco.
  rupicola roo-*pi*-ko-la. Growing on rocks.
  Spain, Portugal.
  tazetta ta-*ze*-ta. Italian name meaning a
  small cup. Mediterranean region.
    italicus ee-*tă*-li-kus. (= *N. canaliculatus*
    hort.). Italian. N and E Mediterranean
    region.
  triandrus tree-*ăn*-drus. With three stamens
  (three are larger than the other three).
  Angel's Tears. Portugal, Spain.
    concolor see *N. triandrus pallidulus*
    pallidulus pa-*lid*-ew-lus. (= *N. triandrus*
    concolor). Rather pale. Golden Angels
    Tears.
  viridiflorus vi-ri-di-*flō*-rus. With green
  flowers. SW Spain, Morocco.
  watieri wo-tee-*e*-ree. After M. Watier,
  Inspector of Woods and Forests in
  Morocco c 1920. Atlas Mountains.

Nasturtium nas-*tur*-tee-um *Cruciferae*. From
L. *nasus tortus* (a twisted nose) referring to
the smell of the leaves. Aquatic perennial
herb.
  officinale o-fi-ki-*nah*-lee. Sold as a herb.
  Watercress. Europe.

Nasturtium see *Tropaeolum majus*
Native's Comb see *Pachycereus pecten-
aboriginum*
Navelwort see *Omphalodes*
  Venus's see *O. linifolia*
Neanthe bella see *Chamaedorea elegans*

Nectaroscordum nek-ta-rō-*skor*-dum
Liliaceae. From Gk. *nektar* (nectar) and
skordon (garlic). Perennial herb.
  siculum sik-ew-lum. (= *Allium siculum*). Of
  Sicily. S Europe, Turkey.

Neillia *neel*-ee-a *Rosaceae*. After Patrick
Neill (1776–1851), Scottish naturalist.

Neillia (continued)
Deciduous shrubs.
  sinensis si-*nen*-sis. Of China.
  thibetica ti-*be*-ti-ka. (= *N. longeracemosa*).
  Of Tibet. China.

Nelumbo ne-*lum*-bō *Nymphaeaceae*. The
Sinhalese name. Aquatic perennial herbs.
Lotus.
  lutea loo-tee-a. Yellow (the flowers).
  American Lotus. E North America.
  nucifera new-*ki*-fe-ra. Nut-bearing. Indian
  Lotus, Sacred Lotus. S Asia to Australia.

Nemesia ne-*me*-see-a *Scrophulariaceae*. From
nemesion the Gk. name for a similar plant.
Annual herbs. S Africa.
  strumosa stroo-*mō*-sa. With cushion-like
  swellings.
  versicolor ver-*si*-ko-lor. Variously coloured.
    'Compacta' kom-păk-ta. (= *N. compacta*
    hort.). Compact.

Nemophila ne-*mo*-fi-la *Hydrophyllaceae*.
From Gk. *nemos* (a glade) and *phileo* (to love),
they grow in shady places. Annual herbs.
Baby Blue Eyes. California.
  maculata măk-ew-*lah*-ta. Spotted (the
  corolla).
  menziesii men-zeez-ee-ee. (= *N. insignis*).
  After Menzies, see *Menziesia*.

Neobuxbaumia polylopha see *Cephalocereus
polylophus*

Neolitsea nee-ō-*lit*-see-a *Lauraceae*. From
Gk. *neos* (new) and *Litsea*, a related genus
(from the Japanese name). Semi-hardy,
evergreen shrub or tree.
  sericea say-*ri*-kee-a. Silky (the young
  growths). Japan, Korea, China.

Neolloydia nee-ō-*loyd*-ee-a *Cactaceae*. After
Francis Ernest Lloyd (1868–1947),
American botanist, the name *Lloydia* having
already been used for a genus of bulbous
herbs.
  grandiflora grănd-i-*flō*-ra. Large-flowered.
  N Mexico.
  conoidea kon-ō-*i*-dee-a. (= *N. texensis*).
  Cone-like. Texas, N Mexico.
  macdowellii mak-*dowl*-ee-ee. (=
  Echinomastus macdowellii). After
  Mcdowell, a plant exporter of Mexico.
  N Mexico.

Neoporteria nee-ō-por-*te*-ree-a *Cactaceae*.
After Carlos Porter, a Chilean entomologist.

Neoporteria (continued)
Chile.
  fusca fus-ka. Dark (the spines).
  subgibbosa sub-gi-*bō*-sa. Somewhat swollen
  on one side.
  villosa vil-*lō*-sa. Softly hairy.

Neoregelia nee-ō-ray-*gel*-ee-a *Bromeliaceae*.
After Eduard Albert von Regel (1815–92).
Tender, evergreen herbs. Brazil.
  carolinae kă-ro-*leen*-ie. (= *N. marechallii*).
  After Carolina. Blushing Bromeliad.
    'Tricolor' *tri*-ko-lor. Three-coloured (the
    leaves).
  marmorata mar-mo-*rah*-ta. Marbled (the
  leaves). Marble Plant.
  sarmentosa sar-men-*tō*-sa. Creeping.
  spectabilis spek-*tah*-bi-lis. Spectacular.
  Finger-nail Plant.

Nepenthes nay-*pen*-theez *Nepenthaceae*. Gk.
name of a plant. Tender herbs. Pitcher
Plant.
  gracilis gră-ki-lis. Graceful. SE Asia.
  khasiana kah-zee-*ah*-na. Of the Khasi Hills,
  Assam.
  maxima mahk-si-ma. Larger. Borneo to
  New Guinea.
  mirabilis mee-*rah*-bi-lis. Wonderful. SE
  Asia.
  rafflesiana răf-alz-ee-*ah*-na. After Sir
  Thomas Stamford Raffles (1781–1826),
  scientific patron and founder of Singapore.
  SE Asia.
  sanguinea sang-*gwin*-ee-a. Blood-red (the
  pitchers). Malaysia.
  ventricosa ven-tri-*kō*-sa. Swollen on one side
  (the pitchers). Philippines.

Nepeta ne-pe-ta *Labiatae*. The L. name.
Perennial herbs.
  cataria ka-*tah*-ree-a. Of cats, which are
  attracted to it. Catmint, Catnip. Europe,
  Asia.
  × *faassenii* fah-*sen*-ee-ee. *N. mussinii* × *N.
  nepetella*. After J. H. Faassen, Dutch
  nurseryman.
  grandiflora grănd-i-*flō*-ra. Large-flowered.
  Caucasus.
  mussinii mu-*sin*-ee-ee. After Puschkin, see
  Puschkinia. Caucasus.
  nepetella ne-pe-*tel*-la. Diminutive of *Nepeta*.
  SW Europe.
  nervosa ner-*vō*-sa. Conspicuously veined.
  Kashmir.

Nephrolepis nef-rō-*lep*-is *Oleandraceae*.
From Gk. *nephros* (a kidney) and *lepis* (a scale)

Nephrolepsis (continued)
referring to the shape of the indusium.
Tender ferns. Sword Fern.
  cordifolia kor-di-*fo*-lee-a. With heart-
  shaped leaves. Tropics and sub-tropics.
  exaltata eks-al-*tah*-ta. Very tall. Tropics.
    'Bostoniensis' bos-ton-ee-*en*-sis. Of
    Boston. Boston Fern.

Nerine nay-*ree*-nay *Amaryllidaceae*. After
Nerine, a sea nymph. Semi-hardy, Bulbous
herbs. S Africa.
  bowdenii bow-*den*-ee-ee. After Mr
  Athelston Bowden who introduced it.
  filifolia fee-li-*fo*-lee-a. With thread-like
  leaves.
  flexuosa fleks-ew-*ō*-sa. Wavy (the perianth
  lobes).
  sarniensis sar-nee-*en*-sis. Of Guernsey
  (Sarnia), where it has long been
  naturalised.
  undulata un-dew-*lah*-ta. Wavy (the perianth
  lobes).

Nerium *nay*-ree-um *Oleaceae*. The Gk.
name. Tender evergreen, poisonous shrub.
  oleander o-lee-*ăn*-der. From *oleandra*, the
  Italian name. Oleander. Mediterranean
  region to E Asia.

Nertera *ner*-te-ra *Rubiaceae*. From Gk.
nerteros (lowly) referring to the dwarf habit.
Tender herb.
  granadensis grăn-a-*den*-sis. (= *N. depressa*).
  Of Granada, Colombia. Bead Plant.
  S America and perhaps the same species in
  New Zealand and Tasmania.

Nettle Tree see *Celtis*
Never-never Plant see *Ctenanthe
oppenheimiana*
New Zealand Burr see *Acaena*
New Zealand Daisy see *Celmisia*
New Zealand Flax see *Phormium tenax*
New Zealand Lilac see *Hebe hulkeana*

Nicandra ni-*kăn*-dra *Solanaceae*. After
Nikander of Colophon, Greek physician and
poet c 137 B.C. Annual herb.
  physalodes fi-sa-*lō*-deez. Like *Physalis*.
  Apple of Peru. Peru.

Nicotiana nee-kō-tee-*ah*-na *Solanaceae*.
After Jean Nicot (1530–1600) who
introduced the tobacco plant to France.
Annual herbs.
  alata ah-*lah*-ta. (= *N. affinis*). Winged (the
  petioles). S America.

Nicotiana (continued)
  glauca glow-ka. Glaucous (the shoots and
  leaves). Bolivia, Argentina.
  × *sanderae* sahn-da-rie. *N. alata* × *N.
  forgetiana*. After Mrs Sander.
  sylvestris sil-*ves*-tris. Of woods. Argentina.
  tabacum ta-*băk*-um. Said to be the
  Caribbean name for a pipe or from Haitian
  taina, a roll of tobacco in a maize leaf.
  Tobacco Plant. Cult.

Nidularium need-ew-*lah*-ree-um
Bromeliaceae. From L. *nidus* (a nest), the
flowers are borne in a nest-like depression in
the centre of a cluster of bracts. Tender,
evergreen herbs. Brazil.
  fulgens ful-gens. Shining (the bracts).
  innocentii in-o-*sent*-ee-ee. Said to be named
  after Pope Innocenti.
    striatum stree-*ah*-tum. (= *N. striatum*).
    Striped (the leaves).
  purpureum pur-*pewr*-ree-um. Purple (the
  leaves).
  rutilans ru-ti-lănz. Reddish (the flowers).

Nierembergia nee-e-ram-*berg*-ee-a
Solanaceae. After Juan Eusebio Nieremberg
(1595–1658), a Spanish Jesuit. Perennial
herbs. Cup Flower.
  hippomanica hip-o-*măn*-i-ka. From Gk.
  hippomanes, a plant that drives horses mad
  or they love to eat. Argentina.
    violacea vee-o-*lah*-kee-a. (= *N. caerulea*).
    Violet (the flowers).
  repens ree-pens. (= *N. rivularis*). Creeping.
  S America.

Nigella ni-*gel*-la *Ranunculaceae*. Diminutive
of L. *niger* (black) referring to the black
seeds. Annual herbs.
  arvensis ar-*ven*-sis. Of fields. N Africa,
  Europe, W Asia.
  damascena dăm-a-*skay*-na. Of Damascus.
  Love-in-a-mist. S Europe, N Africa.
  hispanica his-*pah*-ni-ka. Spanish. Spain,
  Portugal.
  integrifolia in-teg-ri-*fo*-lee-a. With entire
  (lower) leaves. Turkestan.
  orientalis o-ree-en-*tah*-lis. Eastern. SW
  Asia.
  sativa sa-*teev*-a. Cultivated (the seeds are
  used in seasoning). SE Europe, W Asia.

Ninebark see *Physocarpus*
Nirre see *Nothofagus antarctica*

Nolana nō-*lah*-na *Nolanaceae*. From L. *nola*
(a small bell) referring to the shape of the

Nolana (continued)
corolla. Perennial herbs. Chilean Bellflower.
Chile.
acuminata a-kew-mi-*nah*-ta. Long-pointed.
paradoxa pă-ra-*doks*-a. Unusual.

Nomocharis no-mō-*kă*-ris *Liliaceae*. From
Gk. *nomos* (a meadow) and *charis* (grace).
Bulbous perennials.
aperta a-*per*-ta. Closed. China.
farreri fă-ra-ree. After Farrer, see *Viburnum
farreri*. Burma.
mairei mair-ree-ee. After Maire, see
Incarvillea mairei. China.
pardanthina par-dan-*theen*-a. Like
Pardanthina (*Belamcanda*), the spotted
flowers. China.
saluenensis săl-ew-en-*en*-sis. From near the
Nu Jiang (Salween River), W Yunnan.
Burma, China, Tibet.

Nopalxochia nō-pal-*ho*-kee-a *Cactaceae*.
From a Mexican name. Mexico.
ackermannii ă-ker-*măn*-ee-ee. (=
Epiphyllum ackermannii). After Georg
Ackermann who introduced it. Orchid
Cactus.
phyllanthoides fil-lanth-*oi*-deez. Like
Phyllanthus.

Norfolk Island Pine see *Araucaria
heterophylla*

Nothofagus no-thō-*fah*-gus *Fagaceae*. From
Gk. *nothos* (false) and *fagus* (beech) but
notofagus (southern beech) may have been
intended. Deciduous and evergreen trees.
Southern Beech.
antarctica ăn-*tark*-ti-ka. Of Antarctic
regions. Nirre. Southern S America.
betuloides bet-ew-*loi*-deez. Like *Betula*.
S Chile, S Argentina.
dombeyi dom-bee-ee. After Dombey, see
Dombeya. Chile, Argentina.
fusca fus-ka. Brown. New Zealand.
menziesii men-*zeez*-ee-ee. After Menzies
who collected the type specimen, see
Menziesia. New Zealand.
obliqua o-*blee*-kwa. Oblique (the leaf base).
Roblé. Chile, Argentina.
procera prō-*kay*-ra. Tall. Rauli. Chile,
Argentina.
solandri so-*lăn*-dree. After Daniel Carl
Solander (1736–82), a botanist on Cook's
first voyage. New Zealand.

Notocactus no-tō-*kăk*-tus *Cactaceae*. From
Gk. *notos* (southern) and *Cactus* q.v.

Notocactus (continued)
referring to their southerly distribution.
apricus a-*pree*-kus. Sun-loving. Uruguay.
concinnus kon-*kin*-us. Elegant. S Brazil,
Uruguay.
haselbergii hah-zal-*berg*-ee-ee. After Dr von
Haselberg of Stralsind, a cactus grower.
Scarlet Ball Cactus. S Brazil.
leninghausii len-ing-*howz*-ee-ee. After
Leninghaus. Golden Ball Cactus. Brazil.
mammulosus măm-ew-*lō*-sus. Bearing
nipples. S America.
muricatus mew-ri-*kah*-tus. Rough with
spines. S Brazil.
ottonis o-*tō*-nis. After Friedrich Otto
(1782–1856), curator of Berlin Botanic
Garden. SE South America.
scopa skō-pa. Broom-like. Silver Ball
Cactus. S Brazil, Uruguay.
submammulosus sub-măm-ew-*lō*-sus. With
small nipples. N Argentina.
tabularis tăb-ew-*lah*-ris. Flat-topped.
Brazil, Uruguay.

Notospartium no-tō-*spar*-tee-um
Leguminosae. From Gk. *notos* (southern) and
Spartium q.v. Semi-hardy shrub.
carmichaeliae kar-mie-*keel*-ee-ie. From the
resemblance to *Carmichaelia*. New
Zealand.

Nuphar new-far *Nymphaeaceae*. From the
Arabic name. Aquatic perennial herbs.
advena ad-*ven*-a. Adventive. Spatterdock.
E and C United States.
lutea loo-tee-a. Yellow. Brandy Bottle,
Yellow Water Lily. N hemisphere.
pumila pew-mi-la. Dwarf. Europe, Asia.

Nyctocereus nik-tō-*kay*-ree-us *Cactaceae*.
From Gk. *nyktos* (night) and *Cereus* q.v.,
they flower at night.
chontalensis chon-ta-*len*-sis. From the
territory of the Chontal Indians, Oaxaca.
S Mexico.
serpentinus ser-pen-*teen*-us. Snake-like.
Mexico.

Nymphaea nimf-*ie*-a *Nymphaeaceae*. The
classical name after Nymphe, a water
nymph. Aquatic perennial herbs.
alba ăl-ba. White. White Water Lily.
Europe, N Africa, Asia.
caerulea kie-*ru*-lee-a. Deep blue (the
flowers). Blue Egyptian Lotus. N and
C Africa.
candida kăn-di-da. White. N Europe,
N Asia.

Nymphaea (continued)
capensis ka-*pen*-sis. Of the Cape of Good Hope. Cape Blue Water Lily. S Africa.
lotus lō-tus. A Gk. name for several plants. White Egyptian Lotus. Egypt.
× *marliacea* mar-lee-ă-kee-a. After Joseph Latour Marliac (born 1830).
odorata o-dō-*rah*-ta. Scented. E United States.
pygmaea see *N. tetragona*
stellata ste-*lah*-ta. Star-like. S and E Asia.
tetragona tet-ra-*gōn*-a. (= *N. pygmaea*). Four-angled. NE Asia, N America.
tuberosa tew-be-rō-sa. Tuberous. N America.

Nymphoides nimf-*oi*-deez *Gentianaceae*. Like *Nymphaea*. Aquatic perennial herb.
peltata pel-*tah*-ta. With the petiole attached to the lower surface of the leaf blade, literally, shield-like. Europe, Asia.

Nyssa *ni*-sa *Nyssaceae*. After Nyssa (Nysa) a water nymph, the first-described species, *N. aquatica* grows in swamps. Deciduous trees.
sinensis si-*nen*-sis. Of China.
sylvatica sil-*vă*-ti-ka. Of woods. Black Gum, Tupelo. EN America.

O

Oak see *Quercus*
  Black see *Q. velutina*
  Black Jack see *Q. marilandica*
  Common see *Q. robur*
  Cork see *Q. suber*
  Daimio see *Q. dentata*
  Durmast see *Q. petraea*
  Golden, of Cyprus see *Q. alnifolia*
  Holm see *Q. ilex*
  Hungarian see *Q. frainetto*
  Lebanon see *Q. libani*
  Lucombe see *Q.* × *hispanica* 'Lucombeana'
  Pedunculate see *Q. robur*
  Pin see *Q. palustris*
  Red see *Q. rubra*
  Scarlet see *Q. coccinea*
  Sessile see *Q. petraea*
  Shingle see *Q. imbricaria*
  Tanbark see *Lithocarpus densiflorus*
  Turkey see *Q. cerris*
Oat Grass see *Arrhenatherum elatius*
Obedient Plant see *Physostegia virginiana*
Ocean Spray see *Holodiscus discolor*

Ocimum ō-ki-mum *Labiatae*. From *okimon* the Gk. name for an aromatic herb. Annual herbs. SE Asia.
basilicum ba-si-li-kum. The classical name, meaning royal or princely. Basil.
minimum *mi*-ni-mum. Smaller. Bush Basil.

Oconee Bells see *Shortia galacifolia*

× **Odontioda** o-don-tee-ō-da *Orchidaceae*. Intergeneric hybrids, from the names of the parents. *Cochlioda* × *Odontoglossum*. Greenhouse orchids.

× **Odontocidium** o-don-to-*kid*-ee-um *Orchidaceae*. Intergeneric hybrids, from the names of the parents. *Odontoglossum* × *Oncidium*. Greenhouse orchids.

Odontoglossum o-don-to-*glos*-um *Orchidaceae*. From Gk. *odontos* (a tooth) and *glossa* (a tongue) referring to the toothed lip. Greenhouse orchids.
bictoniense bik-ton-ee-*en*-see. Of Bicton, Devon. Mexico, C America.
cervantesii ser-văn-*tes*-ee-ee. After Professor Vincentio Cervantes. Mexico, Guatemala.
citrosmum see *O. pendulum*
cordatum kor-*dah*-tum. Heart-shaped (the lip). Mexico, Guatemala.
crispum *kris*-pum. Wavy-edged (the petals). Lace Orchid. Colombia.
cristatum kris-*tah*-tum. Crested (the disc). Colombia, Ecuador.
grande *grăn*-dee. Large. Tiger Orchid. Mexico, Guatemala.
harryanum hă-ree-*ah*-num. After Sir Harry Veitch. Colombia, Peru.
laeve *lie*-vee. Smooth. Mexico, Guatemala.
nobile nō-bi-lee. (= *O. pescatorei*). Notable. Colombia.
pendulum *pen*-dew-lum. (= *O. citrosmum*). Pendulous (the inflorescence). Mexico, Guatemala.
pescatorei see *O. nobile*
pulchellum pul-*kel*-um. Pretty. Lily of the Valley Orchid. Guatemala.
rossii ros-ee-ee. After John Ross, who collected orchids in Mexico in the 19th century. Mexico, Guatemala.
schlieperianum shlee-pa-ree-*ah*-num. After Adolph Schlieper, an orchid collector. Costa Rica, Panama.
triumphans tree-*um*-fanz. Splendid. Colombia.
uroskinneri ew-rō-*ski*-na-ree. After Skinner, see *Cattleya skinneri*. Guatemala.

× *Odontonia* o-don-*ton*-ee-a *Orchidaceae*.
Intergeneric hybrids, from the names of the
parents. *Miltonia* × *Odontoglossum*.
Greenhouse orchids.

Oenothera oy-nō-*the*-ra *Onagraceae*. From
oinotheras the Gk. name of a plant. Biennial
and perennial herbs.
   acaulis a-*kaw*-lis. (= *O. taraxacifolia*).
   Stemless. Chile.
   berlandieri ber-lăn-dee-*e*-ree. (= *O. speciosa
   childsii*). After J. L. Berlandier (died
   1851), German botanist who explored
   Texas and Mexico. Texas, Mexico.
   biennis bee-*en*-is. Biennial. Evening
   Primrose. E N America.
   caespitosa kie-spi-*tō*-sa. Tufted. W N
   America.
      riparia see *O. tetragona riparia*
   erythosepala e-rith-rō-*se*-pa-la. (= *O.
   lamarckiana*). With red sepals. Cult.
   fruticosa froo-ti-*kō*-sa. Shrubby (which it
   isn't). Sundrops. E United States.
   grandiflora grănd-i-*flō*-ra. Large-flowered.
   Alabama.
   laciniata la-kin-ee-*ah*-ta. Deeply cut (the
   leaves). United States.
   lamarckiana see *O. erythrosepala*
   missouriensis mi-sur-ree-*en*-sis. Of Missouri.
   S Central United States.
   perennis pe-*ren*-is. Perennial. E N America.
   speciosa spe-kee-*ō*-sa. Showy. S Central
   United States.
      childsii see *O. berlandieri*
   stricta strik-ta. Erect. S America.
   taraxacifolia see *O. acaulis*
   tetragona tet-ra-*gō*-na. Four-angled.
   E United States.
      riparia ree-*pah*-ree-a. (= *O. caespitosa
      riparia*). Of river banks. SE United
      States.

Okra see *Abelmoschus esculentus*
Old Maid see *Catharanthus roseus*
Old Man's Beard see *Clematis vitalba*
Old Woman see *Artemisia stelleriana*

Olea o-lee-a *Oleaceae*. The L. name. Semi-
hardy, evergreen tree.
   europaea oy-rō-*pie*-a. European. Olive. SW
   Asia.

Oleander see *Nerium oleander*

Olearia o-lee-*ah*-ree-a *Compositae*. After
Adam Ölschläger (Olearius) (1603–71).

Olearia (continued)
Evergreen shrubs. Daisy Bush.
   avicenniifolia ă-vi-sen-ee-i-*fo*-lee-a. With
   leaves like *Avicennia*. New Zealand.
   chathamica cha-*tăm*-i-ka. Of the Chatham
   Islands.
   frostii *frost*-ee-ee. After Charles Frost.
   Victoria (Australia).
   gunniana see *O. phlogopappa*
   × *haastii* hahst-ee-ee. *O. avicenniifolia* × *O
   moschata*. After Sir Johann Franz Julius
   von Haast (1824–87) who collected the type
   specimen. New Zealand.
   ilicifolia ee-li-ki-*fo*-lee-a. *Ilex*-leaved. New
   Zealand.
   insignis in-*sig*-nis. (= *Pachystegia insignis*).
   Notable. New Zealand.
   macrodonta măk-rō-*don*-ta. With large teeth
   (the leaves). New Zealand.
   × *mollis* mol-lis. *O. ilicifolia* × *O. lacunosa*
   Softly hairy. The plant commonly grown
   under this name is of the parentage *O.
   ilicifolia* × *O. moschata*.
   nummulariifolia num-ew-lah-ree-i-*fo*-lee-a.
   With coin-shaped leaves. New Zealand.
   phlogopappa flog-ō-*pă*-pa. (= *O. gunniana*.
   O. stellulata hort.). With a *Phlox*-like
   pappus. SE Australia, Tasmania.
      Splendens *splen*-denz. (= *O. gunniana*
      'Splendens'. *O. stellulata* 'Splendens').
      Splendid.
   × *scilloniensis* si-lon-ee-*en*-sis. *O. lirata* ×
   O. phlogopappa. Of the Scilly Isles, it was
   raised at Tresco.
   semidentata se-mee-den-*tah*-ta. Half
   toothed, the leaves are toothed in the
   upper half. Chatham Islands.
   stellulata 'Splendens' see *O. phlogopappa*
   Splendens
   traversii tra-*vers*-ee-ee. After W. T. L.
   Travers (1819–1903). Chatham Islands.
   virgata vir-*gah*-ta. Twiggy. New Zealand.
   'Waikariensis' wie-kah-ree-*en*-sis. Of
   Waikari, New Zealand.
   'Zennorensis' zen-o-*ren*-sis. *O. ilicifolia* ×
   O. lacunosa. Of Zennor, Cornwall.

Olive see *Olea europaea*

Omphalodes omf-a-*lō*-deez *Boraginaceae*.
From Gk. *omphalos* (a navel) referring to a
navel-like depression in the seeds. Perennial
herbs. Navelwort.
   cappadocica kăp-a-*do*-ki-ka. Of Cappadocia
   (Turkey). W Asia.
   linifolia lee-ni-*fo*-lee-a. *Linum*-leaved.
   Venus's Navelwort. SW Europe.

Omphalodes (continued)
 luciliae loo-*sil*-ee-ie. After Lucile Boissier.
 Greece, W Asia.
 verna ver-na. Of Spring (flowering). Blue-
 eyed Mary. SE Europe.

Oncidium ong-*kid*-ee-um *Orchidaceae*. From
Gk. *onkos* (a tumour) referring to a swelling
on the lip. Greenhouse orchids.
 altissimum ăl-*tis*-i-mum. Tallest. W Indies.
 aureum ow-ree-um. Golden (the lip). Peru.
 cavendishianum kă-van-dish-ee-*ah*-num.
 After William George Spencer Cavendish,
 6th Duke of Devonshire. Mexico,
 Guatemala.
 cebolleta see *O. longifolium*
 cheirophorum kay-*ro*-fo-rum. Hand-bearing.
 Colombia Buttercup. Costa Rica, Panama.
 concolor kon-ko-lor. Similarly coloured (the
 petals and sepals). Brazil.
 crispum kris-pum. Finely wavy (the petals).
 Brazil.
 cucullatum ku-kew-*lah*-tum. Hood-like.
 Colombia, Ecuador.
 flexuosum fleks-ew-ō-sum. Tortuous.
 Dancing Doll Orchid. Brazil.
 incurvum in-*kur*-vum. Incurved. Mexico.
 longifolium long-gi-*fo*-lee-um. (= *O.
 cebolleta*). Long-leaved. Tropical America.
 longipes long-gi-pays. Long-stalked. Brazil.
 luridum loo-ri-dum. Pale yellow. Tropical
 America.
 macranthum ma-*krănth*-um. Large-
 flowered. Ecuador.
 marshallianum mar-shăl-ee-*ah*-num. After
 Mr W Marshall of Enfield who grew the
 type specimen. Brazil.
 ornithorhyncum or-ni-thō-*ring*-kum. Like a
 bird's beak. Mexico to S America.
 papilio pah-*pi*-lee-ō. A butterfly. Butterfly
 Orchid. S America.
 phalaenopsis fă-lie-*nop*-sis. Moth-like.
 Colombia, Ecuador.
 pulchellum pul-*kel*-um. Pretty. Jamaica,
 Guyana.
 pumilum pew-mi-lum. Dwarf. Brazil.
 pusillum pu-*sil*-um. Dwarf. C and
 S America.
 sarcodes sar-*kō*-deez. Flesh-like. Brazil.
 splendidum splen-di-dum. Splendid.
 Guatemala.
 triquetrum tri-*kwee*-trum. Three-angled (the
 pseudobulbs). Jamaica.
 varicosum vah-ri-*kō*-sum. With dilated
 veins. Brazil.
 wentworthianum went-wurth-ee-*ah*-num.
 After Lord Fitzwilliam. Guatemala.

Onion see *Allium cepa*
 Tree see *A. cepa* Proliferum
 Welsh see *A. fistulosum*

Onoclea o-*nok*-lee-a *Athyriaceae*. From
onokleia, Gk. name for another plant, from
onos (a vessel) and *kleio* (to close), the
pinnules of the fertile fronds curl round the
sori, enclosing them. Fern.
 sensibilis sen-*si*-bi-lis. Sensitive (to early
 frosts). Sensitive Fern. Temperate N
 hemisphere.

Ononis o-*nō*-nis *Leguminosae*. The Gk. name.
Perennial herbs and sub-shrubs.
 aragonensis ă-ra-gon-*en*-sis. Of Aragon, NE
 Spain. Pyrenees, Spain, N Africa.
 fruticosa froo-ti-*kō*-sa. Shrubby. SW
 Europe.
 rotundifolia ro-tun-di-*fo*-lee-a. With
 rounded leaves (leaflets). C and SW
 Europe.

Onopordum o-nō-*por*-dum *Compositae*. From
onopordon, the Gk. name. Biennial herbs.
 acanthium a-kănth-ee-um. Spiny. Scotch
 Thistle, Europe, W Asia.
 nervosum ner-vō-sum. (= *O. arabicum*
 hort.). Veined (the undersides of the leaves).
 Spain, Portugal.
 tauricum tow-ri-kum. Of the Crimea.
 Balkan peninsula, Black Sea region.

Onosma o-*nos*-ma *Boraginaceae*. From Gk.
onos (an ass) and *osme* (smell) referring to the
roots. Perennial herbs.
 alboroseum ăl-bō-*ros*-ee-um. White and
 rose-coloured, the flowers change from
 white to red. W Asia.
 echioides e-kee-*oi*-deez. Like *Echium*. Italy,
 SE Europe.
 stellulatum stel-ew-*lah*-tum. With small
 stars, referring to the star-shaped hairs.
 Yugoslavia.
 tauricum tow-ri-kum. Of the Crimea.
 Golden Drop. SE Europe, W Asia.

Ophiopogon o-fee-ō-*pō*-gon *Liliaceae*. From
Gk. *ophis* (a snake) and *pogon* (a beard).
Perennial herbs with grass-like foliage.
 japonicus ja-*pon*-i-kus. Of Japan. Japan,
 Korea.
 planiscapus plahn-i-*skah*-pus. With a flat
 scape. Japan.
  'Nigrescens' ni-*gres*-enz. Blackish (the
  leaves).

Ophrys *of*-ris *Orchidaceae*. Gk. name for an orchid. Hardy, terrestrial orchids.
　apifera a-*pi*-fe-ra. Bee-bearing, the flowers resemble bees. Bee Orchid. Europe.
　fusca *fus*-ka. Brown. Mediterranean region.
　speculum *spek*-ew-lum. A mirror, from the mirror-like blue spot on the disc. Mediterranean region.

Oplismenus op-*lis*-men-us *Gramineae*. From Gk. *hoplismos* (a weapon) referring to the awns. Tender grass.
　hirtellus hir-*tel*-us. Rather hairy. Basket Grass. Texas to S America.

Opuntia o-*pun*-tee-a *Cactaceae*. Gk. name of a plant that grew near Opus (Opuntis) in Ancient Greece. Prickly Pear.
　articulata ar-tik-ew-*lah*-ta. Jointed (the stem). Argentina.
　　'Diademata' dee-a-day-*mah*-ta. (= *O. diademata*). Crowned.
　basilaris bă-si-*lah*-ris. Basal, it branches from the base. SW United States, N Mexico.
　bergeriana ber-ga-ree-*ah*-na. After Alwyn Berger (1871–1931). Cult.
　bigelovii big-a-*lov*-ee-ee. After Jacob Bigelow (1787–1879). SW United States, N Mexico.
　brasiliensis bra-zil-ee-*en*-sis. (= *Brasiliopuntia brasiliensis*). Of Brazil.
　compressa see *O. humifusa*
　cylindrica si-*lin*-dri-ka. (= *Austrocylindropuntia cylindrica*). Cylindrical (the stem). Cane Cactus. Ecuador, Peru.
　decumbens day-*kum*-benz. Prostrate. Mexico, Guatemala.
　diademata see *O. articulata* 'Diademata'
　ficus-indica *fee*-kus-*in*-di-ka. Fig of India. Indian Fig Cactus. Cult.
　humifusa hum-i-*few*-sa. (= *O. compressa*). Low-growing. United States.
　humilis see *O. tuna*
　imbricata im-bri-*kah*-ta. Densely overlapping. Chain-link Cactus. SW United States. Mexico.
　kleiniae klien-ee-ie. From the resemblance to *Kleinia*. SW United States.
　leucotricha loo-*ko*-tri-ka. With white hairs. Mexico.
　macrorhiza măk-rō-*ree*-za. With large roots. United States.
　microdasys mik-rō-*dăs*-is. From Gk. *mikros* (small) and *dasys* (shaggy) referring to the small areoles. N Mexico.

Opuntia (continued)
　　'Albispina' ăl-bi-*speen*-a. White-spined.
　rufida roo-fi-da. (= *O. rufida*). Reddish (the spines). Cinnamon Cactus. Texas, N Mexico.
　monacantha see *O. vulgaris*
　ovata ō-*vah*-ta. Ovate (the fruit). Andes.
　polyacantha po-lee-a-*kănth*-a. Many-spined. W N America.
　robusta rō-*bus*-ta. Robust. Mexico.
　rufida see *O. microdasys rufida*
　salmiana săl-mee-*ah*-na. After Prince Joseph Salm-Reifferscheid-Dyck (1773–1861). German authority on succulents. Brazil, Argentina.
　scheeri *shear*-ree. After Frederick Scheer (c. 1792–1868) an amateur botanist who grew cacti. Mexico.
　subulata soo-bew-*lah*-ta. Awl-shaped (the leaves). Eve's Pin Cactus. Argentina.
　sulphurea sul-*fu*-ree-a. Sulphur-yellow (the flowers). S America.
　tomentosa tō-men-*tō*-sa. Hairy. Mexico.
　tuna *too*-na. (= *O. humilis*). Mexican name for the *Opuntia* fruit. Jamaica.
　tunicata tun-i-*kah*-ta. Coated, the white, papery sheaths on the spines. Texas to Chile.
　verschaffeltii vair-sha-*felt*-ee-ee. After Verschaffelt. N Bolivia.
　vestita ves-*tee*-ta. Clothed with hairs (the areoles). Cotton-pole Cactus. Bolivia.
　vulgaris vul-*gah*-ris. (= *O. monacantha*). Common. Brazil, Argentina.

Orache see *Atriplex hortensis*
Orange, Bitter see *Citrus aurantium*
　Seville see *C. aurantium*
　Sweet see *C. sinensis*
Orchid, Bee see *Ophrys apifera*
　Black see *Coelogyne pandurata*
　Butterfly see *Oncidium papilio*
　Common Spotted see *Dactylorhiza fuchsii*
　Cradle see *Anguloa*
　Dancing Doll see *Oncidium flexuosum*
　Early Purple see *Orchis mascula*
　Fox-tail see *Aerides*
　Lace see *Odontoglossum crispum*
　Lady's Slipper see *Cypripedium*
　Lily of the Valley see *Odontoglossum pulchellum*
　Meadow see *Dactylorhiza incarnata*
　Moth see *Phalaenopsis*
　Pansy see *Miltonia*
　Ram's Head Lady's Slipper see *Cypripedium arietinum*
　Slipper see *Paphiopedilum*
　Soldier see *Orchis militaris*

Orchid (continued)
  Star of Bethlehem see *Angraecum sesquipedale*
  Tiger see *Odontoglossum grande*
Orchid Bush see *Bauhinia acuminata*
Orchid Tree, Purple see *Bauhinia variegata*

Orchis or-kis *Orchidaceae*. The Gk. name. Hardy terrestrial orchids.
  elata see *Dactylorhiza elata*
  foliosa see *Dactylorhiza foliosa*
  fuchsii see *Dactylorhiza fuchsii*
  incarnata see *Dactylorhiza incarnata*
  maderensis see *Dactylorhiza foliosa*
  mascula mahs-kew-la. Male, compared to less robust 'female' species. Early Purple Orchid. Europe, N Africa, N and W Asia.
  militaris mee-li-*tah*-ris. Like a soldier. Soldier Orchid. Europe. W Asia, Siberia.
  praetermissa see *Dactylorhiza majalis praetermissa*
  spectabilis spek-*tah*-bi-lis. Spectacular. E N America.

Oregon Grape see *Mahonia aquifolium*

Origanum o-ree-*gah*-num *Labiatae*. The Gk. name. Perennial herbs.
  amanum a-*mah*-num. Of the Amanus Mts., S Turkey.
  dictamnus dik-*tăm*-nus. The Gk. name. Cretan Dittany. Crete.
  × *hybridum* *hib*-ri-dum. *O. dictamnus* × *O. sipyleum*. Hybrid.
  libanoticum li-ba-*no*-ti-kum. Of Lebanon.
  marjorana mar-jo-*rah*-na. An old name from the Gk. name *amarakus*. Sweet Marjoram. N Africa, SW Asia.
  onites o-*nee*-teez. Gk. name for a kind of marjoram. Pot Marjoram. Mediterranean region, W Asia.
  pulchellum pul-*kel*-um. Pretty. W Asia.
  rotundifolium ro-tun-di-*fo*-lee-um. Round-leaved. Turkey.
  scabrum *skăb*-rum. Rough. S Greece.
  vulgare vul-*gah*-ree. Common. Wild Marjoram. Europe to C Asia.

Ornithogalum or-ni-*tho*-ga-lum *Liliaceae*. From Gk. *ornis* (a bird) and *gala* (milk). Bulbous, perennial herbs.
  arabicum a-*ră*-bi-kum. Arabian. Mediterranean region.
  balansae ba-*lăn*-zie. After Balansa, see *Crocus balansae*. NE Turkey.
  montanum mon-*tah*-num. Of mountains. SE Europe, Turkey.
  nutans new-tănz. Nodding (the flowers). SE

Ornithogalum (continued)
  Europe, Turkey.
  thyrsoides thur-*soi*-deez. With flowers in a thyrse (a type of inflorescence). Chincherinchee. S Africa.
  umbellatum um-bel-*ah*-tum. The flowers appear to be in umbels. Star of Bethlehem. Europe, N Africa.

Orontium o-*ron*-tee-um *Araceae*. Classical name for a water plant growing in the Syrian river Orontes. Aquatic perennial herb.
  aquaticum a-*kwah*-ti-kum. Growing in water. Golden Club. E United States.

Oroya o-*roy*-a *Cactaceae*. After La Oroya in the Peruvian Andes where the following grows.
  peruviana pe-roo-vee-*ah*-na. Of Peru.

Orris see *Iris germanica florentina*
Osage Orange see *Maclura pomifera*
Osier, Common see *Salix viminalis*
  Purple see *S. purpurea*

Osmanthus os-*mănth*-us *Oleaceae*. From Gk. *osme* (fragrance) and *anthos* (a flower) referring to the fragrant flowers. Evergreen shrubs.
  armatus ar-*mah*-tus. Spiny (the leaves). W China.
  × *burkwoodii* burk-*wud*-ee-ee. *O. delavayi* × *O. decorus*. (= × *Osmarea burkwoodii*). After Burkwood and Skipwith, the raisers.
  decorus de-*kō*-rus. (= *Phillyrea decora*). Beautiful. Lazistan, on the SE coast of the Black Sea.
  delavayi del-a-*vay*-ee. After Delavay who introduced it to France in 1890, see *Abies delavayi*. China.
  × *fortunei* for-*tewn*-ee-ee. *O. fragrans* × *O. heterophyllus*. After Robert Fortune who introduced it in 1862, see *Fortunella*.
  fragrans frah-granz. Fragrant. China.
    aurantiacus ow-răn-tee-*ah*-kus. Orange (the flowers).
  heterophyllus he-te-rō-*fil*-lus. (= *O. aquifolium*. *O. ilicifolius*). With variable leaves. Japan.
  yunnanensis yoo-nan-*en*-sis. Of Yunnan, China.

× **Osmarea burkwoodii** see *Osmanthus* x *burkwoodii*

Osmaronia os-ma-*rō*-nee-a *Rosaceae*. From Gk. *osme* (fragrance) and *Aronia* q.v. a related genus. Deciduous shrub.

Osmaronia (continued)
cerasiformis ke-ra-si-*form*-is. Cherry-shaped (the fruit). Oso Berry. California.

Osmunda os-*mun*-da *Osmundaceae*. Derivation obscure. Ferns.
cinnamomea kin-a-*mō*-mee-a. Cinnamon-coloured (the fronds). Cinnamon Fern. Widely distributed.
claytoniana klay-ton-ee-*ah*-na. After John Clayton (1686–1773), Virginian botanist. Interrupted Fern. N America, Asia.
regalis ray-*gah*-lis. Royal. Royal Fern. Europe, Asia.

Oso Berry see *Osmaronia cerasiformis*

Osteospermum ost-ee-ō-*sperm*-um *Compositae*. From Gk. *osteon* (a bone) and *sperma* (a seed). Semi-hardy sub-shrubs. S Africa.
ecklonis ek-*lon*-is. (= *Dimorphotheca ecklonis*). After Christian Friedrich Ecklon (1795–1868), German apothecary.
jucundum yoo-*kun*-dum. (= *Dimorphotheca barberiae*). Pleasing.

Ostrowskia os-*trov*-skee-a *Campanulaceae*. After Michael Nicholazewitsch von Ostrowsky, a patron of botany. Perennial herb.
magnifica mahg-*ni*-fi-ka. Magnificent. Giant Bellflower. Turkestan.

Ostrya *os*-tree-a *Carpinaceae*. From *ostrys* the Gk. name. Deciduous trees.
carpinifolia kar-peen-i-*fo*-lee-a. With leaves like *Carpinus*. Hop Hornbeam. S Europe, W Asia, Caucasus.
japonica ja-*pon*-i-ka. Of Japan. Japan, China, Korea.
virginiana vir-jin-ee-*ah*-na. Of Virginia. Iron Wood. E N America.

Oswego Tea see *Monarda didyma*
Othonnopsis cheirifolia see *Hertia cheirifolia*

Ourisia ow-*ris*-ee-a *Scrophulariaceae*. After Ouris, a governor of the Falkland Islands where the first species was found. Herbaceous perennials.
alpina ăl-*peen*-a. Alpine. Andes.
coccinea kok-*kin*-ee-a. Scarlet. Andes.
elegans ay-le-gahnz. Elegant. Chile.
macrophylla măk-rō-*fil*-la. Large-leaved. New Zealand.

Our Lady's Milk Thistle see *Silybum marianum*
Our Lord's Candle see *Yucca whipplei*

Oxalis oks-*ah*-lis *Oxalidaceae*. The Gk. name for sorrel, from *oxys* (acid). Hardy and tender herbs.
acetosella a-kay-to-*se*-la. L. name for plants with acid leaves. Wood Sorrel. Europe, N Asia.
adenophylla a-den-o-*fil*-la. With glandular leaves. Chile, Argentina.
articulata ar-tik-ew-*lah*-ta. Jointed. Paraguay.
cernua see *O. pes-caprae*
chrysantha kris-*ănth*-a. With golden flowers. Brazil.
depressa day-*pres*-a. (= *O. inops*). Low-growing. S Africa.
dispar dis-par. Unusual. Guiana.
enneaphylla en-ee-a-*fil*-la. With nine leaflets. Falkland Islands.
hirta hir-ta. Hairy. S Africa.
inops see *O. depressa*
laciniata la-kin-ee-*ah*-ta. Deeply cut (the leaves). Patagonia.
magellanica mă-ge-*lăn*-i-ka. Of the Magellan region. Patagonia, Australia, New Zealand.
oregona o-ree-*gō*-na. Of Oregon. W United States.
ortgiesii ort-*geez*-ee-ee. After Eduard Ortgies (1829–1916). Peru.
pes-caprae pays-*kăp*-rie. (= *O. cernua*). Like a goat's foot (the leaves). S Africa.

Oxlip see *Primula elatior*
Ox-tongue Lily see *Haemanthus coccineus*

Oxycoccus oks-ee-*kok*-us *Ericaceae*. From Gk. *oxys* (acid) and *kokkos* (a round berry). Prostrate, evergreen shrubs. Cranberry.
macrocarpus măk-rō-*kar*-pus. (= *Vaccinium macrocarpon*). With large fruit. American Cranberry. E N America.
palustris pa-*lus*-tris. (= *Vaccinium oxycoccus*). Growing in marshes. Small Cranberry. Europe, N Asia, N America.

Oxydendrum oks-ee-*den*-drum *Ericaceae*. From Gk. *oxys* (acid) and *dendron* (a tree) referring to the acid-tasting leaves. Deciduous tree.
arboreum ar-*bo*-ree-um. Tree-like. Sorrel Tree. E N America.

Oxypetalum oks-ee-*pe*-ta-lum*Asclepiadaceae*.
From Gk. *oxys* (sharp) and *petalum* (a petal).
Tender climber.
  caeruleum kie-*ru*-lee-um. (= *Tweedia*
  caerulea). Deep blue (the flowers). Brazil,
  Uruguay.

Ozothamnus o-zō-*thăm*-nus *Compositae*.
From Gk. *ozo* (a smell) and *thamnos* (a shrub).
Evergreen shrubs.
  coralloides ko-ra-*loi*-deez. (= *Helichrysum
  coralloides*). Coral-like (the shoots). New
  Zealand.
  ledifolius lay-di-*fo*-lee-us. (= *Helichrysum
  ledifolium*). With leaves like *Ledum*.
  Tasmania.
  rosmarinifolius rōs-ma-reen-i-*fo*-lee-us. (=
  Helichrysum rosmarinifolium). With leaves
  like *Rosmarinus*. SE Australia, Tasmania.
  selago se-*lah*-go. (= *Helichrysum selago*).
  Like *Lycopodium selago*. New Zealand.

P

Pachistima see *Paxistima*

Pachycereus pă-kee-*kay*-ree-us *Cactaceae*.
From Gk. *pachys* (thick) and *Cereus* q.v.
referring to the thick shoots. W Mexico.
  pecten-aboriginum pek-ten-ă-bo-*ree*-gi-num.
  As the common name, Native's Comb.
  pringlei *pring*-gal-ee. After G. G. Pringle
  who collected in Mexico c 1887.

Pachyphragma pă-kee-*frăg*-ma *Cruciferae*.
From Gk. *pachys* (thick) and *phragma* (a
partition) referring to the stout-ribbed septum
of the pod. Herbaceous perennial.
  macrophyllum măk-rō-*fil*-lum. Large-
  leaved. Caucasus.

Pachyphytum pă-kee-*fi*-tum *Crassulaceae*.
From Gk. *pachys* (thick) and *phyton* (a plant).
Tender succulents. Mexico.
  amethystinum see *Graptopetalum
  amethystinum*
  bracteosum brăk-tee-ō-sum. With
  conspicuous bracts (on the flower stem).
  glutinicaule gloo-tin-i-*kaw*-lee. (= *P.
  brevifolium* hort.). With sticky stems.
  Sticky Moonstones.
  oviferum ō-*vi*-fe-rum. Egg-bearing,
  referring to the egg-shaped leaves.
  Moonstones.

Pachysandra pă-kis-*ăn*-dra *Buxaceae*. From
Gk. *pachys* (thick) and *andros* (male),
referring to the thick stamens. Evergreen sub-
shrubs.
  procumbens prō-*kum*-benz. Prostrate.
  Allegheny Spurge. SE United States.
  terminalis ter-mi-*nah*-lis. Terminal (the
  flower spikes). Japan.

Pachystachys pă-kee-*stă*-kis *Acanthaceae*.
From Gk. *pachys* (thick) and *stachys* (a spike)
referring to the dense inflorescences. Tender
shrubs.
  coccinea kok-*kin*-ee-a. (= *Jacobinia
  coccinea*). Scarlet. W Indies, S America.
  lutea *loo*-tee-a. Yellow (the bracts).
  Lollipop Plant. Peru.

Pachystegia insignis see *Olearia insignis*

Pachystima see *Paxistima*

× **Pachyveria** pă-kee-*ve*-ree-a *Crassulaceae*.
Intergeneric hybrid, from the names of the
parents. *Echeveria* × *Pachyphytum*. Tender
succulent.
  pachyphytoides pă-kee-fit-oi-deez. *Echeveria
  gibbiflora* × *Pachyphytum bracteosum*. Like
  Pachyphytum.

Paeonia pie-on-ee-a *Paeoniaceae*. From the
Gk. name *paionia*, meaning of Paion,
physician to the gods. Herbaceous perennials
and shrubs. Paeony.
  anomala a-*nom*-a-la. Unusual. W and
  C Asia.
  arietina see *P. mascula arietina*
  clusii *clooz*-ee-ee. After Clusius, see
  Gentiana clusii. Crete.
  delavayi de-la-*vay*-ee. After Delavay, see
  Abies delavayi. W China.
  emodi e-*mō*-dee. Of *Emodi Montes* (the
  Himalaya).
  lactiflora lăk-ti-*flō*-ra. With milky flowers.
  NE Asia.
  × *lemoinei* la-*mwŭn*-ee-ee. *P. lutea* × *P.
  suffruticosa*. After Messrs Lemoine who
  raised some forms.
  lutea *loo*-tea. Yellow. SW China.
    ludlowii lŭd-*lō*-ee-ee. After Frank
    Ludlow (1885–1972) who, with George
    Sherriff, introduced it in 1936. SE Tibet.
  mascula mahs-kew-la. Male, used for
  vigorous species. Europe.
    arietina a-ree-e-*tee*-na. (= *P. arietina*).
    Like a ram's head. E Europe, W Asia.
  mlokosewitschii mlo-ko-sa-*vich*-ee-ee. After

Paeonia (continued)
   Ludwig Franzevich Mlokosewitsch
   (1831–1909), who discovered it. Caucasus.
   obovata ob-ō-*vah*-ta. Obovate (the terminal
   leaflet). E Asia.
   officinalis o-fi-ki-*nah*-lis. Sold as a herb.
   Europe.
   peregrina pe-re-*green*-a. Foreign. S and
   E Europe.
   potaninii po-tah-*nin*-ee-ee. After
   Nicolaevich Potanin (1835–1920), Russian
   explorer and plant collector. China.
      trollioides trol-ee-*oi*-deez. Like *Trollius*.
   × *smouthii* smooth-ee-ee. *P. lactiflora* × *P.
   tenuifolia*. After M. Smouth.
   suffruticosa suf-froo-ti-*kō*-sa. Sub-shrubby.
   Moutan. N China.
   tenuifolia ten-ew-i-*fo*-lee-a. With slender
   leaves (leaf segments). SE Europe,
   Caucasus.
   wittmanniana vit-măn-ee-*ah*-na. After
   Wittmann, who collected in the Caucasian
   Taurus c 1840. Caucasus.

Painted Daisy see *Chrysanthemum carinatum*
Painted Drop-tongue see *Aglaonema crispum*
Painted Feather see *Vriesia carinata*
   Dwarf see *V. psittacina*
Painted Tongue see *Salpiglossis sinuata*
Painted Wood-lily see *Trillium undulatum*
Painter's Palette see *Anthurium andreanum*

Paliurus pă-lee-*ew*-rus *Rhamnaceae*. The Gk.
name. Deciduous shrub or small tree.
   spina-christi speen-a-*kris*-tee. Christ's
   Thorn, it is believed to have been used for
   the crown of thorns. S Europe, W Asia.

Palm, Australian Fan see *Livistona australis*
   Bamboo see *Chamaedorea erumpens*
   Betel Nut see *Areca catechu*
   Burmese Fishtail see *Caryota mitis*
   Canary Island Date see *Phoenix canariensis*
   Chinese Fan see *Livistona chinensis*
   Chusan see *Trachycarpus fortunei*
   Curly Sentry see *Howea belmoreana*
   Date see *Phoenix dactylifera*
   Desert Fan see *Washingtonia filifera*
   Dwarf Fan see *Chamaerops humilis*
   Fan see *Trachycarpus fortunei*
   Fishtail see *Caryota*
   Lady see *Rhapis*
   Miniature Date see *Phoenix roebelinii*
   Paradise see *Howea forsteriana*
   Parlour see *Chamaedorea elegans*
   Sago see *Cycas revoluta*
   Sentry see *Howea*
   Thread see *Washingtonia robusta*

Palm (continued)
   Toddy see *Caryota urens*
   Weddell see *Microcoelum weddellianum*
   Wine see *Caryota urens*
   Yatay see *Butia yatay*
   Yellow see *Chrysalidocarpus lutescens*

Pamianthe păm-ee-*ănth*-ee *Amaryllidaceae*.
After Major Albert Pam (1875–1955) to
whom the following was sent in 1926.
Tender, bulbous herb.
   peruviana pe-roo-vee-*ah*-na. Of Peru.

Pampas Grass see *Cortaderia selloana*

Pancratium păn-*krăt*-ee-um *Amaryllidaceae*.
Gk. name for a bulbous plant. Bulbous
perennials.
   canariense ka-nah-ree-*en*-see. Of the Canary
   Islands.
   illyricum i-*li*-ri-kum. Of Illyria (W
   Yugoslavia). Corsica, Sardinia, Capri etc.
   maritimum ma-*ri*-ti-mum. Growing near the
   sea. Sea Daffodil, Sea Lily. Mediterranean
   region.

Pandanus păn-da-nus *Pandanaceae*. From
pandan the Malayan name. Tender,
evergreen shrubs and trees. Screw Pine.
   baptistii băp-*tist*-ee-ee. After Baptist. New
   Britain Islands.
   candelabrum kăn-day-*lah*-brum. Like a
   candelabra. W Africa.
   pygmaeus pig-*mie*-us. Dwarf. Madagascar.
   sanderi sahn-da-ree. After the Sander
   nursery. Timor.
   veitchii veech-ee-ee. After the Veitch
   nursery. Polynesia.

Panda Plant see *Kalanchoe tomentosa*,
Philodendron bipennifolium

Pandorea păn-*do*-ree-a *Bignoniaceae*. After
Pandora of Gk. mythology. Tender,
evergreen shrubs or trees.
   jasminoides yăs-min-*oi*-deez. Jasmine-like.
   Bower Plant. Australia.
   pandorana păn-do-*rah*-na. After Pandora.
   Wonga-wonga Vine. SE Asia to Australia.

Panicum *pah*-ni-kum *Gramineae*. The L.
name for millet. Annual grasses.
   capillare kă-pi-*lah*-ree. Hair-like. Witch
   Grass. E N America.
   miliaceum mi-lee-*ah*-kee-um. Like *Milium*.
   Asia.
   virgatum vir-*gah*-tum. Wand-like. N and
   C America.

Pansy see *Viola*
 Garden see *V.* × *wittrockiana*
Panther Lily see *Lilium pardalinum*

Papaver pa-*pah*-ver *Papaveraceae*. The L.
name. Annual, biennial and perennial herbs.
Poppy.
 alpinum ăl-*peen*-um. Alpine. Alpine Poppy.
 Origin and identity of cultivated plants
 uncertain.
 bracteatum see *P. orientale*
 commutatum kom-ew-*tah*-tum. Changeable.
 Caucasus, W Asia.
 glaucum glow-kum. Glaucous (the leaves).
 Tulip Poppy. W Asia.
 miyabeanum see *P. nudicaule*
 nudicaule new-di-*kaw*-lee. (= *P.
 miyabeanum*). Bare-stemmed. Arctic
 Poppy, Iceland Poppy. N America,
 Europe, Asia.
 orientale o-ree-en-*tah*-lee. (= *P.
 bracteatum*). Eastern. W Asia.
 pilosum pi-*lō*-sum. Hairy. W Asia.
 rupifragum roo-*pi*-fra-gum. Rock-breaking,
 i.e. growing in rock crevices. Spanish
 Poppy. S Spain.
 somniferum som-*ni*-fe-rum. Sleep-bearing.
 Opium Poppy. SE Europe. W Asia.
 spicatum spee-*kah*-tum. With flowers in
 spikes. W Asia.

Paphiopedilum pă-fee-ō-*pe*-di-lum
Orchidaceae. Gk. Paphos, site of a temple on
Cyprus where Aphrodite was worshipped
and *pedilon* (a slipper). Greenhouse orchids,
sometimes listed under *Cypripedium*. Slipper
Orchid.
 acmodontum ăk-mō-*don*-tum. With a sharp
 tooth (on the front margin of the lip).
 Philippines.
 barbatum bar-*bah*-tum. Bearded (the warts
 on the petals). Malaya.
 bellatulum be-*lah*-tew-lum. Pretty. SE Asia.
 callosum ka-*lō*-sum. Calloused (the petals).
 Thailand.
 concolor kon-ko-lor. Similarly coloured (the
 petals and sepals). SE Asia.
 fairrieanum fair-ree-*ah*-num. After Mr
 Fairrie of Liverpool who bought it in a
 sale of Assam plants.
 hirsutissimum hir-soo-*tis*-i-mum. Very hairy
 (the scape). Assam to S China.
 insigne in-*sig*-nee. Remarkable. Assam.
 niveum niv-ee-um. Snow-white (the lip).
 Thailand, Malaya.
 philippinense fi-li-peen-*en*-see. Of the
 Philippines.

Paphiopedilum (continued)
 purpuratum pur-pew-*rah*-tum. Purplish (the
 flowers). Hong Kong, S China.
 rothschildianum roths-chield-ee-*ah*-num.
 After Rothschild. Borneo.
 spicerianum spie-sa-ree-*ah*-num. After Mr
 Spicer, a tea planter who introduced many
 orchids. Assam, N Burma.
 sukhakulii soo-ka-*koo*-lee-ee. After P.
 Sukhakuli, a Thai nurseryman. NE
 Thailand.
 tonsum tōn-sum. Smooth. Sumatra.
 venustum ve-*nus*-tum. Handsome.
 Himalaya.
 villosum vi-*lō*-sum. Softly hairy (the scape).
 Assam, China, Thailand.

Papyrus see *Cyperus papyrus*

Paradisea pă-ra-*dees*-ee-a *Liliaceae*. After
Count Giovanni Paradisi (1760–1826).
Perennial herb.
 liliastrum lee-lee-*ăs*-trum. Like *Lilium*. St
 Bruno's Lily. S Europe.

Parahebe pă-ra-*hay*-bay *Scrophulariaceae*.
From Gk. *para* (close to) and *Hebe* q.v.
Dwarf, evergreen shrubs related to *Hebe*.
New Zealand.
 catarractae kă-ta-*răk*-tie. (= *Hebe
 catarractae*). Of waterfalls.
 decora de-*kō*-ra. Beautiful.
 lyallii lie-*ăl*-ee-ee. After David Lyall
 (1817–95), naval surgeon and naturalist
 who collected the type specimen.

Paris Daisy see *Chrysanthemum frutescens*

Parkinsonia par-kin-*son*-ee-a *Leguminosae*.
After John Parkinson (1567–1650), London
apothecary and botanical author. Tender,
evergreen tree.
 aculeata a-kew-lee-*ah*-ta. Prickly.
 Jerusalem Thorn. Tropical America.

Parnassia par-*năs*-ee-a *Saxifragaceae*. From
the 16th century name *Gramen Parnassi*,
referring to Mt Parnassus, Greece. Bog
garden perennial herb.
 palustris pa-*lus*-tris. Of marshes. Grass of
 Parnassus. N temperate regions.

Parochetus pa-*ro*-ke-tus *Leguminosae*. From
Gk. *para* (near) and *ochetus* (a brook)
referring to the habitat. Perennial herb.
 communis kom-*ew*-nis. Common. Shamrock
 Pea. Himalaya, E Africa.

Parodia pa-rō-dee-a *Cactaceae*. After Lorenzi
Raimondo Parodi (1895–1966), Argentine
botanist. N Argentina.
  aureispina ow-ree-i-*speen*-a. With golden
  spines. Golden Tom Thumb.
  chrysacanthion kris-a-*kănth*-ee-on. With
  golden spines.
  maassii mahs-ee-ee. After W. Maass. Also
  Bolivia.
  microsperma mik-rō-*sperm*-a. With small
  seeds.
  mutabilis mew-*tah*-bi-lis. Changeable.
  nivosa ni-*vō*-sa. Snow-white (the spines).
  sanguiniflora sang-gwin-i-*flō*-ra. With blood-
  red flowers.
  scopaoides skō-pa-*oi*-deez. Broom-like.

Parrotia pa-*rōt*-ee-a *Hamamelidaceae*. After
F. W. Parrot (1792–1841), a German
naturalist. Deciduous tree.
  persica *per*-si-ka. Of Iran (Persia). Persian
  Ironwood. SW Caspian Sea region.

Parrotiopsis pa-rōt-ee-*op*-sis
Hamamelidaceae. From *Parrotia* q.v. and
Gk. *-opsis* indicating resemblance. Deciduous
tree.
  jacquemontiana zhahk-a-mont-ee-*ah*-na.
  After Victor Jacquemont (1801–32),
  French naturalist. W Himalaya.

Parsley see *Petroselinum crispum*
Parsley Vine see *Vitis vinifera* 'Apiifolia'
Parsnip see *Pastinaca sativa*

Parthenocissus par-then-ō-*kis*-us *Vitaceae*.
From Gk. *parthenos* (a virgin) and *kissos* (ivy)
referring to the common name, Virginia
Creeper. Deciduous climbers.
  henryana hen-ree-*ah*-na. After Augustine
  Henry who discovered it, see *Illicium
  henryi*. C China.
  himalayana him-a-lay-*ah*-na. Of the
  Himalaya.
  inserta in-*ser*-ta. Inserted, it needs to be
  attached to walls. N America.
  quinquefolia kwing-kwee-*fo*-lee-a. With five
  leaves (leaflets). Virginia Creeper.
  N America.
  thomsonii tom-*son*-ee-ee. After Thomson,
  see *Aster thomsonii*. China, Assam.
  tricuspidata tri-kus-pi-*dah*-ta. Three-
  pointed (the leaves). Boston Ivy. Japan,
  China.

Partridge Berry see *Mitchella repens*
Pasque Flower see *Pulsatilla*

Passiflora pă-si-*flō*-ra *Passifloraceae*. From L.
passio (passion) and *flos* (a flower). The parts
of the flower have been compared with
various aspects of the crucifixion of Christ.
Evergreen, mainly tender, climbers. Passion
Flower.
  × *allardii* a-*lard*-ee-ee. *P. caerulea*
  'Constance Elliott' × *P. quadrangularis*.
  After Edgar John Allard (c 1877–1918),
  gardener and hybridist.
  antioquiensis ăn-tee-ō-kee-*en*-sis. Of
  Antioquia, Colombia.
  caerulea kie-*ru*-lee-a. Blue (the flowers).
  Blue Passion Flower. S Brazil.
  × *caponii* ka-*pon*-ee-ee. *P. quadrangularis* ×
  P. racemosa. After Mr W. J. Capon who
  raised it in 1953.
  coccinea kok-*kin*-ee-a. Scarlet (the flowers).
  Red Granadilla. N S America.
  edulis e-*dew*-lis. Edible (the fruit).
  Granadilla. Brazil.
  × *exoniensis* eks-ō-nee-*en*-sis. *P.
  antioquiensis* × *P. mollisima*. Of Exeter.
  incarnata in-kar-*nah*-ta. Flesh-coloured (the
  flowers). May Apple, Maypop. S United
  States.
  laurifolia low-ri-*fo*-lee-a. *Laurus*-leaved.
  Yellow Granadilla. W Indies, S America.
  manicata măn-i-*kah*-ta. Long-sleeved.
  N S America.
  mixta miks-ta. Mixed. N S America.
  mollissima mol-*lis*-i-ma. Very softly hairy
  (the shoots and undersides of the leaves).
  N S America.
  quadrangularis kwod-rang-gew-*lah*-ris.
  Four-angled (the shoots). Giant
  Granadilla. S America.
  racemosa ră-kay-*mō*-sa. With flowers in
  racemes. Brazil.
  umbilicata um-bi-lee-*kah*-ta. With a navel.
  S America.

Passion Flower see *Passiflora*
  Blue see *P. caerulea*

Pastinaca păs-ti-*nah*-ka *Umbelliferae*. The L.
name for parsnip and carrot, from *pastus*
(food). Biennial herb.
  sativa sa-*tee*-va. Cultivated. Parsnip.
  Europe, Asia.

Paulownia pow-lō-nee-a *Scrophulariaceae*.
After Anna Paulowna (1795–1865), daughter
of Czar Paul I of Russia. Deciduous trees.
China.
  lilacina li-la-*keen*-a. (= *P. fargesii* hort.).
  Lilac (the flowers).
  tomentosa tō-men-*tō*-sa. Hairy (the leaves).

Pavonia pa-*von*-ee-a *Malvaceae*. After Jose Antonio Pavon (1754–1840), Spanish botanist. Tender, evergreen shrub.
  multiflora mul-tee-*flō*-ra. Many-flowered. Brazil.

Pawpaw see *Asimina triloba*

Paxistima pǎks-*i*-sti-ma *Celastraceae*. (= *Pachistima. Pachystima*). From Gk. *pachys* (thick) and *stigma*. Evergreen shrubs.
  canbyi kǎn-bee-ee. After its discoverer William Marriot Canby (1831–1904). E United States.
  myrtifolia mur-ti-*fo*-lee-a. *Myrtus*-leaved. NW North America.

Pea, Garden see *Pisum sativum*
Pea Tree see *Caragana arborescens*
Peach see *Prunus persica*
Peacock Plant see *Calathea makoyana*
Peacock Tiger Flower see *Tigridia pavonia*
Pear see *Pyrus*
  Common see *P. communis*
  Willow-leaved see *P. salicifolia*
Pearl Fruit see *Margyricarpus pinnatus*
Pearl Grass see *Briza maxima*
Pearl Plant see *Haworthia margaritifera*
Pearlwort see *Sagina*
Pearly Everlasting see *Anaphalis*

Pedilanthus pe-di-*lǎnth*-us *Euphorbiaceae*. From Gk. *pedilon* (a slipper) and *anthos* (a flower) referring to the shape of the involucre. Tender succulents.
  tithymaloides ti-thee-mah-*loi*-deez. Like *Tithymalus* (= *Euphorbia*). Ribbon Cactus, Slipper Flower. C and N South America.
  smallii smawl-ee-ee. Jacob's Ladder. After John Kunkel Small (1869–1938), American botanist. Florida, Cuba.

Pelargonium pe-lar-*gon*-ee-um *Geraniaceae*. From Gk. *pelargos* (a stork) referring to the storksbill-like fruit. Tender perennial herbs and shrubs. Geranium. S Africa.
  crispum kris-pum. Finely wavy (the leaves). Lemon-scented Pelargonium.
  denticulatum den-tik-ew-*lah*-tum. Toothed (the leaves).
  × *domesticum* do-*mes*-ti-kum. Cultivated. Regal Pelargonium.
  × *fragrans* frah-granz. *P. exstipulatum* × *P. odoratissimum*. Fragrant (the leaves). Nutmeg Pelargonium.

Pelargonium (continued)
  fulgidum ful-gi-dum. Shining (the scarlet flowers).
  graveolens gra-*vee*-o-lenz. Aromatic. Rose Pelargonium.
  × *hortorum* hor-*to*-rum. Of gardens. Zonal Pelargonium.
  odoratissimum o-dō-ra-*tis*-i-mum. Highly scented (the leaves). Apple Pelargonium.
  peltatum pel-*tah*-tum. Peltate (the leaves). Ivy-leaved Pelargonium.
  quercifolium kwer-ki-*fo*-lee-um. *Quercus*-leaved. Oak-leaved Pelargonium.
  tetragonum tet-ra-*gō*-num. Four-angled (the stems).
  tomentosum tō-men-*tō*-sum. Hairy. Peppermint-scented Pelargonium.
  triste tris-tee. Sad, the sombre-coloured flowers.
  zonale zō-*nah*-lee. Zoned (the leaves). Mainly grown as hybrids.

Pellaea pe-*lie*-a *Sinopteridaceae*. From Gk. *pellaios* (dark) referring to the often dark stalks. Ferns.
  atropurpurea aht-rō-pur-*pewr*-ree-a. Deep purple (the stalks). Purple Cliff Brake. N America.
  rotundifolia ro-tund-i-*fo*-lee-a. With round leaves (pinnae). Button Fern. New Zealand.
  viridis vi-ri-dis. Green. Green Cliff Brake. Africa.

Pellionia pe-lee-*on*-ee-a *Urticaceae*. After Alphonse Odet Pellion (1796–1868), a French Admiral. Tender, evergreen herbs.
  daveauana dǎ-vō-*ah*-na. After Jules Daveau (1852–1929), a director of Lisbon Botanical Garden. Trailing Watermelon Begonia. SE Asia.
  pulchra pul-kra. Pretty. Rainbow Vine. Vietnam.

Peltiphyllum pel-ti-*fil*-lum *Saxifragaceae*. From Gk. *pelte* (a shield) and *phyllon* (a leaf) referring to the shield-like leaves. Perennial herb.
  peltatum pel-*tah*-tum. Peltate (the leaves). Umbrella Plant. Oregon, California.

Pennisetum pen-i-*say*-tum *Gramineae*. From L. *penna* (a feather) and *seta* (a bristle) referring to the feathery bristles around the spikelets. Perennial grasses.
  alopecuroides ǎ-lō-pek-ew-*roi*-deez. Like *Alopecurus*. Asia.

Pennisetum (continued)
orientalis o-ree-en-*tah*-lis. Eastern.
Abyssinia.
setaceum say-*tah*-kee-um. Bristly. Fountain
Grass. Africa.
villosum vi-*lō*-sum. Softly hairy. Abyssinian
Feathertop. Africa.

Pennyroyal see *Mentha pulegium*

Penstemon pen-*stay*-mon *Scrophulariaceae*.
From Gk. *pente* (five) and *stemon* (a stamen)
referring to the five stamens. Perennial herbs
and shrubs.
alpinus ăl-*peen*-us. Alpine. W United
States.
barbatus bar-*bah*-tus. (= *Chelone barbata*).
Bearded (the lower lip of the corolla). SW
United States, Mexico.
barretiae ba-*ret*-ee-ie. After Mrs Barret who
discovered it. Oregon.
cordifolius kor-di-*fo*-lee-us. With heart-
shaped leaves. S California.
davidsonii day-vid-*son*-ee-ee. (= *P.
menziesii*). After Davidson. W N America.
× *edithae* ee-dith-ie. *P. barretiae* × *P.
rupicola*. After Edith Hardin English.
fruticosus froo-ti-*kō*-sus. Shrubby.
W N America.
scouleri skool-a-ree. (= *P. scouleri*). After
Dr John Scouler (1804–71), who
collected with David Douglas.
hartwegii hart-*weg*-ee-ee. After Carl
Theodore Hartweg (1812–71). Mexico.
heterophyllos he-te-rō-*fil*-lus. With variable
leaves. California.
menziesii see *P. davidsonii*
newberryi new-*be*-ree-ee. After J. S.
Newberry who discovered it. W United
States.
humilior hu-*mil*-ee-or. (= *P. roezlii*
hort.). Low-growing.
ovatus ō-*vah*-tus. Ovate (the leaves).
W N America.
pinifolius peen-i-*fo*-lee-us. With leaves like
Pinus. SW United States, Mexico.
roezlii hort. see *P. newberryi humilior*
rupicola roo-*pi*-ko-la. Growing on rocks.
W N America.
scouleri see *P. fruticosus scouleri*
virens vi-renz. Green. W United States.

Pentapterygium serpens see *Agapetes serpens*

Pentas *pen*-tas *Rubiaceae*. From Gk. *pentas*
(a series of five), it differs from related
genera in having the floral parts in fives.

Pentas (continued)
Tender sub-shrub.
lanceolata lăn-kee-o-*lah*-ta. Lance-shaped
(the leaves). Egyptian Star Cluster.
E Africa, Arabia.

Pen Wiper see *Kalanchoe marmorata*

Peperomia pe-pe-*rom*-ee-a *Piperaceae*. From
Gk. *peperi* (pepper) and *homoios*
(resembling), it is closely related to the
pepper plant. Tender, evergreen, succulent
herbs.
argyreia ar-gi-*ree*-a. Silvery (the leaves).
Watermelon Peperomia, Rugby Football
Plant. Tropical S America.
caperata kǎ-pe-*rah*-ta. Wrinkled (the
leaves). Brazil.
fraseri fray-za-ree. After Fraser who
collected the type specimen. Mignonette
Peperomia. Ecuador.
glabella gla-*bel*-la. Rather glabrous. Wax
Privet. Tropical America.
griseoargentea gri-see-ō-ar-*gen*-tee-a. Grey-
silver (the leaves). Ivy-leaf Peperomia.
Brazil.
magnoliifolia mǎg-nol-ee-i-*fo*-lee-a.
Magnolia-leaved. Desert Privet. W Indies,
S America.
obtusifolia ob-tew-si-*fo*-lee-a. Blunt-leaved.
Baby Rubber Plant. Tropical America.
orba or-ba. (= *P. 'Princess Astrid'*). L. for
orphan, it is of unknown origin.
scandens skăn-denz. Climbing. Identity and
origin of cultivated plants uncertain.
verticillata ver-ti-ki-*lah*-ta. Whorled (the
leaves). W Indies.

Pepper, Chilli see *Capsicum frutescens*
Christmas see *C. annuum*
Sweet see *C. annuum*
Peppermint, Black see *Mentha × piperita*
White see *M. × piperita officinalis*

Pereskia pe-*res*-kee-a *Cactaceae*. After
Nicholas Claude Fabre de Peiresc
(1580–1637), French naturalist. Leafy Cacti.
aculeata a-kew-lee-*ah*-ta. Prickly. Barbados
Gooseberry. Tropical America.
grandifolia grănd-i-*fo*-lee-a. (= *P.
grandiflora*). Large-leaved. Brazil.

Perilla pe-*ril*-la *Labiatae*. Derivation
obscure. Annual herb.
frutescens froo-*tes*-enz. Shrubby. Himalaya,
E Asia.

Peristrophe pe-*ri*-stro-fay *Acanthaceae*.
From Gk. *peri* (around) and *strophe* (turning)
referring to the twisted corolla tube. Tender
perennial herbs.
  hyssopifolia hi-sō-pi-*fo*-lee-a. (= *P.*
  angustifolia. *P. salicifolia*). Hyssopus-
  leaved. Java.
  speciosa spe-kee-*ō*-sa. Showy. India.

Periwinkle see *Vinca*
  Greater see *V. major*
  Lesser see *V. minor*
  Madagascar see *Catharanthus roseus*

Pernettya per-*net*-ee-a *Ericaceae*. After
Antoine Joseph Pernetty (1716–1801).
Evergreen shrubs.
  leucocarpa loo-kō-*kar*-pa. White-fruited.
  Chile, Argentina.
  mucronata mew-kron-*ah*-ta. Mucronate (the
  leaves). Chile, Argentina.
  prostrata pros-*trah*-ta. Prostrate. Andes.
    pentlandii pent-*länd*-ee-ee. After Joseph
    Barclay Pentland (1797–1873) who
    collected the type specimen. S America.

Perovskia pe-*rof*-skee-a *Labiatae*. After V.
A. Perovsky (1794–c 1857). Sub-shrubs.
  abrotanoides a-bro-ta-*noi*-deez. Like
  Artemisia abrotanum. Russia, W Asia.
  atriplicifolia ă-tri-pli-ki-*fo*-lee-a. *Atriplex*-
  leaved. W Himalaya, Afghanistan.

Persea *per*-see-a *Lauraceae*. Gk. name of a
tree. Tender, evergreen tree.
  americana a-me-ri-*kah*-na. (= *P.*
  gratissima). American. Avocado Pear.
  C America.

Persian Ironwood see *Parrotia persica*
Persian Shield see *Strobilanthes dyerianus*
Persian Violet see *Exacum affine*
Persimmon see *Diospyros virginiana*
  Chinese see *D. kaki*

Petasites pe-ta-*see*-teez *Compositae*. From
Gk. *petasos* (a hat) referring to the large
leaves. Perennial herbs.
  fragrans *frah*-granz. Fragrant. Winter
  Heliotrope. W Mediterranean region.
  japonicus ja-*pon*-i-kus. Of Japan. Korea,
  China, Japan.

Petrea *pet*-ree-a *Verbenaceae*. After Lord
Robert James Petre (1713–43), botanical and
horticultural patron. Tender climber.
  volubilis vol-*ew*-bi-lis. Twining. Purple

Petrea (continued)
  Wreath, Queen's Wreath. Mexico,
  C America.

Petrocoptis pet-ro-*kop*-tis *Caryophyllaceae*.
From Gk. *petros* (a rock) and *kopto* (to
break), from the habit of growing in rock
crevices. Perennial herb.
  glaucifolia glow-ki-*fo*-lee-a. (= *P. lagascae*.
  Lychnis lagascae). With glaucous leaves.
  Pyrenees, N Spain.

Petrorhagia pet-ro-*rah*-gee-a
Caryophyllaceae. From Gk. *petros* (a rock)
and *rhagas* (a chink), the following grows in
cracks in rocks. Perennial herb.
  saxifraga săks-*if*-ra-ga. (= *Tunica*
  saxifraga). The L. version of the Gk. name
  above. C and S Europe to C Asia.

Petroselinum pet-ro-se-*leen*-um
Umbelliferae. From Gk. *petros* (a rock) and
selinon (parsley). Biennial herb.
  crispum *kris*-pum. Finely wavy (the leaves).
  Parsley. Europe, W Asia.

Petunia pe-*tewn*-ee-a *Solanaceae*. From
petun, the Brazilian name for tobacco.
Annual herbs.
  × *hybrida* *hib*-ri-da. Hybrid. Common
  Petunia.

Peyote see *Lophophora williamsii*

Pfeiffera *fie*-fa-ra *Cactaceae*. After Louis Carl
Georg Pfeiffer (1805–77). German botanist.
  ianothele ee-ahn-ō-*thay*-lay. With violet
  nipples. N Argentina.

Phacelia fa-*kel*-ee-a *Hydrophyllaceae*. From
Gk. *phakelos* (a bundle) referring to the
clustered flowers. Annual herbs.
  campanularia kăm-păn-ew-*lah*-ree-a. Like
  Campanula. California Bluebell. W United
  States.
  tanacetifolia tăn-a-set-i-*fo*-lee-a. *Tanacetum*-
  leaved. SW United States, Mexico.
  viscida *vis*-ki-da. Sticky. California.

Phaius *fie*-us *Orchidaceae*. From Gk. *phaius*
(dusky) referring to the flowers of one species.
Greenhouse orchid.
  tankervilliae tang-ka-*vil*-ee-ie. (= *P.*
  wallichii). After Emma Lady Tankerville
  (c 1750–1836). Himalaya.

Phalaenopsis fă-lie-*nop*-sis *Orchidaceae*.
From Gk. *phalaina* (a moth) and *-opsis*

Phalaenopsis (continued)
indicating resemblance. Greenhouse orchids.
Moth Orchid.
  amabilis a-*mah*-bi-lis. Beautiful. SE Asia to
  N Australia.
  esmeralda see *Doritis pulcherrima*
  lueddemanniana loo-da-măn-ee-*ah*-na. After
  Lueddemann, a Paris orchid grower.
  Philippines.
  parishii pă-*rish*-ee-ee. After the Rev.
  Charles Samuel Parish, an army surgeon
  who collected orchids in Burma. Burma.
  sanderiana sahn-da-ree-*ah*-na. After the
  Sander orchid nursery. Philippines.
  schilleriana shi-la-ree-*ah*-na. After Herr
  Schiller of Hamburg, an orchid grower.
  Philippines.
  stuartiana stew-art-ee-*ah*-na. After Stuart
  Low, see *Cymbidium lowianum*.
  Philippines.

Phalaris fa-*lah*-ris *Gramineae*. Gk. name for
a grass. Annual and perennial grasses.
  arundinacea a-run-di-*nah*-kee-a. Reed-like.
  Reed Canary Grass. Europe, Asia,
  N America.
    'Picta' *pik*-ta. Painted (the leaves).
    Gardener's Garters.
  canariensis ka-nah-ree-*en*-sis. Of the Canary
  Islands. Canary Grass. Canary Islands,
  N Africa.

Phaseolus fa-*see*-o-lus *Leguminosae*. From
phaselos the Gk. name for another bean.
Annual herbs. Tropical America.
  coccineus kok-*kin*-ee-us. Scarlet. Scarlet
  Runner Bean.
  vulgaris vul-*gah*-ris. Common. French
  Bean, Runner Bean.

Pheasant's Eye see *Adonis autumnalis*

Phellodendron fe-lo-*den*-dron *Rutaceae*.
From Gk. *phellos* (cork) and *dendron* (a tree)
referring to the corky bark. Deciduous trees.
  amurense ăm-ew-*ren*-see. Of the Amur
  region. NE Asia.

Philadelphus fil-a-*del*-fus *Philadelphaceae*.
The Gk. name. Deciduous shrubs.
Sometimes referred to as Syringa q.v.
  'Burfordensis' bur-fad-*en*-sis. Of Burford
  Court where it was raised.
  coronarius ko-rō-*nah*-ree-us. Used in
  garlands. Mock Orange. SE Europe,
  W Asia.
  delavayi de-la-*vay*-ee. After Delavay who
  discovered and introduced it in 1887, see

Philadelphus (continued)
  Abies delavayi. W China, SE Tibet,
  N Burma.
  microphyllus mik-rō-*fil*-lus. Small-leaved.
  SW United States.

Philesia fi-*leez*-ee-a *Liliaceae*. From Gk.
phileo (to love) referring to the attractive
flowers. Evergreen shrub.
  magellanica mă-ge-*lăn*-i-ka. Of the region
  of the Magellan Straits. S Chile.

Phillyrea fi-*li*-ree-a *Oleaceae*. The Gk. name.
Evergreen shrubs and trees.
  angustifolia ang-gus-ti-*fo*-lee-a. Narrow-
  leaved. S Europe, N Africa.
    'Rosmarinifolia' rōs-ma-reen-i-*fo*-lee-a.
    With leaves like *Rosmarinus*.
  decora see *Osmanthus decorus*
  latifolia lah-tee-*fo*-lee-a. Broad-leaved.
  Mediterranean region.

Philodendron fi-lo-*den*-dron *Araceae*. From
Gk. *phileo* (to love) and *dendron* (a tree)
referring to their tree-climbing habit. Tender,
evergreen climbers.
  angustisectum ang-gus-ti-*sek*-tum. With
  narrow divisions (the leaves). Colombia.
  bipennifolium bi-pen-i-*fo*-lee-um. With
  bipinnate leaves. Panda Plant. Brazil.
  bipinnatifidum bi-pi-nah-ti-*fi*-dum. (= *P.
  panduriforme* hort.). Bipinnately divided
  (the leaves). Brazil.
  domesticum do-*mes*-ti-kum. (= *P. hastatum*
  hort.). Cultivated. Cult.
  erubescens e-roo-*bes*-enz. Blushing, the rosy
  stipules and young leaves. Colombia.
  hastatum hort. see *P. domesticum*
  imbe im-bee. The native name. Brazil.
  laciniatum see *P. pedatum*
  melanochrysum me-la-*no*-kris-um. Black-
  gold, from the colour of the leaves.
  Colombia.
  panduriforme hort. see *P. bipinnatifidum*
  pedatum pe-*dah*-tum. (= *P. laciniatum*).
  Like a bird's foot (the leaves). S America.
  pinnatifidum pin-ah-ti-*fi*-dum. Pinnately
  divided (the leaves). Venezuela, Trinidad.
  sagittifolium să-gi-ti-*fo*-lee-um. With arrow-
  shaped leaves. S Mexico.
  scandens skăn-denz. Climbing. Tropical
  America.
  selloum se-*lō*-um. After Sello, see *Feijoa
  sellowiana*. S Brazil.

Phlomis *flo*-mis *Labiatae*. Gk. name for a
plant. Perennial herbs and evergreen shrubs.
  chrysophylla kris-o-*fil*-la. Golden-leaved

Phlomis (continued)
(referring to dried specimens). W Asia.
fruticosa froo-ti-*kō*-sa. Shrubby. Jerusalem
Sage. Mediterranean region.
italica ee-*tāl*-i-ka. Italian. Balearic Islands.
russelliana rū-sel-ee-*ah*-na. After Russell, a
physician at Aleppo. W Asia.
samia sah-mee-a. Of the island of Samos in
the Aegean. Greece.

Phlox floks *Polemoniaceae*. From Gk. *phlox*
(a flame). Annual and perennial herbs.
adsurgens ăd-*sur*-genz. Erect, the flowering
stems from prostrate shoots. Oregon,
California.
amoena a-*moy*-na. Pleasant. SE United
States.
bifida bi-fi-da. Divided into two (the
petals). E United States.
divaricata di-vah-ri-*kah*-ta. Spreading.
E N America.
douglasii dūg-*lăs*-ee-ee. After Douglas, see
Iris douglasii. W United States.
drummondii drū-*mond*-ee-ee. After Thomas
Drummond (c 1790–1835), who collected
in N America. Pride of Texas. Texas.
maculata măk-ew-*lah*-ta. Spotted (the
stems). E United States.
nana nah-na. Dwarf. SW United States,
N Mexico.
paniculata pa-nik-ew-*lah*-ta. With flowers
in panicles. E United States.
× *procumbens* prō-*kum*-benz. *P. stolonifera*
× *P. subulata*. Prostrate.
stolonifera sto-lō-*ni*-fe-ra. Bearing stolons.
E United States.
subulata soob-ew-*lah*-ta. Awl-shaped (the
leaves).

Phoenix fee-niks *Palmae*. The Gk. name.
Tender palms.
canariensis ka-nah-ree-*en*-sis. Of the Canary
Islands. Canary Island Date Palm.
dactylifera dăk-ti-*li*-fe-ra. Finger-bearing.
Date Palm. W Asia, N Africa.
roebelinii rur-be-*lin*-ee-ee. After M. Robelin
who collected for Sander's in SE Asia.
Miniature Date Palm. Laos.

Phormium *for*-mee-um *Agavaceae*. From Gk.
phormion (a mat), fibre is produced from the
leaves of *P. tenax*. Evergreen perennial herbs.
New Zealand.
cookianum kuk-ee-*ah*-num. (= *P. colensoi*).
After Captain Cook.
tenax ten-ahks. Tough. New Zealand Flax.

Photinia fō-*tin*-ee-a *Rosaceae*. From Gk.
photos (light) referring to the shining leaves
of some species. Deciduous and evergreen
shrubs and trees.
beauverdiana bō-ver-dee-*ah*-na. After
Gustave Beauverd (1867–1942), botanist
and artist. W China.
davidiana dă-vid-ee-*ah*-na. (= *Stranvaesia
davidiana*). After David who discovered it
in 1869, see *Davidia*. China.
× *fraseri* fray-za-ree. *P. glabra* × *P.
serrulata*. After the Fraser nurseries,
Alabama, who raised it.
glabra glăb-ra. Glabrous (the leaves).
Japan, China.
serrulata se-ru-*lah*-ta. With small teeth (the
leaves). China.
villosa vi-*lō*-sa. Softly hairy (the young
leaves and shoots). Japan, China, Korea.
laevis lie-vis. Smooth.

Phuopsis foo-*op*-sis *Rubiaceae*. From Gk.
phou (a kind of valerian) and -*opsis* indicating
resemblance. Perennial herb.
stylosa sti-*lō*-sa. With a prominent style.
Caucasus.

Phygelius foo-*gay*-lee-us *Scrophulariaceae*.
Probably from Gk. *phyge* (flight) and *helios*
(the sun). Semi-hardy sub-shrubs. S Africa.
aequalis ie-*kwah*-lis. Equal, the corolla
lobes, which are not in *P. capensis*.
capensis ka-*pen*-sis. Of the Cape of Good
Hope.
'Coccineus' kok-*kin*-ee-us. Scarlet.

Phyla fi-la *Verbenaceae*. From Gk. *phyla* (a
tribe) referring to the compound flower heads.
Perennial herb.
nodiflora nō-di-*flō*-ra. (= *Lippia nodiflora*).
With flowers borne from the nodes.
Widely distributed.

× **Phylliopsis** fil-lee-*op*-sis *Ericaceae*.
Intergeneric hybrid, from the names of the
parents. *Kalmiopsis* × *Phyllodoce*. Evergreen
shrub.
hillieri hi-lee-a-ree. *Kalmiopsis leachiana* ×
Phyllodoce breweri. After Hillier's in whose
nursery it was found.

Phyllitis fi-*li*-tis *Aspleniaceae*. From Gk.
phyllon (a leaf). Evergreen fern.
scolopendrium sko-lo-*pen*-dree-um. From
Gk. *skolopendra* (a millipede) referring to
the arrangement of the sori. Hart's-tongue
Fern. Europe.

Phyllocladus fi-*lo*-kla-dus *Podocarpaceae*.
From Gk. *phyllon* (a leaf) and *klados* (a
branch) referring to the flattened, leaf-like
shoots. Semi-hardy conifer.
   alpinus ăl-*peen*-us. Alpine. Celery Pine.
New Zealand.

Phyllodoce fi-*lo*-do-kee *Ericaceae*. After
Phyllodoce, a sea nymph. Dwarf, evergreen
shrubs.
   aleutica a-*loo*-ti-ka. Of the Aleutian Islands.
Japan to Alaska.
   breweri broo-a-ree. After Brewer, who
discovered it in 1862. California.
   caerulea kie-*ru*-lee-a. Deep blue (the
corolla). Arctic and alpine N hemisphere.
   empetriformis em-pet-ri-*form*-is. Like
Empetrum. W N America.
   × *intermedia* in-ter-*med*-ee-a. *P.*
empetriformis × *P. glanduliflora*.
Intermediate (between the parents).
W N America.
   nipponica ni-*pon*-i-ka. Of Japan.

Phyllostachys fi-*lo*-sta-kis *Gramineae*. From
Gk. *phyllon* (a leaf) and *stachys* (a spike)
referring to the leafy inflorescences.
Bamboos. China.
   aurea ow-ree-a. Golden (the canes).
   bambusoides băm-bew-*soi*-deez. Like
Bambusa, a related genus.
   flexuosa fleks-ew-*ō*-sa. Zig-zag (the stems).
   nigra nig-ra. Black (the old stems). Black
Bamboo.

Physalis *fi*-sa-lis *Solanaceae*. From Gk. *physa*
(a bladder) referring to the bladder-like
fruits. Perennial herbs.
   alkekengi ăl-ke-*ken*-jee. (= *P. franchetii*).
From *al kakendi* the Arabic name. Chinese
Lantern. SE Europe to Japan.
   peruviana pe-roo-vee-*ah*-na. Of Peru. Cape
Gooseberry. S America.

Physocarpus fi-sō-*kar*-pus *Rosaceae*. From
Gk. *physa* (a bladder) and *karpon* (a fruit)
referring to the inflated fruits. Deciduous
shrubs. Ninebark.
   opulifolius op-ew-li-*fo*-lee-us. With leaves
like *Viburnum opulus*. E N America.

Physoplexis fi-sō-*pleks*-is *Campanulaceae*.
From Gk. *physa* (a bladder) and *plexis*
(plaiting). Perennial herb.
   comosa ko-*mō*-sa. (= *Phyteuma comosum*).
Tufted. Devil's Claw. S Alps.

Physostegia fi-sō-*stee*-gee-a *Labiatae*. From
Gk. *physa* (a bladder) and *stege* (covering)
referring to the inflated calyx which covers
the fruit. Herbaceous perennial.
   virginiana vir-jin-ee-*ah*-na. Of Virginia.
Obedient Plant. E N America.

Phyteuma fi-*tew*-ma *Campanulaceae*. The
Gk. name for *Reseda phyteuma*. Perennial
herbs.
   comosum see *Physoplexis comosa*
   hemisphaericum he-mi-*sfie*-ri-kum.
Hemispherical (the flower heads).
S Europe.
   orbiculare or-bik-ew-*lah*-ree. Orbicular (the
flower heads). Rampion. Europe.
   spicatum spee-*kah*-tum. With flowers in
spikes. Spiked Rampion. Europe.

Phytolacca fi-tō-*lă*-ka *Phytolaccaceae*. From
Gk. *phyton* (a plant) and L. *lacca* referring
to the lac insect *Laccifer lacca* from which a
dye is obtained. A red dye is obtained from
the berries. Perennial herb.
   americana a-me-ri-*kah*-na. American.
Pokeweed. United States, Mexico.

Picea *pi*-kee-a *Pinaceae*. L. name for a pitch-
producing pine, from *pix* (pitch). Evergreen
conifers. Spruce.
   abies ă-bee-ayz. L. name for fir (*Abies*).
Norway Spruce. Europe.
      'Acrocona' ăk-rō-*kō*-na. With terminal
cones.
      'Clanbrassiliana' klăn-bra-sil-ee-*ah*-na.
After Lord Clanbrassil who grew the
original plant.
      'Nidiformis' nee-di-*form*-is. Nest-
shaped.
   asperata ă-spe-*rah*-ta. Rough (the foliage).
Dragon Spruce. W China.
   brachytyla bră-kee-*tee*-la. With short
swellings. China.
   breweriana broo-a-ree-*ah*-na. After William
Henry Brewer (1828–1910), American
botanist who discovered it. Brewer Spruce.
California, Oregon.
   glauca glow-ka. Glaucous (the leaves).
White Spruce. N America.
   jezoensis ye-zō-*en*-sis. Of Hokkaido (Yezo),
Japan. Yezo Spruce. NE Asia.
      hondoensis hon-dō-*en*-sis. Of Honshu
(Hondo), Japan. Hondo Spruce.
   koyamae koy-*ah*-mie. After Mitsuo Koyama
(1885–1935) who discovered it in 1911.
Japan, Korea.
   likiangensis li-kee-ang-*gen*-sis. Of Likiang,

Picea (continued)
Yunnan. W China, Tibet.
  purpurea see *P. purpurea*
mariana mă-ree-*ah*-na. Of Maryland. Black
Spruce. N America.
omorika o-mo-ri-ka. The native name.
Serbian Spruce. Yugoslavia.
orientalis o-ree-en-*tah*-lis. Eastern. Turkey,
Caucasus.
pungens *pung*-genz. Sharp-pointed (the
leaves). Colorado Spruce. W United States.
  Glauca *glow*-ka. Glaucous (the leaves).
purpurea pur-*pewr*-ree-a. (= *P. likiangensis
purpurea*). Purple (the young cones). China.
sitchensis sit-*ken*-sis. Of Sitka, Alaska. Sitka
Spruce. N California to Alaska.
smithiana smith-ee-*ah*-na. After Sir James
Edward Smith (1759–1828). W Himalaya.

Pickerel Weed see *Pontederia cordata*

Picrasma pi-*krăs*-ma *Simaroubaceae*. From
Gk. *pikra* (a bitter taste) referring to the
bitter leaves and wood. Deciduous tree.
quassioides kwah-see-*oi*-deez. Like *Quassia
amara*. Himalaya, E Asia.

Pieris *pee*-e-ris *Ericaceae*. From *Pierides* a
name of the Muses (goddesses of the arts).
Evergreen shrubs.
floribunda flŏ-ri-*bun*-da. Profusely
flowering. SE United States.
formosa for-*mō*-sa. Beautiful. Himalaya,
China.
  forrestii fo-*rest*-ee-ee. After Forrest who
  introduced it, see *Abies delavayi forrestii*.
  W China.
japonica ja-*pon*-i-ka. Of Japan.
taiwanensis tie-wahn-*en*-sis. Of Taiwan.

Piggy-back Plant see *Tolmiea menziesii*

Pilea *pee*-lee-a *Urticaceae*. From L. *pileus* (a
cap), referring to the female flowers. Tender,
evergreen herbs.
cadierei kă-dee-*e*-ree-ee. After R. P. Cadiere
who collected it in 1938. Aluminium
Plant. Vietnam.
involucrata in-vo-loo-*krah*-ta. With an
involucre, the leaves are clustered beneath
the flowers. Friendship Plant. C and S
America.
microphylla mik-rō-*fil*-la. Small-leaved.
Artillery Plant. Tropical America.
nummulariifolia num-ew-lah-ree-i-*fo*-lee-a.
With coin-shaped leaves. Creeping
Charlie. C and S America.
repens ree-penz. Creeping. W Indies.

Pilea (continued)
spruceana sproo-see-*ah*-na. After Richard
Spruce (1817–93), who collected in
S America.

Pileostegia pee-lee-ō-*stee*-gee-a
Hydrangeaceae. From Gk. *pilos* (a cap) and
stege (a covering) referring to the corolla.
Evergreen climber.
viburnoides vee-burn-*oi*-deez. Like
Viburnum (the flower heads). NE India,
China, Taiwan.

Pilosocereus palmeri see *Cephalocereus
palmeri*

Pimelea pi-*me*-lay-a *Thymelaeaceae*. From
Gk. *pimele* (fat) referring to the high oil
content of the seeds. Tender, evergreen
shrubs.
ferruginea fe-roo-*gin*-ee-a. Rusty.
W Australia.
longiflora long-gi-*flō*-ra. Long-flowered.
W Australia.
prostrata pros-*trah*-ta. Prostrate. New
Zealand.
rosea ro-see-a. Rose-coloured (the flowers).
W Australia.
spectabilis spek-*tah*-bi-lis. Spectacular.
W Australia.

Pimpernel, Bog see *Anagallis tenella*
  Scarlet see *A. arvensis*
Pincushion Flower see *Hakea laurina*
Pine see *Pinus*
  Aleppo see *P. halepensis*
  Arolla see *P. cembra*
  Austrian see *P. nigra*
  Balkan see *P. peuce*
  Beach see *P. contorta*
  Bhutan see *P. wallichiana*
  Big-cone see *P. coulteri*
  Bishop see *P. muricata*
  Bosnian see *P. heldreichii leucodermis*
  Bristle-cone see *P. aristata*
  Chinese see *P. tabuliformis*
  Chinese White see *P. armandii*
  Corsican see *P. nigra maritima*
  Digger see *P. sabiniana*
  Dwarf Mountain see *P. mugo*
  Eastern White see *P. strobus*
  Himalayan see *P. wallichiana*
  Japanese Black see *P. thunbergii*
  Japanese Red see *P. densiflora*
  Japanese White see *P. parviflora*
  Knobcone see *P. attenuata*
  Lacebark see *P. bungeana*
  Limber see *P. flexilis*

Pine (continued)

Lodgepole see *P. contorta latifolia*
Macedonian see *P. peuce*
Maritime see *P. pinaster*
Mexican White see *P. ayacahuite*
Monterey see *P. radiata*
Northern Pitch see *P. rigida*
Scots see *P. sylvestris*
Stone see *P. pinea*
Sugar see *P. lambertiana*
Umbrella see *P. pinea*
Western White see *P. monticola*
Western Yellow see *P. ponderosa*
Weymouth see *P. strobus*
Whitebark see *P. albicaulis*
Yunnan see *P. yunnanensis*
Pineapple see *Ananas comosus*
Wild see *A. bracteatus*

Pinguicula ping-*gwi*-kew-la *Lentibulariaceae*. From L. *pinguis* (fat) referring to the greasy appearance of the leaves. Carnivorous herbs. Butterwort.
alpina ăl-*peen*-a. Alpine. Europe.
caudata kaw-*dah*-ta. (= *P. bakeriana*). With a tail, referring to the long, slender spur. Mexico.
grandiflora grănd-i-*flō*-ra. Large-flowered. W Europe.
gypsicola gip-*si*-ko-la. Lime-loving. Mexico.
vulgaris vul-*gah*-ris. Common. N America, Europe, Asia.

Pink see *Dianthus*
Cheddar see *D. gratianopolitanus*
Clove see *D. caryophyllus*
Fringed see *D. superbus*
Glacier see *D. glacialis*
Indian see *D. chinensis*
Maiden see *D. deltoides*
Pink Sand Verbena see *Abronia umbellata*
Pink Siris see *Albizia julibrissin*

Pinus *pee*-nus *Pinaceae*. The L. name. Evergreen conifers. Pine.
albicaulis ăl-bi-*kaw*-lis. White-stemmed. Whitebark Pine. W N America.
aristata ă-ris-*tah*-ta. Awned (the slender cone bristles). Bristlecone Pine. SW United States.
armandii ar-*mond*-ee-ee. After Armand David who discovered it, see *Davidia*. Chinese White Pine. W and C China.
attenuata a-ten-ew-*ah*-ta. Drawn out (the cones). Knobcone Pine. W United States, N Mexico.
ayacahuite ie-a-ka-*hee*-tee. The Mexican

Pinus (continued)
name, Mexican White Pine. Mexico, Guatemala.
bungeana bung-gee-*ah*-na. After Alexander von Bunge (1803–90), Russian botanist who collected the type specimen. Lacebark Pine. China.
cembra *kem*-bra. The Italian name. Arolla Pine. Alps to the Carpathians.
contorta kon-*tor*-ta. Twisted (the young shoots). Beach Pine. W N America (coastal).
latifolia lah-tee-*fo*-lee-a. Broad-leaved. Lodgepole Pine. Rocky Mountains.
coulteri *kool*-ta-ree. After its discoverer. Thomas Coulter (1793–1843), an Irish physician. Big-cone Pine. California.
densiflora den-si-*flō*-ra. Densely flowered. Japanese Red Pine. Japan, Korea, China.
'Oculis-draconis' *ok*-ew-lis-dra-*kō*-nis. Dragon's-eye.
'Umbraculifera' um-brahk-ew-*li*-fe-ra. Umbrella-bearing.
flexilis *fleks*-i-lis. Flexible (the young shoots). Limber Pine. W N America.
halepensis hă-le-*pen*-sis. Of Aleppo (Syria). Aleppo Pine. S Europe, W Asia.
heldreichii hel-*driek*-ee-ee. After Theodor von Heldreich (1822–1902). SE Europe.
leucodermis loo-ko-*derm*-is. (= *P. leucodermis*). White-skinned, the bark of old trees. Bosnian Pine. Yugoslavia, Albania.
jeffreyi *jef*-ree-ee. After John Jeffrey (1826–54), a gardener at the Edinburgh Botanic Garden, who discovered it. W United States, N Mexico.
lambertiana lăm-bert-ee-*ah*-na. After Aylmer Bourke Lambert (1761–1842). Sugar Pine. W United States.
leucodermis see *P. heldreichii leucodermis*
montezumae mon-tee-*zoo*-mie. After Montezuma, an Aztec Emperor. S and C Mexico.
monticola mon-*ti*-ko-la. Growing on mountains. Western White Pine. W N America.
mugo *mew*-gō. An old Tyrolese name. Dwarf Mountain Pine. C and SE Europe.
muricata mew-ri-*kah*-ta. Rough with spines (the cone). Bishop Pine. California.
nigra *nig*-ra. Black (the bark). Austrian Pine. SC and SE Europe.
maritima ma-*ri*-ti-ma. Growing near the sea. Corsican Pine. S Italy, Sicily, Corsica, N Africa.
'Hornibrookiana' horn-i-bruk-ee-*ah*-na. After Murray Hornibrook, an authority

Pinus (continued)
on dwarf conifers.
parviflora par-vi-*flō*-ra. Small-flowered.
Japanese White Pine. Japan.
peuce poy-kay. Classical name of a pine.
Macedonian Pine, Balkan Pine.
Yugoslavia.
pinaster peen-*ăs*-ter. L. name meaning
inferior to the cultivated pine, *P. pinea*.
Maritime Pine. SW Europe, N Africa.
pinea pee-nee-a. L. name for pine-nuts,
which in this tree are edible. Stone Pine,
Umbrella Pine. S Europe, W Asia.
ponderosa pon-de-*rō*-sa. Heavy (the wood).
Western Yellow Pine. W N America.
radiata ră-dee-*ah*-ta. Radiating (lines on the
cone scales). Monterey Pine. Monterey,
California.
rigida ri-gi-da. Rigid (the leaves). Northern
Pitch Pine. E N America.
sabiniana sa-been-ee-*ah*-na. After Sabine.
Digger Pine. California.
strobus stro-bus. L. name of a gum-yielding
tree. Eastern White Pine, Weymouth
Pine. E N America.
sylvestris sil-*ves*-tris. Of woods. Scots Pine.
N Europe, N Asia.
'Beuvronensis' burv-ron-*en*-sis. Of
Beuvron, near Orléans, France.
tabuliformis tăb-ew-li-*form*-is. Flat-topped.
Chinese Pine. China.
yunnanensis see *P. yunnanensis*
thunbergii thun-*berg*-ee-ee. After Thunberg,
see *Thunbergia*. Japanese Black Pine.
Japan.
wallichiana wo-lik-ee-*ah*-na. After
Nathaniel Wallich (1786–1854). Danish
surgeon and botanist with the E India Co.
Bhutan Pine, Himalayan Pine. Himalaya.
yunnanensis yoo-nan-*en*-sis. (= *P.
tabuliformis yunnanensis*). Of Yunnan.
Yunnan Pine. W China.

Piptanthus pip-*tănth*-us *Leguminosae*. From
Gk. *pipto* (to fall) and *anthos* (a flower), the
flowers fall intact. Evergreen shrub.
nepalensis ne-pa-*len*-sis. (= *P.
laburnifolius*). Of Nepal. Himalaya.

Pistacia pis-*tah*-kee-a *Anacardiaceae*. From
Gk. *pistake*, pistachio nut, obtained from
another species. Deciduous shrub or tree.
chinensis chi-*nen*-sis. Of China.

Pistia pis-tee-a *Araceae*. From Gk. *pistos*
(water) referring to its habitat. Floating,
tender perennial herb.
stratiotes stră-tee-*ō*-teez. The Gk. name.

Pistia (continued)
Water Lettuce. Tropics and subtropics.

Pisum *pee*-sum *Leguminosae*. The L. name.
Annual herb.
sativum sa-*teev*-um. Cultivated. Garden
Pea. Europe, Asia.

Pitcher Plant see *Nepenthes, Sarracenia*
California see *Darlingtonia californica*
Northern see *Sarracenia purpurea*
Yellow see *S. flava*

Pittosporum pi-*tos*-po-rum *Pittosporaceae*.
From Gk. *pitta* (pitch) and *sporum* (a seed)
referring to the sticky seeds. Tender and
semi-hardy, evergreen trees and shrubs.
adaphniphylloides a-daf-ni-fil-*oi*-deez. Not
P. daphniphylloides which it was originally
thought to be. China.
crassifolium krăs-i-*fo*-lee-um. Thick-leaved.
New Zealand.
dallii dăl-ee-ee. After J. Dall (died 1912),
who discovered it in 1905. New Zealand.
eugenioides ew-jeen-ee-*oi*-deez. Like
Eugenia. New Zealand.
'Garnettii' gar-*net*-ee-ee. *P. ralphii* × *P.
tenuifolium*. After Garnett.
ralphii rălf-ee-ee. After Dr Ralph who
collected the type specimen. New Zealand.
tenuifolium ten-ew-i-*fo*-lee-um. With thin
leaves. New Zealand.
tobira to-*bi*-ra. The native name. E Asia.
undulatum un-dew-*lah*-tum. Wavy-edged
(the leaves). E Australia.

Plagianthus plă-gee-*ănth*-us *Malvaceae*.
From Gk. *plagios* (oblique) and *anthos* (a
flower) referring to the asymmetrical flowers.
Deciduous shrubs and trees. New Zealand.
betulinus bet-ew-*leen*-us. Like *Betula*.
divaricatus di-vah-ri-*kah*-tus. Spreading.

Plane see *Platanus*
London see *P.* × *acerifolia*
Oriental see *P. orientalis*
Plantain Lily see *Hosta*

Platanus plă-ta-nus *Platanaceae*. From
platanos the Gk. name for *P. orientalis*.
Deciduous trees. Plane.
acerifolia a-ke-ri-*fo*-lee-a. (= *P. hispanica*).
Possibly *P. occidentalis* × *P. orientalis*.
With leaves like *Acer*. London Plane. Cult.
orientalis o-ree-en-*tah*-lis. Eastern. Oriental
Plane. SE Europe.

Platycarya plă-tee-*kă*-ree-a *Juglandaceae*.
From Gk. *platys* (broad) and *karyon* (a nut)
referring to the fruit. Deciduous tree.
  strobilacea stro-bi-*lah*-kee-a. Cone-like (the
  fruit). E Asia.

Platycerium plă-tee-*ke*-ree-um
Polypodiaceae. From Gk. *platys* (broad) and
keras (a horn) referring to the flattened, horn-
like fronds. Tender Ferns.
  bifurcatum bi-fur-*kah*-tum. (= *P. alcicorne*
  hort.). Forked into two. Elk's-horn Fern,
  Stag's-horn Fern. Australia, Polynesia.
  grande grănd-ee. Large. Australia.

Platycodon plă-tee-*kō*-don *Campanulaceae*.
From Gk. *platys* (broad) and *kodon* (a bell)
referring to the shape of the corolla. Perennial
herbs.
  grandiflorus grănd-i-*flō*-rus. Large-flowered.
  Balloon Flower. E Asia.
    mariesii ma-*reez*-ee-ee. After Maries, see
    Davallia mariesii.

Platystemon plă-tee-*stay*-mon *Papaveraceae*.
From Gk. *platys* (broad) and *stemon* (a
stamen) referring to the broad stamens.
Annual herb.
  californicus kăl-i-*forn*-i-kus. Of California.
  Cream Cups. S California, N Mexico.

Plecostachys ple-*ko*-sta-kis *Compositae*.
From Gk. *plecos* (to plait) and *stachys* (a
spike) referring to the inflorescence. Annual
herb or semi-hardy sub-shrub.
  serpyllifolia ser-pi-li-*fo*-lee-a. (=
  Helichrysum microphyllum hort.). Thyme-
  leaved.
  S Africa.

Plectranthus plek-*trănth*-us *Labiatae*. From
Gk. *plectron* (a spur) and *anthos* (a flower)
referring to the spurred flowers of the type
species. Tender perennial herbs.
  australis ow-*strah*-lis. Southern. Swedish
  Ivy. SE Australia.
  coleoides ko-lee-*oi*-deez. Like *Coleus*. SW
  India.
  nummularius num-ew-*lah*-ree-us. Coin-like
  (the leaves). S Africa.
  oertendahlii ur-tan-*dahl*-ee-ee. After
  Oertendahl. Candle Plant. Cult.

Pleione *play*-o-nay *Orchidaceae*. After
Pleione, wife of Atlas and mother of the
Pleiades. Semi-hardy or cool greenhouse
orchids.
  bulbocodioides bul-bō-kō-dee-*oi*-deez. (= *P.*

Pleione (continued)
  formosana). Like *Bulbocodium*. Tibet,
  China, Taiwan.
  hookeriana huk-a-ree-*ah*-na. After Sir
  Joseph Hooker. Himalaya, Assam, Tibet.
  humilis *hu*-mi-lis. Low-growing. Himalaya,
  Burma, Tibet.
  praecox *prie*-koks. Early (flowering).
  Himalaya to W China and Thailand.
  pricei *pries*-ee-ee. After William Robert
  Price (1886–1975), who collected in
  Taiwan. Taiwan.

Pleiospilos play-*os*-pi-los *Aizoaceae*. From
Gk. *pleios* (many) and *spilos* (a spot) referring
to the spotted leaves. Tender succulent.
  bolusii bō-*lus*-ee-ee. After Harry Bolus
  (1834–1911). S Africa.

Pleomele reflexa see *Dracaena reflexa*
Plover Eggs see *Adromischus cooperi*
Plum see *Prunus domestica*
Plum Yew see *Cephalotaxus*

Plumbago plum-*bah*-gō *Plumbaginaceae*.
The L. name from *plumbum* (lead). Tender
climbers. Leadwort.
  auriculata ow-rik-ew-*lah*-ta. (= *P.*
  capensis). With auricles (the leaves). Cape
  Leadwort. S Africa.
  indica *in*-di-ka. Indian. Scarlet Leadwort.
  SE Asia.

Plume Bush see *Calomeria amaranthoides*

Plumeria ploo-*me*-ree-a *Apocynaceae*. After
Charles Plumier (1646–1704), French monk
and botanist. Tender tree.
  rubra rub-ra. Red (the flowers). Frangipani.
  Mexico, C America.

Poached Egg Flower see *Limnanthes douglasii*

Podocarpus pod-o-*kar*-pus *Podocarpaceae*.
From Gk. *podos* (a foot) and *karpos* (a fruit)
referring to the stalked fruit. Evergreen
conifers.
  alpinus ăl-*peen*-us. Alpine. Tasmania, New
  South Wales.
  andinus ăn-*deen*-us. Of the Andes. Chile,
  Argentina.
  macrophyllus măk-rō-*fil*-lus. Large-leaved.
  Japan.
  nivalis ni-*vah*-lis. Growing near snow. New
  Zealand.
  nubigenus new-bi-*gen*-us. From among the
  clouds. Chile, Argentina.
  salignus sa-*lig*-nus. Willow-like (the leaves).
  Chile.

Podocarpus (continued)
totara tō-*tah*-ra. The native name. New Zealand.

Podophyllum pod-o-*fil*-lum *Berberidaceae*. From *anapodophyllum*, Gk. *anas* (a duck), *podos* (a foot) and *phyllon* (a leaf), referring to the leaves of *P. peltatum*. Perennial herbs.
hexandrum heks-*ăn*-drum. (= *P. emodi*). With six stamens. Himalaya, W China.
peltatum pel-*tah*-tum. Peltate (the leaves). E N America.

Poinsettia see *Euphorbia pulcherrima*
Annual see *E. cyathophora*
Poke Weed see *Phytolacca americana*

Polemonium po-lee-*mō*-nee-um *Polemoniaceae*. From *polemonion* the Gk. name of a plant. Perennial herbs.
caeruleum kie-*ru*-lee-um. Deep blue (the flowers). Jacob's Ladder. Europe, Asia.
carneum *kar*-nee-um. Flesh-coloured (the flowers). Oregon, California.
foliosissimum fo-lee-ō-*sis*-i-mum. Very leafy. W United States.
pauciflorum paw-si-*flō*-rum. Few-flowered. SW United States, Mexico.
reptans *rep*-tănz. Creeping. E United States.

Polianthes po-lee-*ănth*-eez *Agavaceae*. From Gk. *polios* (grey) and *anthos* (a flower). Tender perennial herb.
tuberosa tew-be-*rō*-sa. Tuberous. Tuberose. Cult.

Policeman's Helmet see *Impatiens glandulifera*
Polka-dot Plant see *Hypoestes phyllostachya*
Polyanthus see *Primula* Polyantha

Polygala po-*li*-ga-la *Polygalaceae*. The classical name from Gk. *polys* (much) and *gala* (milk), they were thought to enhance milk secretion. Herbs and shrubs.
calcarea kăl-*kah*-ree-a. Growing on chalk. Europe.
chamaebuxus kă-mie-*buks*-us. Dwarf *Buxus*. Europe.
vayredae vay-*ree*-die. After Vayreda who rediscovered it in 1877. Pyrenees.

Polygonatum po-li-go-*nah*-tum *Liliaceae*. From *polygonaton*, the Gk. name, from *polys* (many) and *gony* (a knee), referring to the jointed rhizome. Perennial herbs. Solomon's Seal.
biflorum bi-*flō*-rum. Two-flowered. E United States.

Polygonatum (continued)
commutatum kom-ew-*tah*-tum. Changeable. United States.
hookeri hu-ka-ree. After Sir Joseph Hooker. Himalaya, W China.
× *hybridum* hib-ri-dum. *P. multiflorum* × *P. odoratum*. Hybrid.
multiflorum mul-tee-*flō*-rum. Many-flowered. Europe, Asia.
odoratum o-dō-*rah*-tum. Scented. Europe, Asia.
verticillatum ver-ti-ki-*lah*-tum. Whorled (the leaves). Europe to the Himalaya.

Polygonum po-*li*-go-num *Polygonaceae*. From Gk. *polys* (many) and *gony* (a knee) referring to the jointed stems. Perennial herbs and climbers.
affine a-*fee*-nee. Related to. Afghanistan, Himalaya, Tibet.
amphibium ăm-*fi*-bee-um. Growing in water or on land. Widely distributed.
amplexicaule ăm-pleks-i-*kaw*-lee. With stem-clasping leaves. Afghanistan, Himalaya, China.
aubertii ō-*bair*-tee-ee. After the French missionary Aubert who introduced it. China.
baldschuanicum băld-shoo-*ăn*-i-kum. Of Balzhuan, C Asia. S Tadzhikstan.
bistorta bis-*tor*-ta. A medieval name, from L. *bis* (twice) and *tortus* (twisted) referring to the twisted roots. Bistort. Europe to C Asia.
campanulatum kăm-păn-ew-*lah*-tum. Bell-shaped (the corolla). Himalaya.
capitatum kă-pi-*tah*-tum. In a dense head (the flowers). Himalaya to W China.
macrophyllum măk-rō-*fil*-lum. Large-leaved. Himalaya, China.
multiflorum mul-tee-*flō*-rum. Many-flowered. Japan.
sphaerostachyum sfie-rō-*stăk*-ee-um. With spherical flower heads. Himalaya.
tenuicaule ten-ew-i-*kaw*-lee. Slender-stemmed. Japan.
vacciniifolium va-keen-ee-i-*fo*-lee-um. *Vaccinium*-leaved. Himalaya, Tibet.

Polypodium po-lee-*pod*-ee-um *Polypodiaceae*. From Gk. *polys* (many) and *podos* (a foot) referring to the branched rhizome. Ferns.
aureum *ow*-ree-um. Golden (the sori). Hare's-foot Fern. Tropical America.
vulgare vul-*gah*-ree. Common. Common Polypody. Europe, Asia.

Polypody, Common see *Polypodium vulgare*
  Limestone see *Gymnocarpium robertianum*

Polyscias po-*lis*-kee-as *Araliaceae*. From Gk.
polys (many) and *skias* (an umbel) referring
to the numerous umbels in the inflorescence.
Tender, evergreen shrubs.
  balfouriana bål-for-ree-*ah*-na. After Sir
  Isaac Bayley Balfour (1853–1922), who
  collected in Socotra. New Caledonia.
  guilfoylei gil-*foyl*-ee-ee. After W. R.
  Guilfoyle (1840–1912), who collected in
  the S Pacific Islands. Polynesia.

Polystichum po-*li*-sti-kum *Aspidiaceae*.
From Gk. *polys* (many) and *stichos* (a row)
referring to the arrangement of the sori.
Ferns.
  acrostichoides a-kro-sti-*koi*-deez. Like
  Acrostichum, a related genus. Christmas
  Fern, Holly Fern. E N America.
  aculeatum a-kew-lee-*ah*-tum. Prickly. Hard
  Shield Fern. Widely distributed.
  falcatum fål-*kah*-tum. (= *Cyrtomium
  falcatum*). Sickle-shaped (the pinnae).
  Holly Fern. E Asia.
  munitum mew-*nee*-tum. Armed (with teeth).
  American Sword Fern. W N America.
  setiferum say-*ti*-fe-rum. Bristly. Soft Shield
  Fern. Europe.
  tsu-simense tsoo-see-*men*-see. Of Tsu-shima,
  Japan. E Asia.

Pomegranate see *Punica granatum*

Poncirus pon-*si*-rus *Rutaceae*. From the
French name of a kind of citron. Deciduous
shrub.
  trifoliata tri-fo-lee-*ah*-ta. With three-leaves
  (leaflets). China.

Pond Cypress see *Taxodium ascendens*

Pontederia pon-te-*de*-ree-a *Pontederiaceae*.
After Guilo Pontedera (1688–1757), a
professor of botany at Padua. Aquatic
perennial herb.
  cordata kor-*dah*-ta. Heart-shaped (the
  leaves). Pickerel Weed. E N America.

Poor Man's Orchid see *Schizanthus*
Poplar see *Populus*
  Balsam see *P. balsamifera*
  Black see *P. nigra*
  Grey see *P. canescens*
  Lombardy see *P. nigra* 'Italica'
  White see *P. alba*

Poppy see *Papaver*
  Alpine see *Papaver alpinum*
  Arctic see *P. nudicaulis*
  Californian see *Eschscholzia californica*
  Celandine see *Stylophorum diphyllum*
  Crested see *Argemone platyceras*
  Harebell see *Meconopsis quintuplinervis*
  Himalayan Blue see *Meconopsis
  betonicifolia*
  Iceland see *Papaver nudicaule*
  Mexican Tulip see *Hunnemannia
  fumariifolia*
  Opium see *Papaver somniferum*
  Plume see *Macleaya microcarpa*
  Prickly see *Argemone mexicana*
  Red Horned see *Glaucium corniculatum*
  Spanish see *Papaver rupifragum*
  Tree see *Romneya*
  Tulip see *Papaver glaucum*
  Welsh see *Meconopsis cambrica*
  Yellow Horned see *Glaucium flavum*

Populus *pō*-pu-lus *Salicaceae*. The L. name.
Deciduous trees. Poplar.
  alba ål-ba. White (the undersides of the
  leaves). Abele, White Poplar. Europe,
  N Africa to C Asia.
  balsamifera bål-sa-*mi*-fe-ra. Balsam-
  bearing. Balsam Poplar. N America.
  × *berolinensis* be-ro-leen-*en*-sis. *P. laurifolia*
  × *P. nigra* 'Italica'. Of Berlin, where it
  originated.
  candicans kån-di-kånz. White (the
  undersides of the leaves). Balm of Gilead.
  Cult.
  canescens kah-*nes*-enz. Grey-hairy (the
  leaves). Grey Poplar. Europe.
  lasiocarpa lå-see-ō-*kar*-pa. With woolly
  fruits. China.
  nigra nig-ra. Black (the bark). Black Poplar.
  Europe, W Asia.
    betulifolia bet-ew-li-*fo*-lee-a. With leaves
    like *Betula*. W Europe.
    'Italica' ee-*tål*-i-ka. Italian. Lombardy
    Poplar.
    'Robusta' rō-*bus*-ta. Robust.
    'Serotina' se-*ro*-ti-na. Late (into leaf).
  tremula *trem*-ew-la. Trembling (the leaves).
  Aspen. Europe, N Africa, N Asia.
  trichocarpa tri-kō-*kar*-pa. With hairy fruits.
  W N America.
  wilsonii wil-*son*-ee-ee. After Wilson who
  introduced it in 1907, see *Magnolia
  wilsonii*. China.

Portugal Laurel see *Prunus lusitanica*

Portulaca por-tew-*lah*-ka *Portulacaceae*. The L. name for *P. oleracea*. Succulent annuals.
grandiflora gränd-i-*flō*-ra. Large-flowered. Sun Plant. S America.
oleracea o-le-*rah*-kee-a. Vegetable-like. Purslane, India.

Potamogeton po-ta-mo-*gay*-ton *Potamogetonaceae*. From the classical name, from Gk. *potamos* (a river) and *geiton* (a neighbour). Aquatic herbs. Widely distributed.
crispus kris-pus. Wavy (the leaves).
pectinatus pek-ti-*nah*-tus. Comb-like (the arrangement of the leaves).
perfoliatus per-fo-lee-*ah*-tus. With the leaf base surrounding the stem.

Potato see *Solanum tuberosum*

Potentilla po-ten-*til*-la *Rosaceae*. From L. *potens* (powerful) referring to medicinal properties. Perennial herbs and shrubs.
alba ăl-ba. White (the flowers). Europe, Caucasus.
alchemilloides ăl-ke-mil-*oi*-des. Like *Alchemilla*. Pyrenees.
arbuscula ar-*bus*-kew-la. Like a small tree. Himalaya, China.
 'Beesii' *beez*-ee-ee. After Messrs Bees who distributed it.
argyrophylla ar-gi-rō-*fil*-la. With silvery leaves. Himalaya.
 atrosanguinea aht-rō-sang-*gwin*-ee-a. (= *P. atrosanguinea*). Deep red (the flowers).
atrosanguinea see *P. argyrophylla atrosanguinea*
aurea ow-ree-a. Golden (the flowers). Europe.
 chrysocraspeda kris-ō-*krăs*-pe-da. (= *P. ternata*). Golden-fringed. SE Europe.
calabra ka-*lăb*-ra. Of Calabria. Italy, Sicily.
cinerea ki-*ne*-ree-a. (= *P. tommasiniana*). Grey-hairy. Alps.
crantzii *krănts*-ee-ee. After H. J. N. von Crantz (1722–99) a botanical writer. Europe, W Asia.
eriocarpa e-ree-ō-*kar*-pa. With woolly fruits. Himalaya.
fragiformis frah-gi-*form*-is. Strawberry-like. NE Asia.
fruticosa froo-ti-*kō*-sa. Shrubby. N temperate regions.
nepalensis ne-pa-*len*-sis. Of Nepal. Himalaya.
nitida *ni*-ti-da. Shining (the flowers). Alps.

Potentilla (continued)
recta *rek*-ta. Erect. Europe, N Africa.
rupestris roo-*pes*-tris. Growing on rocks. Europe, Asia, W N America.
tabernaemontani tă-ber-nie-mon-*tah*-nee. (= *P. verna*). After Tabernaemontanus, see *Amsonia tabernaemontani*. Europe.
ternata see *P. aurea chrysocraspeda*
tommasiniana see *P. cinerea*
verna see *P. tabernaemontani*

× **Potinara** po-ti-*nah*-ra *Orchidaceae*. Intergeneric hybrid, after M. Potin, president of the French orchid society in 1922. *Brassavola* × *Cattleya* × *Laelia* × *Sophronitis*. Greenhouse orchids.

Powder Puff see *Mammillaria bocasana*

Pratia prah-tee-a *Lobeliaceae*. After Ch. L. Prat-Bernon, a French naval officer. Perennial herb.
angulata ang-gew-*lah*-ta. (= *P. treadwellii*). Angled. New Zealand.

Prayer Plant see *Maranta leuconeura*
Prickly Moses see *Acacia verticillata*
Prickly Pear see *Opuntia*
Prickly Thrift see *Acantholimon*
Pride of Texas see *Phlox drummondii*
Primrose see *Primula vulgaris*
 Fairy see *P. malacoides*

Primula preem-ew-la *Primulaceae*. From L. *primus* (first), referring to the early flowers. Perennial herbs.
acaulis see *P. vulgaris*
allionii ă-lee-ō-nee-ee. After Carlo Allioni (1705–1804), Italian botanist. Maritime Alps.
alpicola ăl-*pi*-ko-la. Growing on mountains. Tibet.
amoena a-*moy*-na. Pleasant. Caucasus.
aurantiaca ow-răn-tee-*ah*-ka. Orange (the flowers). China.
auricula ow-*rik*-ew-la. An old name, from L. *auricula* (an ear), the leaves have been likened to bear's ears. Auricula. Europe.
beesiana beez-ee-*ah*-na. After Bees Nursery. China.
bracteata brăk-tee-*ah*-ta. With bracts. W China.
bulleyana bu-lee-*ah*-na. After A. K. Bulley, see *Iris bulleyana*. China.
burmanica bur-*măn*-i-ka. Of Burma. China, Burma.
capitata kă-pi-*tah*-ta. In a dense head (the flowers). Himalaya.

Primula (continued)

chionantha kee-on-*ănth*-a. With snow-white flowers. China.

clarkei klark-ee-ee. After C. B. Clarke who discovered it in 1876. Kashmir.

cockburniana kō-burn-ee-*ah*-na. After H. Cockburn of the Consular Service, Chungking and the Rev. G. Cockburn who assisted the collector, A. E. Pratt. China.

cortusoides kor-tew-*soi*-deez. Like *Cortusa*. W Siberia.

denticulata den-tik-ew-*lah*-ta. Slightly toothed. Afghanistan, Himalaya, Burma.

edgeworthii ej-*werth*-ee-ee. After Edgeworth, see *Edgeworthia*. Himalaya.

elatior ay-*lah*-tee-or. Taller. Oxlip. Europe, W Asia.

florindae flo-*rin*-die. Discovered by Kingdon-Ward and named after his first wife Florinda. SE Tibet.

frondosa fron-*dō*-sa. Leafy. Bulgaria.

helodoxa he-lō-*doks*-a. Glory of the marsh. China.

hirsuta hir-*soo*-ta. (= *P. rubra*). Hairy. Alps, Pyrenees.

japonica ja-*pon*-i-ka. Of Japan.

× *kewensis* kew-*en*-sis. *P. floribunda* × *P. verticillata*. Of Kew where it was raised in 1898.

littoniana see *P. vialii*

malacoides mă-la-*koi*-deez. Mallow-like. Fairy Primrose. China.

marginata mar-gi-*nah*-ta. Margined (the leaves). SW Alps.

minima mi-ni-ma. Smaller. E Europe.

nutans new-tănz. Nodding (the flowers). China.

obconica ob-*kō*-ni-ka. Like an inverted cone (the calyx). China.

Polyantha po-lee-*ănth*-a. Many-flowered. Polyanthus.

polyneura po-lee-*new*-ra. With many veins. China.

prolifera prō-*li*-fe-ra. Proliferous. Assam.

pulverulenta pul-ve-ru-*len*-ta. Mealy. China.

rosea ro-see-a. Rose-coloured. Himalaya.

rubra see *P. hirsuta*.

secundiflora se-kun-di-*flō*-ra. With flowers on one side of the stalk. Himalaya.

sieboldii see-*bōld*-ee-ee. After Siebold, see *Acanthopanax sieboldii*. Japan.

sikkimensis si-kim-*en*-sis. Of Sikkim. Himalaya, W China.

sinensis si-*nen*-sis. Of China.

spectabilis spek-*tah*-bi-lis. Spectacular. S Alps.

veris ve-ris. Of spring. Cowslip. Europe.

vialii vee-*ahl*-ee-ee. (= *P. littoniana*). After

Primula (continued)

Père Vial. China.

vulgaris vul-*gah*-ris. (= *P. acaulis*). Common. Primrose. Europe.

yargongensis yar-gong-*en*-sis. Of the Yargong gorge, Tibet. W China, Burma, Tibet.

Prince Albert's Yew see *Saxegothaea conspicua*

Prince's Feather see *Amaranthus hybridus*

Princess of the Night see *Selenicereus pteranthus*

Princess Vine see *Cissus sicyoides*

Privet see *Ligustrum*

Common see *L. vulgare*

Proboscidea pro-bos-*ki*-dee-a *Martyniaceae*. From Gk. *proboskis* (an elephant's trunk) referring to the long, curved beak of the fruit. Annual and perennial herbs.

fragrans frah-granz. Fragrant. Mexico.

louisianica loo-eez-ee-*ah*-ni-ka. (= *Martynia louisianica*). Of Louisiana. S United States.

Promenaea pro-men-*ie*-a *Orchidaceae*. After Promeneia, a priestess of the temple of Dodona. Greenhouse Orchids.

stapelioides sta-pel-ee-*oi*-deez. Like *Stapelia*. Brazil.

Propeller Plant see *Crassula falcata*

Prophet Flower see *Arnebia pulchra*

Prostanthera pros-tănth-*e*-ra *Labiatae*. From Gk. *prosthema* (an appendage) and *anthera* (an anther) referring to the spurred anthers. Semi-hardy, evergreen shrubs.

ovalifolia ō-vah-li-*fo*-lee-a. With oval leaves. New South Wales.

rotundifolia ro-tun-di-*fo*-lee-a. Round-leaved. SE Australia.

Protea *prō*-tee-a *Proteaceae*. After Proteus a Gk. sea god with the power of prophecy. Tender, evergreen shrubs. S Africa.

barbigera bar-*bi*-ge-ra. Bearded.

cynaroides si-nah-*roi*-deez. Like *Cynara*. King Protea.

eximea eks-*i*-mee-a. Distinguished.

grandiceps grănd-i-keps. Large-headed.

neriifolia nay-ree-i-*fo*-lee-a. With leaves like *Nerium*.

speciosa spe-kee-*ō*-sa. Showy.

Prunella proo-*nel*-la *Labiatae*. Possibly from L. *prunum* (purple) referring to the flowers

Prunella (continued)
or from German *Braüne* (quinsy) which they
were said to cure. Perennial herbs. Self Heal.
grandiflora gränd-i-*flō*-ra. Large-flowered.
Europe.
× *webbiana* web-ee-*ah*-na. After Webb.

Prunus *proo*-nus *Rosaceae*. The L. name for
the plum tree. Deciduous trees, deciduous
and evergreen shrubs.
× *amygdalo-persica* a-*mig*-da-lō-*per*-si-ka. *P.*
dulcis (= *P. amygdalus*) × *P. persica*.
From the names of the parents.
   'Pollardii' po-*lard*-ee-ee. After Mr
   Pollard, the raiser.
armeniaca ar-men-ee-*ah*-ka. Of Armenia.
Apricot. N China.
avium ä-vee-um. Of birds. Gean, Mazzard.
Europe.
× *blireana* bli-ree-*ah*-na. Of Bléré, France.
cerasifera ke-ra-*si*-fe-ra. Cherry-bearing.
Cherry Plum, Myrobalan. Cult.
   'Pissardii' pee-*sard*-ee-ee. After M.
   Pissard, gardener to the Shah of Persia,
   who sent it to France in 1886.
cerasus ke-ra-sus. L. name for cherry. Sour
Cherry. Cult.
× *cistena* sis-*teen*-a. *P. cerasifera* 'Pissardii'
× *P. pumila*. The Sioux name for baby,
from the dwarf habit.
conradinae kon-ra-*deen*-ie. Name by
Koehne after his wife Conradine. China.
davidiana dä-vid-ee-*ah*-na. After David
who introduced it to France in 1865, see
Davidia. China.
domestica do-*mes*-ti-ka. Cultivated. Plum.
Cult.
dulcis dul-kis. Sweet. Almond. SE Europe,
N Africa, W Asia.
glandulosa glän-dew-*lō*-sa. Glandular. NE
Asia.
   'Hillieri' *hi*-lee-a-ree. After Hillier's who
   raised it.
incisa in-*kee*-sa. Deeply cut (the leaves).
Fuji Cherry. Japan.
   'Praecox' *prie*-koks. Early (flowering).
laurocerasus low-rō-*ke*-ra-sus. As the
common name, Cherry Laurel. SE
Europe, W Asia.
   'Magnoliifolia' mäg-nol-ee-i-*fo*-lee-a.
   Magnolia-leaved.
   'Schipkaensis' ship-ka-*en*-sis. Of the
   Schipka Pass, Bulgaria.
   'Zabeliana' za-bel-ee-*ah*-na. After Zabel.
lusitanica loo-si-*tah*-ni-ka. Of Portugal.
Portugal Laurel. Spain, Portugal.
mume mew-mee. From *ume* the Japanese
name. Japanese Apricot. SW China.

Prunus (continued)
padus pä-dus. Gk. name of a wild cherry.
Bird Cherry. N Europe, N Asia.
persica *per*-si-ka. Of Persia. Peach. China.
pumila pew-mi-la. Dwarf. NE United
States.
sargentii sar-*jent*-ee-ee. After Charles
Sprague Sargent (1841–1927), first
director of the Arnold Arboretum.
N Japan.
serrula se-ru-la. Saw-toothed (the leaves).
W China.
serrulata se-ru-*lah*-ta. With small teeth (the
leaves). Cult.
spinosa spee-*nō*-sa. Spiny. Sloe, Blackthorn.
Europe, N Asia.
subhirtella sub-hir-*tel*-la. Somewhat hairy.
Cult.
   'Autumnalis' ow-tum-*nah*-lis. Of autumn
   (flowering).
tenella te-*nel*-la. Dainty. Dwarf Russian
Almond. E Europe, SW Russia.
triloba tri-*lō*-ba. Three-lobed (the leaves).
China.
   'Multiplex' *mul*-ti-pleks. Much-folded
   (the double flowers).
× *yedoensis* ye-dō-*en*-sis. Of Tokyo (Yedo).
Yoshino Cherry.
   'Ivensii' ie-*venz*-ee-ee. After Arthur J.
   Ivens of Hillier's, who raised it.

Pseuderanthemum soo-de-*ränth*-e-mum
Acanthaceae. From Gk. *pseudo* (false) and
Eranthemum, a related genus. Tender shrubs.
Polynesia.
atropurpureum aht-rō-pur-*pewr*-ree-um.
Deep purple (the leaves).
reticulatum ray-tik-ew-*lah*-tum. Net-veined
(the leaves).
tricolor tri-ko-lor. Three-coloured (the
leaves).

Pseudolarix soo-dō-*lä*-riks *Pinaceae*. From
Gk. *pseudo* (false) and *Larix* q.v. Deciduous,
larch-like conifer.
amabilis a-*mah*-bi-lis. Beautiful. Golden
Larch. E China.

Pseudopanax soo-dō-*pän*-äks *Araliaceae*.
From Gk. *pseudo* (false) and *Panax* a related
genus, see *Acanthopanax*. Semi-hardy,
evergreen shrubs or trees.
crassifolius kräs-i-*fo*-lee-us. Thick-leaved.
New Zealand.
davidii dä-*vid*-ee-ee. After David who
introduced it, see *Davidia*. China.
ferox fe-*rōks*. Spiny (the leaves). New
Zealand.

Pseudotsuga soo-dō-tsoo-ga *Pinaceae*. From Gk. *pseudo* (false) and *Tsuga* q.v. Evergreen conifer.
  menziesii men-*zeez*-ee-ee. After Menzies, see *Menziesia*. Douglas Fir. W N America.
  'Fletcheri' *flech*-a-ree. After Fletcher's nursery who distributed it.

Pseudowintera soo-dō-win-*te*-ra *Winteraceae*. From Gk. *pseudo* (false) and *Wintera*. Originally named *Wintera*, a name which had already been used for *Drimys*, see *D. winteri*. Evergreen shrub.
  colorata ko-lo-*rah*-ta. Coloured (the leaves). New Zealand.

Psidium si-dee-um *Myrtaceae*. From *psidion* the Gk. name for a pomegranate. Tender, evergreen shrub.
  guajava gwah-*hah*-va. The native name. Guava. Tropical America.

Psylliostachys si-lee-*o*-sta-kis *Plumbaginaceae*. From Gk. *psyllion* (a type of plantain) and *stachys* (a spike). Annual herbs. W and C Asia.
  spicata spee-*kah*-ta. (= *Limonium spicatum*). With flowers in spikes.
  suworowii soo-vo-*rov*-ee-ee. (= *Limonium suworowii*). After Ivan Petrowitch Suworow, a medical inspector in Turkestan c 1886.

Ptelea *tel*-ee-a *Rutaceae*. Gk. name for an elm, the winged fruits are similar. Deciduous tree.
  trifoliata tri-fo-lee-*ah*-ta. With three leaves (leaflets). Hop Tree. S Canada. E United States.

Pteris *te*-ris *Pteridaceae*. From Gk. *pteris* (a fern). Tender ferns.
  cretica kray-ti-ka. Of Crete. Cretan Brake. Tropics.
  'Albo-lineata' ăl-bō-li-nee-*ah*-ta. White-lined (the fronds).
  ensiformis ayn-si-*form*-is. Sword-shaped (the pinnae). Sword Brake. E Asia to Australia.
  multifida mul-ti-*fi*-da. Divided many times. E Asia.
  quadriaurita kwod-ree-ow-*ree*-ta. Four-eared (the base of the fronds). Tropics.
  'Argyraea' ar-gi-*ree*-a. Silvery.
  tremula trem-ew-la. Trembling. Australian Brake. Australia, New Zealand.

Pterocactus te-rō-*kăk*-tus *Cactaceae*. From Gk. *pteron* (a wing) and *Cactus* q.v. referring to the winged seeds.
  tuberosus tew-be-*rō*-sus. Tuberous (the root). Argentina.

Pterocarya te-rō-*kă*-ree-a *Juglandaceae*. From Gk. *pteron* (a wing) and *karyon* (a nut) referring to the winged fruit. Deciduous trees. Wing-nut.
  fraxinifolia fraks-i-ni-*fo*-lee-a. *Fraxinus*-leaved. Caucasian Wing-nut. Caucasus, N Iran.
  × *rehderiana* ray-da-ree-*ah*-na. *P. fraxinifolia* × *P. stenoptera*. After Rehder, see *Rehderodendron*.

Pterocephalus te-rō-*kef*-a-lus *Dipsacaceae*. From Gk. *pteron* (a wing) and *kephale* (a head) referring to the feathery fruiting heads. Perennial herb.
  perennis pe-*ren*-is. (= *P. parnassi*). Perennial. S and E Greece.

Pterostyrax te-rō-*sti*-răks *Styracaceae*. From Gk. *pteron* (a wing) and *Styrax* q.v. Deciduous shrub.
  hispida his-pi-da. Bristly (the fruit). Japan, China.

Ptilotrichum ti-*lo*-tri-kum *Cruciferae*. From Gk. *ptilon* (downy) and *trichos* (hair). Dwarf shrub.
  spinosum spee-*nō*-sum. (= *Alyssum spinosum*). Spiny. S France, Spain.

Pulmonaria pul-mon-*ah*-ree-a *Boraginaceae*. From L. *pulmo* (the lung), the leaves were used to treat bronchial infections. Herbaceous perennials. Lungwort.
  angustifolia ang-gus-ti-*fo*-lee-a. Narrow-leaved. C and E Europe.
  longifolia long-gi-*fo*-lee-a. Long-leaved. W Europe.
  officinalis o-fi-ki-*nah*-lis. Sold as a herb. Europe.
  rubra rub-ra. Red (the flowers). SE Europe.
  saccharata să-ka-*rah*-ta. Appearing sprinkled with sugar (the leaves). SE France, N Italy.

Pulsatilla pul-sa-*til*-la *Ranunculaceae*. From L. *pulso* (to strike). Perennial herbs. Pasque Flower.

Pulsatilla (continued)
alpina ăl-*peen*-a. (= *Anemone alpina*).
Alpine. C and S Europe.
halleri hăl-a-ree. (= *Anemone halleri*). After
Albrecht von Haller (1708–77) a Swiss
botanist. Alps.
rubra rub-ra. Red (the flowers). France,
Spain.
vernalis ver-*nah*-lis. (= *Anemone vernalis*).
Of spring. Europe.
vulgaris vul-*gah*-ris. (= *Anemone pulsatilla*).
Common. Europe.

Pummelo see *Citrus maxima*
Pumpkin see *Cucurbita maxima*

Punica pew-ni-ka *Punicaceae*. The L. name.
Semi-hardy evergreen shrub or tree.
granatum grah-*nah*-tum. Many-seeded.
Iran, Afghanistan. Pomegranate.

Purple Bell Vine see *Rhodochiton volubile*
Purple Heart see *Setcreasia pallida*
Purple Moor Grass see *Molinia caerulea*
Purple Passion Vine see *Gynura aurantiaca*
Purple Top see *Verbena bonariensis*
Purple Wreath see *Petrea volubilis*
Purslane see *Portulaca oleracea*
  Rock see *Calandrinia*
  Tree see *Atriplex halimus*
Pussy Ears see *Cyanotis somaliensis*,
Kalanchoe tomentosa

Puschkinia push-*kin*-ee-a *Liliaceae*. After
Count Apollos Mussin-Puschkin (died 1815),
who collected in the Caucasus in 1802.
Bulbous perennial.
scilloides skil-*loi*-deez. (= *P. libanotica*).
Like *Scilla*. Caucasus, W Asia.

Pyracantha pi-ra-*kănth*-a *Rosaceae*. From
Gk. *pyr* (fire) and *akantha* (a thorn) referring
to the spiny shoots and red berries.
angustifolia ang-gus-ti-*fo*-lee-a. Narrow-
leaved. W China.
atalantioides ă-ta-lăn-tee-*oi*-deez. Like
Atalantia (Rutaceae). C China.
coccinea kok-*kin*-ee-a. Scarlet (the fruits).
S Europe, W Asia.
  'Lalandei' la-*lond*-ee-ee. After M.
  Lalande who raised it in France c 1874.
rogersiana ro-jerz-ee-*ah*-na. After G. L.
Coltman-Rogers of Stanage Park who first
exhibited it. W China.

Pyrethrum see *Chrysanthemum coccineum*

Pyrola *pi*-ro-la *Pyrolaceae*. Diminutive of
Pyrus, from the similar leaves. Perennial
herbs. Wintergreen.
asarifolia a-sah-ri-*fo*-lee-a. *Asarum*-leaved.
N America.
elliptica e-*lip*-ti-ka. Elliptic (the leaves).
N America, Japan.
media me-dee-a. Intermediate. Europe,
W Asia.
rotundifolia ro-tun-di-*fo*-lee-a. Round-
leaved. Europe, W Asia.

Pyrus *pi*-rus *Rosaceae*. The L. name.
Deciduous trees. Pear.
calleryana ka-le-ree-*ah*-na. After J. Callery,
a French missionary who collected the
type specimen. China.
communis kom-*ew*-nis. Common. Common
Pear. Europe.
salicifolia să-li-ki-*fo*-lee-a. *Salix*-leaved.
Willow-leaved Pear. Caucasus, Turkey,
NW Iran.

Q

Quaking Grass see *Briza media*
Quamoclit coccinea see *Ipomoea coccinea*
  lobata see *Mina lobata*
  pennata see *Ipomoea quamoclit*
Queen of the Night see *Selenicereus
grandiflorus*
Queen of the Prairie see *Filipendula rubra*
Queen's Wreath see *Petrea volubilis*
Queensland Umbrella Tree see *Schefflera
actinophylla*

Quercus *kwer*-kus *Fagaceae*. The L. name.
Deciduous and evergreen trees. Oak.
acutissima ă-kew-*tis*-i-ma. Very sharply
pointed (the leaves). E Asia.
agrifolia ăg-ri-*fo*-lee-a. With spiny leaves.
California, N Mexico.
alnifolia ăl-ni-*fo*-lee-a. *Alnus*-leaved.
Golden Oak of Cyprus. Cyprus.
canariensis ka-nah-ree-*en*-sis. Of the Canary
Islands. N Africa, S Portugal, Spain.
castaneifolia kas-tăn-ee-i-*fo*-lee-a. *Castanea*-
leaved. N Iran, SW Russia.
cerris ke-ris. The L. name. Turkey Oak.
Europe, Turkey.
coccifera kok-*kif*-e-ra. Berry-bearing,
referring to the kermes insects which
breed on the tree. Kermes Oak.
W Mediterranean region, N Africa.

Quercus (continued)

coccinea kok-*kin*-ee-a. Scarlet (autumn colour). Scarlet Oak. E N America.

dentata den-*tah*-ta. Toothed (the leaves). Daimio Oak. Japan, NE Asia.

frainetto fray-*net*-ō. In error for *farnetto*, the Italian name. Hungarian Oak. SE Europe.

× *hispanica* his-*pah*-ni-ka. *Q. cerris* × *Q. suber*. Spanish, it was originally thought to come from Spain.

　'Ambrozyana' ăm-brō-zee-*ah*-na. After Count Ambrozy of the Arboretum Mlynany, who raised it.

　'Lucombeana' lū-kom-bee-*ah*-na. After Lucombe, an Exeter nurseryman, who raised it. Lucombe Oak.

ilex ee-leks. The L. name. Holm Oak. Mediterranean region.

imbricaria im-bri-*kah*-ree-a. From L. *imbrex* (a tile) the wood was used for roof tiles. Shingle Oak. E United States.

× *kewensis* kew-*en*-sis. *Q. cerris* × *Q. wislizenii*. Of Kew, where it was raised.

libani li-ba-nee. Of Lebanon. Lebanon Oak. W Asia.

macranthera ma-*krănth*-e-ra. With large anthers. Caucasus, N Iran.

macrolepis măk-rō-*lep*-is. With large scales (on the acorn). SE Europe, Turkey.

marilandica mă-ri-*lănd*-i-ka. Of Maryland. Black Jack Oak. E United States.

myrsinifolia mur-si-ni-*fo*-lee-a. *Myrsine*-leaved. E Asia.

palustris pa-*lus*-tris. Of swamps. Pin Oak. E United States.

petraea pe-*trie*-a. Of rocky places. Durmast Oak, Sessile Oak, Europe.

phillyreoides fi-li-ree-*oi*-deez. Like *Phillyrea*. China, Japan.

pyrenaica pi-ray-*nah*-i-ka. Of the Pyrenees. SW Europe, Morocco.

robur rō-bur. The L. name for the oak and its wood. Common Oak, Pedunculate Oak. Europe, Caucasus.

rubra rub-ra. Red (autumn colour). Red Oak. E N America.

suber soo-ber. The L. name. Cork Oak. W Mediterranean region.

× *turneri turn*-a-ree. *Q. ilex* × *Q. robur*. After Mr Spencer Turner (c 1728–76) in whose Essex nursery it was raised.

velutina vel-ew-*teen*-a. Velvety (the buds). Black Oak. E N America.

wislizenii wiz-li-*zen*-ee-ee. After A. Wislizenius. California.

Quince see *Cydonia oblonga*
　Ornamental see *Chaenomeles*

R

Rabbit's Foot see *Maranta leuconeura kerchoveana*
Radish see *Raphanus sativus*
Rainbow Star see *Cryptanthus bromelioides*
Rainbow Vine see *Pellionia pulchra*

Ramonda ra-*mon*-da *Gesneriaceae*. After Louis Francis Ramond (1753–1827), French botanist. Perennial herbs.

myconi mi-*kō*-nee. After Franciso Mico (born 1528), Spanish physician and botanist. Pyrenees, N Spain.

nathaliae na-*tah*-lee-ie. After Queen Nathalia, wife of King Milan. S Yugoslavia, N Greece.

serbica ser-bi-ka. Of Serbia. SE Europe.

Rampion see *Phyteuma orbiculare*
　Spiked see *P. spicatum*

Ranunculus rah-*nun*-kew-lus *Ranunculaceae*. The L. name from *rana* (a frog) as many grow in wet places. Annual and perennial herbs. Buttercup.

aconitifolius ă-kon-ee-ti-*fo*-lee-us. *Aconitum*-leaved. White Bachelors Buttons. C and S Europe.

　'Flore Pleno' *flō*-ree-*play*-nō. With double flowers. Fair Maids of France.

acris ah-kris. Sharp-tasting. Europe, Asia.

　'Flore Pleno' *flō*-ree-*play*-nō. With double flowers. Yellow Bachelors Buttons.

alpestris al-*pes*-tris. Of the lower mountains. Alps, Pyrenees.

amplexicaulis ăm-pleks-i-*kaw*-lis. With leaves clasping the stem. SW Europe.

aquatilis a-*kwah*-ti-lis. Growing in water. Water Crowfoot. Europe.

asiaticus ah-see-*ah*-ti-kus. Asian. Persian Buttercup. Crete, SW Asia.

bulbosus bul-*bō*-sus. Bulbous. Europe.

calandrinioides kă-lan-dree-nee-*oi*-deez. Like *Calandrinia*. Morocco.

ficaria fee-*kah*-ree-a. Like *Ficus*, the fig-like tubers. Lesser Celandine. Europe, W Asia.

glacialis glă-kee-*ah*-lis. Of icy regions. Europe, Greenland.

gramineus grah-*min*-ee-us. Grass-like (the leaves).

lingua ling-gwa. Tongue-like (the leaves). Greater Spearwort. Europe.

Ranunculus (continued)
montana mon-*tah*-na. Of mountains.
Europe, Caucasus.
parnassifolius par-năs-i-*fo*-lee-us. With
leaves like *Parnassia*. Pyrenees, Alps.
pyrenaeus pi-ray-*nie*-us. Of the Pyrenees.
Alps, Pyrenees, Corsica.

Raoulia *rowl*-ee-a *Compositae*. After Edward
Raoul (1815–52), French surgeon who
studied New Zealand plants. Perennial herbs
and sub-shrubs. New Zealand.
australis ow-*strah*-lis. (= *R. lutescens*).
Southern.
hookeri huk-a-ree. After Sir Joseph Hooker.
tenuicaulis ten-ew-i-*kaw*-lis. Slender-
stemmed.

Rape see *Brassica napus*

Raphanus *ră*-fa-nus *Cruciferae*. The L.
name. Biennial herb.
sativus sa-*teev*-us. Cultivated. Radish. Cult.

Raphiolepis see *Rhaphiolepis*
Raspberry see *Rubus idaeus*
Rattlesnake Plant see *Calathea lancifolia*
Rauli see *Nothofagus procera*

Rebutia ra-*bew*-tee-a *Cactaceae*. After P.
Rebut, a French cactus dealer. Crown
Cactus. Unless otherwise stated, Argentina.
aureiflora ow-ree-i-*flō*-ra. Golden-flowered.
deminuta day-mi-*new*-ta. (= *Aylostera
deminuta*). Small.
fiebrigii fee-*brig*-ee-ee. After Dr C. Fiebrig
of the Museum and Garden, Asuncion,
Paraguay. Bolivia.
kupperiana ku-pa-ree-*ah*-na. After
Professor W. Kupper of Munich.
minuscula mi-*nus*-kew-la. Rather small. Red
Crown Cactus.
pseudodeminuta soo-dō-day-mi-*new*-ta.
False *R. deminuta*.
pygmaea see *Lobivia pygmaea*
senilis se-*nee*-lis. An old man, referring to
the white, bristle-like spines. Fire Crown
Cactus.
spegazziniana spe-ga-zeen-ee-*ah*-na. After
Professor Carlos Spegazzini (1858–1926),
an Argentine botanist.
spinosissima speen-ō-*sis*-i-ma. Very spiny.
violaciflora vee-o-lah-ki-*flō*-ra. With violet
flowers.
xanthocarpa zănth-ō-*kar*-pa. With yellow
fruits.

Rechsteineria cardinalis see *Sinningia
cardinalis*
cyclophylla see *S. macropoda*
leucotricha see *S. leucotricha*
Red Cape Tulip see *Haemanthus*
Red Ivy see *Hemigraphis alternata*
Red-hot Cat's Tail see *Acalypha hispida*
Red Hot Poker see *Kniphofia*
Red Nodding Bells see *Streptocarpus dunnii*
Red Ribbons see *Clarkia concinna*
Redbud see *Cercis canadensis*
Redwood, Coast see *Sequoia sempervirens*
Giant see *Sequoiadendron giganteum*
Reed Canary Grass see *Phalaris arundinacea*
Reed Grass see *Glyceria maxima*
Reed-mace see *Typha*

Rehderodendron ray-da-rō-*den*-dron
Styracaceae. After Alfred Rehder
(1863–1949), American botanist, of the
Arnold Arboretum, and Gk. *dendron* (a tree).
Deciduous tree.
macrocarpum măk-rō-*kar*-pum. Large-
fruited. Omei Shan, China.

Rehmannia ray-*mahn*-ee-a *Gesneriaceae*.
After Joseph Rehmann (1753–1831),
German physician. Perennial herb.
elata ay-*lah*-ta. (= *R. angulata* hort.). Tall.
Chinese Foxglove. China.

Reineckia rie-*nek*-ee-a *Liliaceae*. After
Johann Heinrich Julius Reinecke
(1799–1871). Perennial herb.
carnea kar-nee-a. Flesh-pink (the flowers).
China, Japan.

Reinwardtia rien-*vart*-ee-a *Linaceae*. After
Caspar Reinwardt (1773–1854), founder of
Bogar Botanic Garden, Java. Tender sub-
shrub.
indica in-di-ka. (= *R. tetragyna. R. trigyna*).
Indian. Yellow Flax. Himalaya, E and SE
Asia.

Renanthera ray-nan-*the*-ra *Orchidaceae*.
From L. *renes* (kidneys) and *anthera* (an
anther) referring to the kidney-shaped
pollinia. Greenhouse orchids. SE Asia.
coccinea kok-*kin*-ee-a. Scarlet (the flowers).
imschootiana im-shoot-ee-*ah*-na. After A.
van Imschoot of Ghent.

Reseda re-*say*-da *Resedaceae*. The L. name
from *resedo* (to heal) referring to medicinal
properties. Annual herb.

Reseda (continued)
odorata o-dō-*rah*-ta. Fragrant (the flowers).
Mignonette. N Africa.

Resurrection Lily see *Lycoris squamigera*
Resurrection Plant see *Selaginella lepidophylla*
Rex-begonia Vine see *Cissus discolor*

Rhamnus *răm*-nus *Rhamnaceae*. The Gk.
name of a shrub. Deciduous and evergreen,
trees and shrubs.
alaternus ă-la-*tern*-us. The L. name.
S Europe.
cathartica ka-*thar*-ti-ka. Purging. Common
Buckthorn. Europe, Asia.
frangula frang-gew-la. The L. name. Alder
Buckthorn. Europe, N Africa, Asia.

Rhaphiolepis răf-ee-ō-*lep*-is *Rosaceae*.
(*Raphiolepis*). From Gk. *rhaphis* (a needle)
and *lepis* (a scale) referring to the narrow
bracteoles of the inflorescence. Semi-hardy,
evergreen shrubs.
× *delacourii* de-la-*kour*-ree-ee. R. indica ×
R. umbellata. After M. Delacour, the
raiser.
indica in-di-ka. Indian. Indian Hawthorn.
S China.
umbellata um-bel-*ah*-ta. The flowers appear
to be in umbels. Japan, Korea.

Rhapis *ră*-pis *Palmae*. From Gk. *rhapis* (a
needle) referring to the slender leaf
segments. Tender, suckering palms. Lady
Palm. S China.
excelsa eks-*kel*-sa. Tall.
humilis hu-mi-lis. Low-growing.

Rhazya rah-zee-a *Apocynaceae*. After Abu
Bekr-el-Rasi (Rhazes) an Arabian physician
(died c 932). Perennial herb.
orientalis o-ree-en-*tah*-lis. Eastern. SE
Europe, W Asia.

Rheum ray-um *Polygonaceae*. From *rheon*,
the Gk. name for rhubarb. Perennial herbs.
alexandrae ă-leks-*ahn*-drie. After Queen
Alexandra, wife of Edward VII.
Himalaya.
palmatum pahl-*mah*-tum. Lobed like a hand
(the leaves). NE Asia.
rhabarbarum rah-*bar*-ba-rum. Literally the
rhubarb of foreigners. *Rha* being an old
name of the River Volta along which *R.
officinale*, a herbal plant, was brought from
China. Rhubarb. NE Asia.

Rhipsalidopsis rip-sa-li-*dops*-is *Cactaceae*.
From *Rhipsalis* q.v. and Gk. -*opsis* indicating
resemblance. Brazil.
gaertneri gairt-na-ree. (= *Schlumbergera
gaertneri*). After J. Gärtner (1732–91), a
Stuttgart physician. Easter Cactus.
rosea ro-see-a. Rose-coloured (the flowers).

Rhipsalis *rip*-sa-lis *Cactaceae*. From Gk.
rhips (wicker-work) referring to the supple
shoots.
baccifera bah-*ki*-fe-ra. (= *R. cassutha*).
Berry-bearing. Mistletoe Cactus.
S America, Ceylon, Africa.
capilliformis ka-pi-li-*form*-is. Thread-like
(the stems). Brazil.
cereuscula kay-ree-*us*-kew-la. Like a small
Cereus. Brazil, Uruguay.
crispata kris-*pah*-ta. Wavy-edged (the
stems). Brazil.
houlletiana hoo-lay-tee-*ah*-na. After M.
Houllet (1811–90), assistant curator of the
Jardin des Plantes, Paris. Snowdrop
Cactus. Brazil.
paradoxa pă-ra-*doks*-a. Unusual. Chain
Cactus. Brazil.
pilocarpa see *Erythrorhipsalis pilocarpa*
tonduzii ton-*duz*-ee-ee. After J. F. A.
Tonduz (1862–1921), a Swiss botanist.
Costa Rica.
warmingiana var-ming-gee-*ah*-na. After
Professor Johannes Eugenius Bülow
Warming (1841–1924), a Danish botanist.
Brazil.

Rhodanthe manglesii see *Helipterum manglesii*

Rhodiola ro-*dee*-o-la *Crassulaceae*. From Gk.
rhodon (a rose) referring to the rose-scented
root of *R. rosea*. Perennial herbs.
fastigiata fa-stig-ee-*ah*-ta. (= *Sedum
fastigiatum*). Upright. Himalaya, Tibet,
China.
heterodonta he-te-rō-*don*-ta. (= *Sedum
heterodontum*). Variably toothed (the
leaves). Himalaya, Tibet.
rosea ro-see-a. (= *Sedum rhodiola. Sedum
rosea*). Rose-like, see above. Roseroot.
Arctic and alpine N hemisphere.

Rhodochiton ro-*do*-ki-ton *Scrophulariaceae*.
From Gk. *rhodo-* (red) and *chiton* (a cloak)
referring to the enveloping calyx. Semi-hardy
climber.
volubile vol-*ew*-bi-lee. (= *R.*

Rhodochiton (continued)
atrosanguineum). Twining. Purple Bell
Vine. S Mexico.

Rhodohypoxis ro-do-hi-*poks*-is
Hypoxidaceae. From Gk. *rhodo-* (red) and
Hypoxis q.v. Semi-hardy perennial herbs.
baurii bow-ree-ee. After the Rev. R. Baur
who collected *Rhodohypoxis* in S Africa.
S Africa.

Rhodotypos ro-do-*ti*-pos *Rosaceae*. From
Gk. *rhodon* (a rose) and *typos* (type) referring
to the rose-like flowers. Deciduous shrub.
scandens skán-denz. Climbing. China,
Japan.

Rhododendron ro-do-*den*-dron *Ericaceae*.
The Gk. name for *Nerium oleander* from
rhodo- (red) and *dendron* a tree. Evergreen and
deciduous shrubs. Azaleas are marked (Az).
aberconwayi ă-ba-*kon*-way-ee. After Lord
Aberconway. E Yunnan.
albrechtii ăl-*brekt*-ee-ee (Az). After Dr
Albrecht of the Russian Consul at
Hakodate, who collected the type
specimen. C and N Japan.
arborescens ar-bo-*res*-enz (Az). Becoming
tree-like. E N America.
arboreum ar-*bo*-ree-um. Tree-like.
Himalaya.
atlanticum ăt-*lăn*-ti-kum (Az). Of the
Atlantic coast (of N America). SE United
States.
augustinii aw-gus-*tin*-ee-ee. After Augustine
Henry who discovered it in 1886, see
Illicium henryi. C and W China.
chasmanthum kăs-*mănth*-um. With
gaping flowers.
auriculatum ow-rik-ew-*lah*-tum. Auricled
(the leaves). China.
barbatum bar-*bah*-tum. Bearded (the shoots
of some forms). Himalaya.
bullatum see *R. edgeworthii*
bureavii bew-*reev*-ee-ee. After Edouard
Bureau (1831–1918), a French botanist.
NW Yunnan.
calendulaceum ka-len-dew-*lah*-kee-um (Az).
Like *Calendula* (the flower colour).
E N America.
callimorphum kăl-i-*morf*-um. Beautifully
shaped. W Yunnan, Upper Burma.
calophytum kăl-o-*fi*-tum. Beautiful plant.
W Sichuan.
calostrotum kăl-o-strō-tum. With a beautiful
covering (the leaves). Upper Burma.
keleticum kay-*lay*-ti-cum. (= *R.*

Rhododendron (continued)
keleticum). Charming. Tibet.
radicans rah-di-kănz. (= *R. radicans*).
With rooting stems.
caloxanthum see *R. campylocarpum*
caloxanthum
campylocarpum kăm-pi-lō-*kar*-pum. With a
curved fruit. Himalaya.
caloxanthum kă-loks-*ănth*-um. (= *R.
caloxanthum*). A beautiful yellow.
Upper Burma.
campylogynum kăm-pi-*lo*-gi-num. With a
curved ovary. Himalaya, Upper Burma,
W China.
catawbiense ka-taw-bee-*en*-see. From near
the Catawba River. SE United States.
charitopes kă-ri-*tō*-pays. With a graceful
stalk. Upper Burma.
tsangpoense tsang-pō-*en*-see. (= *R.
tsangpoense*). From near the Tsangpo
River, Tibet.
ciliatum ki-lee-*ah*-tum. Fringed with hairs
(the leaves). Himalaya.
cinnabarinum ki-na-ba-*reen*-um. Cinnabar
red. Himalaya.
Concatenans kon-ka-*ten*-anz. (= *R.
concatenans*). Joined together (the
corolla lobes).
xanthocodon zănth-o-*kō*-don. (= *R.
xanthocodon*). A yellow bell.
concatenans see *R. cinnabarinum*
Concatenans
concinnum kon-*kin*-um. Elegant.
W Sichuan.
pseudoyanthinum soo-dō-yăn-*theen*-um.
False *R. yanthinum*.
crassum see *R. maddenii crassum*
decorum de-*kō*-rum. Beautiful. China.
discolor see *R. fortunei discolor*
edgeworthii ej-*werth*-ee-ee. (= *R. bullatum*).
After Edgeworth, see *Edgeworthia*.
Himalaya, W China.
falconeri fawl-*kon*-a-ree. After Hugh
Falconer (1808–65), Scottish doctor and
botanist with the E India Co. Himalaya.
fargesii see *R. oreodoxa fargesii*
fastigiatum fa-stig-ee-*ah*-tum. With upright
branches. Yunnan.
ferrugineum fe-roo-*gi*-nee-um. Rusty (the
shoots and undersides of the leaves).
S Europe (mountains).
fictolacteum see *R. rex fictolacteum*
forrestii fo-*rest*-ee-ee. After Forrest, see
Abies delavayi forrestii, W China, Upper
Burma.
repens ree-pens. Creeping.
fortunei for-*tewn*-ee-ee. After Fortune who
discovered it. See *Fortunella*. E China.

Rhododenron (continued)

discolor *dis*-ko-lor. (= *R. discolor*). Two-coloured (the leaves). C China.

fulvum ful-vum. Tawny (the lower leaf surface). E Himalaya, W China.

glaucophyllum glow-kō-*fil*-lum. With glaucous leaves (the undersides). Himalaya.

haematodes hie-ma-*tō*-deez. Blood-red. Yunnan.

hanceanum hǎns-ee-*ah*-num. After Henry Fletcher Hance (1827–56), British Consul in China. SW Sichuan.

hippophaeoides hi-po-fa-ee-*oi*-deez. Like *Hippophae*. W China.

hirsutum hir-*soo*-tum. Hairy. Alps, Yugoslavia.

impeditum im-pe-*dee*-tum. Tangled. Yunnan.

imperator see *R. uniflorum imperator*

indicum in-di-kum (Az). Indian. S Japan.

insigne in-*sig*-nee. Distinguished. W Sichuan.

kaempferi kempf-a-ree (Az). After Engelbert Kaempfer (1651–1716), German physician. Japan.

keleticum see *R. calostrotum keleticum*

kiusianum kee-oo-see-*ah*-num (Az). Of Kyushu, Japan.

lepidostylum le-pi-dō-*sti*-lum. With a scaly style. W China.

leucaspis loo-*kǎs*-pis. A white shield, referring to the white, saucer-shaped flowers. E Himalaya.

Loderi-*lō*-da-ree. *R. fortunei* × *R. griffithianum*. After Sir Edmund Loder who raised it.

lutescens loo-*tes*-enz. Yellowish (the flowers). W China.

luteum loo-tee-um (Az). (= *Azalea pontica*). Yellow (the flowers). W Caucasus, Turkey.

macabeanum ma-kay-bee-*ah*-num. After Mr M'Cabe, Deputy Commissioner for the Naga Hills. Assam, Manipur.

maddenii ma-*den*-ee-ee. After Major Madden of the Bengal Civil Service. Himalaya.

  crassum krǎs-um. (= *R. crassum*). Thick (the leaves). NW Yunnan, Upper Burma.

moupinense moo-pin-*en*-see. Of Moupin. W Sichuan.

mucronulatum mew-kron-ew-*lah*-tum. With a short point (the leaves). NE Asia.

nakaharae nǎ-ka-*hah*-rie. After G. Nakahara, a Japanese botanist. Taiwan.

neriiflorum nay-ree-i-*flō*-rum. *Nerium*-

Rhododendron (continued)

flowered. E Himalaya, W China.

obtusum ob-*tew*-sum (Az). Blunt (the leaves). Cult.

  'Amoenum' a-*moy*-num. Pleasant.

occidentale ok-ki-den-*tah*-lee (Az). Western. W N America.

orbiculare or-bik-ew-*lah*-ree. Rounded (the leaves). W Sichuan.

oreodoxa o-ree-ō-*doks*-a. Glory of the mountain. W Sichuan.

  fargesii far-*jee*-zee-ee. (= *R. fargesii*). After Farges who discovered it, see *Decaisnea fargesii*. Hubei, Sichuan.

oreotrephes o-ree-o-tre-feez. Growing on mountains. W China, Himalaya.

pemakoense pe-ma-kō-*en*-see. Of Pemako, Tibet.

polycladum po-lee-*klǎd*-um. With many branches. China.

  Scintillans *skin*-ti-lǎnz. (= *R. scintillans*). Gleaming.

ponticum pon-ti-kum. Of Pontus (NE Turkey). W Asia, Balkans.

poukhanense see *R. yedoense poukhanense*

pruniflorum proon-i-*flō*-rum. (= *R. tsangpoense pruniflorum*). *Prunus*-flowered. Upper Burma, Assam.

pseudochrysanthum soo-dō-kris-*ǎnth*-um. False *R. chrysanthum*. Taiwan.

quinquefolium kwing-kwee-*fo*-lee-um (Az). Five-leaved, the leaves are in terminal clusters of five. Japan.

racemosum rǎ-kay-*mō*-sum. With flowers in racemes. W China.

radicans see *R. calostrotum radicans*

reticulatum ray-tik-ew-*lah*-tum (Az). Net-veined (the lower surface of the leaves). Japan.

rex reks. King. W China.

  fictolacteum fik-tō-*lak*-tee-um. (= *R. fictolacteum*). False *R. lacteum*. W China, SE Tibet.

rubiginosum roo-bi-gi-*nō*-sum. Rusty (the lower leaf surface). W China, SE Tibet.

russatum rus-*ah*-tum. Russet, the lower leaf surface. W China.

saluenense sǎl-ew-en-*en*-see. Of the Nu Jiang (Salween River) W Yunnan. W China, SE Tibet.

sargentianum sar-jent-ee-*ah*-num. After Sargent, see *Prunus sargentii*. W Sichuan.

schlippenbachii shlip-an-*bahk*-ee-ee (Az). After Baron Schlippenbach who discovered it in 1854. Korea, E Russia, E China.

scintillans see *R. polycladum* Scintillans

simsii simz-ee-ee. After John Sims

Rhododendron (continued)
(1749–1831). S and C China, Taiwan,
Burma.
sinogrande si-nō-*grăn*-dee. The Chinese *R.
grande*. Burma, Assam, Tibet, W China.
souliei soo-lee-ee. After the French
missionary Jean André Soulie (1858–1903)
who discovered it. W Sichuan.
sutchuenense sūch-wen-*en*-see. Of Sichuan
(Szechwan). W Hubei, E Sichuan.
tephropeplum tef-rō-*pep*-lum. Grey-cloaked.
E Himalaya, Upper Burma.
thomsonii tom-*son*-ee-ee. After Thomson,
see *Aster thomsonii*. Himalaya.
tsangpoense see *R. charitopes tsangpoense*
pruniflorum see *R. pruniflorum*
uniflorum ew-ni-*flō*-rum. One-flowered. SE
Tibet.
   imperator im-*pe*-ra-tor. (= *R. imperator*).
   Emperor, it was originally named *R.*
   'Purple Emperor'.
valentinianum vă-len-tin-ee-*ah*-num. After
Père S. P. Valentin, missionary in China
c 1919. Yunnan.
vaseyi vay-zee-ee (Az). After George R.
Vasey (1822–93) who discovered it in 1878.
N Carolina.
viscosum vis-*kō*-sum (Az). Sticky (the
flowers). E N America.
wardii word-ee-ee. After Kingdon Ward
who collected the type specimen, see
Cassiope wardii. W China, Tibet.
williamsianum wil-yămz-ee-*ah*-num. After
J. C. Williams of Caerhays. W Sichuan.
xanthocodon see *R. cinnabarinum
xanthocodon*
yakushimanum ya-koo-shee-*mah*-num. Of
Yakushima, an island of S Japan.
yedoense ye-dō-*en*-see (Az). Of Tokyo
(Yedo). Cult.
   poukhanense poo-ka-*nen*-see. (= *R.
   poukhanense*). Of Poukhan-san, Korea.
   Korea.
yunnanense yoo-nan-*en*-see. Of Yunnan.
W China, Tibet, Upper Burma.

Rhoeo rō-ee-ō *Commelinaceae*. Derivation
obscure. Tender perennial herb.
spathacea spa-*thah*-kee-a. (= *R. discolor*).
Spathe-like (the large bracts). Boat Lily.
W Indies, C America.

Rhoicissus rō-i-*kis*-us *Vitaceae*. From L.
rhoicus (of *Rhus*) and *Cissus* q.v. Tender,
evergreen climber.
capensis ka-*pen*-sis. (= *Cissus capensis*). Of
the Cape of Good Hope. S Africa.
rhomboidea hort. see *Cissus rhombifolia*

Rhombophyllum rom-bō-*fil*-lum *Aizoaceae*.
From Gk. *rhombos* (a rhombus) and *phyllon* (a
leaf) referring to the shape of the leaves.
Tender succulents. S Africa.
nelii nel-ee-ee. After G. C. Nel
(1885–1960). Elk's Horns.
rhomboideum rom-*boi*-dee-um. Diamond-
shaped (the leaves).

Rhubarb see *Rheum rhabarbarum*

Rhus rus *Anacardiaceae*. The L. name of *R.
coriaria*. Deciduous shrubs and trees.
Sumach.
cotinoides see *Cotinus obovatus*
cotinus see *Cotinus coggygria*
glabra glăb-ra. Glabrous. Smooth Sumach.
N America.
   'Laciniata' la-kin-ee-*ah*-ta. Deeply cut
   (the leaflets).
typhina tee-*fee*-na. Like *Typha*. Stag's-Horn
Sumach. E N America.
   'Dissecta' di-*sek*-ta. (= 'Laciniata').
   Deeply cut (the leaflets).

Ribbon Gum see *Eucalyptus viminalis*

Ribes rie-beez *Grossulariaceae*. From Arabic
or Persian *ribas* (acid-tasting), referring to
the fruit. Deciduous and evergreen shrubs.
alpinum ăl-*peen*-um. Alpine. Mountain
Currant. Europe.
gayanum gay-*ah*-num. After C. Gay, a
French botanist. Chile.
laurifolium low-ri-*fo*-lee-um. *Laurus*-leaved.
W China.
nigrum nig-rum. Black (the fruit). Black
Currant. Europe.
odoratum o-dō-*rah*-tum. Fragrant (the
flowers). Buffalo Currant. C United
States.
rubrum rub-rum. Red (the fruits). Red
Currant. W Europe.
sanguineum sang-*gwin*-ee-um. Blood-red.
Flowering Currant. W N America.
speciosum spe-kee-ō-sum. Showy.
California.
uva-crispa oo-va-*kris*-pa. A medieval name
meaning crisp grape. Gooseberry. Europe,
N Africa.

Richea ree-shee-a *Epacridaceae*. After C.-A.-
G. Riche, a French naturalist (died 1791).
Evergreen shrub.
scoparia skō-*pah*-ree-a. Broom-like.
Tasmania.

Ricinus *ri*-ki-nus *Euphorbiaceae*. The L. name from *ricinus* (a tick), the seeds resemble ticks. Tender tree or shrub grown as an annual.
  communis kom-*ew*-nis. Common. Castor-oil Plant. Tropical Africa.

Robinia ro-*bin*-ee-a *Leguminosae*. After Jean Robin (1550–1629), herbalist to Henry IV of France. Deciduous trees and shrubs.
  hispida his-pi-da. Bristly (the shoots). Rose Acacia. SE United States.
  kelseyi kel-see-ee. After Mr Harlan P. Kelsey who introduced it to cultivation. E United States.
  pseudacacia sood-a-kay-see-a. False *Acacia*. Locust, False Acacia. E United States.
    'Frisia' *free*-see-a. Of Friesland.
    'Umbraculifera' um-brah-kew-*li*-fe-ra. (= 'Inermis'). Umbrella-bearing (the shape of the head). Mop-headed Acacia.
  × *slavinii* sla-*vin*-ee-ee. *R. kelseyi* × *R. pseudoacacia*. After B. H. Slavin, who collected the seed from which it was raised.
    'Hillieri' *hi*-lee-a-ree. (*R.* × *hillieri*). After Hillier's who raised it.

Roblé see *Nothofagus obliqua*

Rochea rosh-ee-a *Crassulaceae*. After Daniel de la Roche (1743–1813), a Swiss physician. Tender succulents. S Africa.
  coccinea kok-*kin*-ee-a. Scarlet (the flowers).
  falcata see *Crassula falcata*
  × *versicolor* ver-si-ko-lor. (*R. coccinea* × *R. subulata*). Variously coloured (the flowers).

Rock Jasmine see *Androsace*
Rock Lily see *Arthropodium*
Rock Rose see *Cistus*

Rodgersia ro-*jerz*-ee-a *Saxifragaceae*. After Rear Admiral John Rodgers (1812–82). Perennial herbs.
  aesculifolia ie-skew-li-*fo*-lee-a. With leaves like *Aesculus*. China.
  pinnata pi-*nah*-ta. Pinnate (the leaves). China.
  podophylla po-dō-*fil*-la. With stoutly stalked leaves. China, Japan.
  sambucifolia săm-bew-ki-*fo*-lee-a. *Sambucus*-leaved. China.
  tabularis tăb-ew-*lah*-ris. Table-like (the leaves). N China, Korea.

Rohdea rō-dee-a *Liliaceae*. After Michael

Rohdea (continued)
Rohde (1782–1812). Tender perennial herb.
  japonica ja-*pon*-i-ka. Of Japan. Lily of China. China, Japan.

Romneya rom-*nee*-a *Papaveraceae*. After Dr Thomas Romney Robinson (1792–1882), Irish astronomer. Sub-shrubs. Tree Poppy.
  coulteri kool-ta-ree. After Dr Thomas Coulter who discovered it in 1833. S California.
  trichocalyx tri-kō-*kă*-liks. With a hairy calyx. S California, N Mexico.

Romulea rom-*ew*-lee-a *Iridaceae*. After Romulus, founder and first King of Rome. Cormous herbs.
  bulbocodium bul-bō-*kō*-dee-um. Like *Bulbocodium*. Mediterranean region, N Africa, W Asia.
  clusiana kloo-zee-*ah*-na. After Clusius, see *Gentiana clusii*. Spain.
  requienii rek-wee-*en*-ee-ee. After Requien, see *Mentha requienii*. Mediterranean region.
  rosea ro-see-a. Rose-coloured (the flowers). S Africa.

Rooksbya euphorbioides see *Carnegiea euphorbioides*

Rosa ro-sa *Rosaceae*. The L. name. Shrubs and climbers. Rose.
  × *alba* ăl-ba. White (the flowers). White Rose.
  banksiae banks-ee-ie. After Lady Dorothea Banks, wife of Sir Joseph Banks. Banksian Rose. China.
  brunonii broo-*non*-ee-ee. After Robert Brown (1773–1858), Scottish botanist. Himalayan Musk Rose. Himalaya, W China.
  canina ka-*neen*-a. Of dogs. Dog Rose. Europe, SW Asia.
  centifolia kent-i-*fo*-lee-a. With 100 leaves (petals). Holland Rose, Provence Rose. Cult.
  chinensis chin-*en*-sis. Of China. Cult.
    'Mutabilis' see *R.* × *odorata* 'Mutabilis'
    'Viridiflora' vi-ri-di-*flō*-ra. With green flowers.
  damascena dă-ma-*skay*-na. Of Damascus. Damask Rose. Cult.
    'Trigintipetala' tree-gin-ti-*pe*-ta-la. With thirty petals. Kazanlik Rose.
    versicolor ver-*si*-ko-lor. Variously coloured. York and Lancaster Rose.
  ecae ee-kie. After Mrs E. C. Aitchinson

Rosa (continued)
(E.C.A.) whose husband introduced it to
cultivation. C Asia.
eglanteria eg-lan-*te*-ree-a. (= *R. rubiginosa*).
From Provencal *aiglent*, originally,
probably from L. *aculeatus* (prickly).
Eglantine, Sweet Briar. Europe, N Africa.
elegantula ay-le-*gănt*-ew-la. (= *R. farreri*).
Elegant. China.
'Persetosa' per-say-*tō*-sa. Very bristly.
Threepenny-bit Rose.
farreri see *R. elegantula*
filipes fee-li-pays. Slender-stalked (the
flowers). W China.
foetida foy-ti-da. Foetid (the flowers).
C Asia.
'Bicolor' *bi*-ko-lor. Two-coloured.
Austrian Copper Briar.
'Persiana' per-see-*ah*-na. Of Persia.
gallica gă-li-ka. Of France. Red Rose.
Europe, W Asia.
glauca glow-ka. (= *R. rubrifolia*). Glaucous
(the bloom on the stem and young leaves).
S Europe.
× *harisonii* hă-ri-*son*-ee-ee. *R. foetida* × *R.
pimpinellifolia*. After George Folliot
Harison (died 1846), of New York, who
raised it.
helenae he-len-ie. After Ernest Wilson's
wife, Ellen. China.
hugonis hew-*gō*-nis. After its introducer
Father Hugh Scallan (Pater Hugo). NW
China.
moyesii moyz-ee-ee. After the Rev. J.
Moyes, a missionary in W China.
W Sichuan.
'Fargesii' far-*jeez*-ee-ee. After Farges, see
Decaisnea fargesii
nitida ni-ti-da. Shining (the leaves).
E N America.
× *odorata* o-dō-*rah*-ta. *R. chinensis* × *R.
gigantea*. Scented.
'Mutabilis' mew-*tah*-bi-lis. (= *R.
chinensis* 'Mutabilis'). Changeable (the
flower colour).
omeiensis see *R. sericea omeiensis
pteracantha* see *R. sericea pteracantha*
pimpinellifolia pim-pi-nel-i-*fo*-lee-a. (= *R.
spinosissima*). Pimpinella-leaved. Burnet
Rose. Europe to C Asia.
'Grandiflora' grănd-i-*flō*-ra. (= *altaica*
hort.). Large-flowered.
rubiginosa see *R. eglanteria*
rubrifolia see *R. glauca*
rugosa roo-*gō*-sa. Wrinkled (the leaves). NE
Asia.
sericea say-*ri*-kee-a. Silky-hairy. Himalaya,
China.

Rosa (continued)
omeiensis ō-may-*en*-sis. (= *R. omeiensis*).
Of the Omei Shan, China.
pteracantha te-ra-*kănth*-a. (= *R. omeiensis
pteracantha*). With winged spines.
spinosissima see *R. pimpinellifolia*
wichuraiana vi-kewr-ra-ee-*ah*-na. After
Max Wichura, a Prussian diplomat who
collected the type specimen. Japan, Korea.
xanthina zăn-*theen*-a. Yellow (the flowers).
China, Korea.

Rosary Vine see *Ceropegia woodii*

Roscoea ros-*kō*-ee-a *Zingiberaceae*. After
William Roscoe of Liverpool (1753–1831).
Semi-hardy perennial herbs.
alpina ăl-*peen*-a. Alpine. Himalaya, Tibet.
cautleyoides kawt-lee-*oi*-deez. Like
Cautleya. W China.
humeana hewm-ee-*ah*-na. After David
Hume of Edinburgh Botanic Garden.
W China.
purpurea pur-*pewr*-ree-a. Purple (the
flowers). Himalaya.

Rose see *Rosa*
  Banksian see *R. banksiae*
  Burnet see *R. pimpinellifolia*
  Damask see *R. damascena*
  Dog see *R. canina*
  Himalayan Musk see *R. brunonii*
  Holland see *R. centifolia*
  Kazanlik see *R. damascena* 'Trigintipetala'
  Provence see *R. centifolia*
  Threepenny-bit see *R. elegantula*
  'Persetosa'
  White see *R.* × *alba*
  York and Lancaster see *R. damascena
  versicolor*
Rose Acacia see *Robinia hispida*
Rose of China see *Hibiscus rosa-sinensis*
Rose of Heaven see *Silene coeli-rosea*
Rose Pincushion see *Mammillaria
zeilmanniana*
Rose of Sharon see *Hypericum calycinum*
Roseroot see *Rhodiola rosea*

Rosmarinus rōs-ma-*reen*-us *Labiatae*. The L.
name from *ros* (dew) and *marinus* (of the sea).
Evergreen shrub.
officinalis o-fi-ki-*nah*-lis. Sold as a herb.
Rosemary. Mediterranean region.
'Prostratus' pros-*trah*-tus. (= *R.
lavendulaceus* hort.). Prostrate.

Rough Bindweed see *Smilax aspera*

Rowan see *Sorbus aucuparia*
Royal Red Bugler see *Aeschynanthus pulcher*
Royal Paint Brush see *Haemanthus magnificus*
Royal Nodding Bells see *Streptocarpus wendlandii*
Rubber Plant see *Ficus elastica*

Rubus *rub*-us *Rosaceae*. The L. name for the blackberry. Deciduous and evergreen shrubs.
calycinoides kă-li-ki-*noi*-deez. Like *R. calycinus*. Himalaya.
cockburnianus kō-burn-ee-*ah*-nus. After Cockburn, see *Primula cockburniana*. China.
deliciosus day-li-kee-*ō*-sus. Delightful (the flowers). W N America.
idaeus ee-*die*-us. Of Mt. Ida. Raspberry. Europe, N Asia.
laciniatus la-kin-ee-*ah*-tus. Deeply cut (the leaves). Cut-leaved Bramble. Cult.
loganobaccus lō-ga-nō-*bah*-kus. After James Harvey Logan (1841–1928), an American judge who raised it, and L. *baccus* (a berry). Loganberry. Cult.
nepalensis ne-pa-*len*-sis. Of Nepal. Himalaya.
odoratus o-dō-*rah*-tus. Scented. E N America.
phoenicolasius foy-nee-ko-*lah*-see-us. With purple hairs (on the shoots). Wineberry. E Asia.
spectabilis spek-*tah*-bi-lis. Spectacular. Salmonberry. W N America.
thibetanus ti-bet-*ah*-nus. Of Tibet. W China.
tricolor tri-ko-lor. Three-coloured. W China.
ulmifolius ul-mi-*fo*-lee-us. *Ulmus*-leaved. Europe.
'Bellidiflorus' be-li-di-*flō*-rus. *Bellis*-flowered.

Rudbeckia rud-*bek*-ee-a *Compositae*. After Olof Rudbeck the elder (1630–1702) and the younger (1660–1740). Annual, biennial and perennial herbs. Coneflower.
bicolor bi-ko-lor. Two-coloured. United States.
fulgida *ful*-gi-da. Shining. United States.
  deamii *deem*-ee-ee. After its discoverer, Charles Clemon Deam (1865–1953).
hirta *hir*-ta. Hairy. Black-eyed Susan. United States.
laciniata la-kin-ee-*ah*-ta. Deeply cut (the leaves). N America.
maxima *mahks*-i-ma. Larger. United States.
purpurea see *Echinacea purpurea*

Rudbeckia (continued)
subtomentosa sub-tō-men-*tō*-sa. Somewhat hairy. N America.
triloba tri-*lō*-ba. Three-lobed (the lower leaves). N America.

Rue see *Ruta graveolens*

Ruellia roo-*el*-ee-a *Acanthaceae*. After Jean Ruel (1474–1537), French herbalist. Tender herbs and shrubs.
ciliosa ki-lee-*ō*-sa. Fringed with hairs (the calyx). SE United States.
devosiana da-vos-ee-*ah*-na. After Cornelius de Vos (1806–95). Brazil.
macrantha ma-*krănth*-a. Large-flowered. Christmas Pride. Brazil.
makoyana mă-koy-ah-na. After Jacob Makoy, a Belgian nurseryman. Trailing Velvet Plant. Brazil.
portellae por-*tel*-ie. After Francisco Portella of Rio de Janeiro who sent plants to Kew c 1880. Brazil.

Rugby Football Plant see *Peperomia argyreia*

Rumex *ru*-meks *Polygonaceae*. The L. name for *R. acetosa*. Perennial herbs.
acetosa a-kay-*tō*-sa. Old name for plants with acid leaves. Common Sorrel. Europe, Asia, N America.
scutatus skoo-*tah*-tus. Shield-bearing. French sorrel. C and S Europe.

Runner Bean see *Phaseolus vulgaris*
Scarlet see *P. coccineus*
Rupturewort see *Herniaria glabra*

Ruscus *rus*-kus *Liliaceae*. The L. name. Evergreen shrubs.
aculeatus a-kew-lee-*ah*-tus. Prickly. Butcher's Broom. Europe, N Africa, W Asia.
hypoglossum hi-pō-*glos*-um. Beneath the tongue, the flowers are borne under a tongue-like bract. S Europe, Turkey.
racemosus see *Danae racemosa*

Russelia rūs-*el*-ee-a *Scrophulariaceae*. After Dr Alexander Russel (1715–68). Tender shrubs.
equisetiformis ek-wi-say-ti-*form*-is. (= *R. juncea*). Like *Equisetum*. Coral Plant. Mexico.
sarmentosa sar-men-*tō*-sa. Twiggy. C America.

Ruta *roo*-ta *Rutaceae*. The L. name.

Ruta (continued)
Evergreen sub-shrub.
graveolens gra-*vee*-o-lenz. Strong-smelling.
Rue. SE Europe.

Rutabaga see *Brassica napus* Napobrassica

S

Saffron see *Crocus sativus*
Saffron Spike see *Aphelandra squarrosa*
Sage see *Salvia*
    Cardinal see *S. fulgens*
    Common see *S. officinalis*
    Gentian see *S. patens*
    Mealy-cup see *S. farinacea*
    Pineapple see *S. rutilans*
Sage Brush see *Artemisia tridentata*

Sagina sa-*geen*-a *Caryophyllaceae*. From L.
sagina (fodder), sheep were fed on a related
plant. Perennial herbs.
glabra gläb-ra. (= *Minuartia verna* hort.).
Glabrous. Pyrenees, Alps.
    'Aurea' *ow*-ree-a. Golden.
subulata soob-ew-*lah*-ta. Awl-shaped (the
leaves). Europe.

Sagittaria sä-gi-*tah*-ree-a *Alismataceae*. From
L. *sagitta* (an arrow) referring to the shape
of the leaves. Aquatic perennial herbs.
Arrowhead.
latifolia lah-tee-*fo*-lee-a. Broad-leaved.
N America.
sagittifolia sä-gi-ti-*fo*-lee-a. With arrow-
shaped leaves. Europe, Asia.
    'Flore Pleno' *flō*-ree-*play*-nō. (= *S.
japonica* 'Flore Pleno'). With double
flowers.
trifolia tri-*fo*-lee-a. Three-leaved, the leaves
are three-lobed. N hemisphere.

Saguaro see *Carnegiea gigantea*

Saintpaulia saynt-*pawl*-ee-a *Gesneriaceae*.
After Baron Walter von Saint Paul-Illaire
(1860–1910), who discovered *S. ionantha*.
Tender perennial herbs. African Violet.
Tanzania.
confusa kon-*few*-sa. Confused (with another
species).
grandifolia gränd-i-*fo*-lee-a. Large-leaved.
grotei grō-tee-ee. After Grote.
ionantha ee-on-*änth*-a. With violet flowers.
shumensis shoom-*en*-sis. Of Shume,
Tanzania.

Salal see *Gaultheria shallon*

Salix sä-liks *Salicaceae*. The L. name.
Deciduous trees and shrubs. Willow.
acutifolia a-kew-ti-*fo*-lee-a. With pointed
leaves. Russia.
aegyptiaca ie-gip-tee-*ah*-ka. Egyptian.
Musk Willow. W Asia.
alba äl-ba. White (the leaves). White
Willow. Europe, W Asia.
    Argentea ar-*gen*-tee-a. Silvery (the
leaves). Silver Willow.
    'Britzensis' brits-*en*-sis. Of Britz, near
Berlin.
    'Caerulea' kie-*ru*-lee-a. Blue (the leaves).
Cricket Bat Willow.
    Vitellina vi-te-*leen*-a. Egg-yolk yellow
(the shoots). Golden Willow.
    'Vitellina Pendula' see *S.* 'Chrysocoma'
'Boydii' *boyd*-ee-ee. After William Brack
Boyd (1831–1918), Scottish naturalist who
discovered it.
caprea käp-ree-a. Of goats (the foliage was
used for goat fodder). Goat Willow.
Europe, NW Asia.
'Chrysocoma' kris-*o*-ko-ma. (= *S. alba*
'Vitellina Pendula'). Golden-haired,
referring to the weeping, yellow shoots.
Weeping Willow.
daphnoides däf-*noi*-deez. Laurel-like. Violet
Willow. Europe.
elaeagnos e-lee-*äg*-nos. Gk. name of a
willow, see *Elaeagnus*. Europe, W Asia.
exigua eks-ig-ew-a. Small. W N America.
fragilis frä-gi-lis. Fragile (the shoots). Crack
Willow. Europe, Russia.
hastata häs-*tah*-ta. Spear-shaped (the
leaves). Europe, Asia.
    'Wehrhahnii' vair-*hahn*-ee-ee. After H.
R. Wehrhahn.
lanata lah-*nah*-ta. Woolly. Woolly Willow.
N Europe, N Asia.
matsudana mät-soo-*dah*-na. After Matsuda.
Peking Willow. N China.
    'Pendula' *pen*-dew-la. Weeping.
    'Tortuosa' tort-ew-ō-sa. Twisted (the
shoots). Dragon's Claw Willow.
'Melanostachys' me-la-*no*-sta-kis. With
black spikes.
pentandra pen-*tän*-dra. With five stamens.
Bay Willow. Europe, Asia.
purpurea pur-*pewr*-ree-a. Purple (the
shoots). Purple Osier. Europe, N Asia.
repens ree-penz. Creeping. Creeping
Willow. Europe, Asia.
reticulata ray-tik-ew-*lah*-ta. Net-veined (the
leaves). Arctic and alpine N hemisphere.

Salix (continued)
sachalinensis sa-kah-lin-*en*-sis. Of Sakhalin. NE Asia.
viminalis vee-min-*ah*-lis. With long, slender shoots. Common Osier. Europe, Asia.

Salmon Blood Lily see *Haemanthus multiflorus*
Salmonberry see *Rubus spectabilis*

Salpiglossus sal-pi-*glos*-us *Solanaceae*. From Gk. *salpinx* (a trumpet) and *glossa* (a tongue). Tender herb.
sinuata sin-ew-*ah*-ta. Wavy-edged (the leaves). Painted Tongue. Chile.

Salsify see *Tragopogon porrifolius*
Salt Tree see *Halimodendron halodendron*

Salvia *săl*-vee-a *Labiatae*. The L. name, from *salvus* (safe) referring to medicinal properties. Annual and perennial herbs and shrubs. Sage.
ambigens see *S. caerulea*
argentea ar-*gen*-tee-a. Silvery (the leaves). S Europe.
azurea a-*zew*-ree-a. Deep blue (the flowers). C United States.
  grandiflora gránd-i-*flō*-ra. (= *pitcheri*). Large-flowered.
blepharophylla blef-a-rō-*fil*-la. With fringed leaves. Mexico.
caerulea kie-*ru*-lee-a. (= *S. ambigens. S. guaranitica*). Deep blue. Brazil.
farinacea fă-ree-*nah*-kee-a. Mealy. Mealy-cup Sage. SW United States.
fulgens *ful*-genz. Shining (the flowers). Cardinal Sage. Mexico.
greggii greg-ee-ee. After Dr John Gregg who discovered it in 1848. Texas, Mexico.
guaranitica see *S. caerulea*
haematodes see *S. pratensis*
horminum see *S. viridis*
involucrata in-vo-loo-*krah*-ta. With bracts around the flowers. Mexico.
  'Bethellii' be-*thel*-ee-ee. After Mr Bethell who raised it.
microphylla mik-rō-*fil*-la. Small-leaved. Mexico.
nemorosa ne-mo-*rō*-sa. Of woods. Europe.
officinalis o-fi-ki-*nah*-lis. Sold as a herb. Common Sage. S Europe.
  'Icterina' ik-*te*-ri-na. Jaundice-yellow (the leaves).
patens *pă*-tenz. Spreading (the flowers). Gentian Sage. Mexico.
pratensis prah-*tayn*-sis. (= *S. haematodes*). Of meadows. Europe.
rutilans *roo*-ti-lănz. Reddish (the flowers).

Salvia (continued)
  Pineapple Sage. Mexico.
sclarea *sklah*-ree-a. From L. *clarus* (clear) the seeds and leaves are used in eye lotions. Clary. S Europe.
splendens *splen*-denz. Splendid. Brazil.
uliginosa ew-li-gi-*nō*-sa. Of marshes. S America.
viridis *vi*-ri-dis. (= *S. horminum*). Green (the bracts of some forms). S Europe.

Salvinia săl-*veen*-ee-a *Salviniaceae*. After Professor Antonio Maria Salvini (1633–1729). Tender, floating ferns.
auriculata ow-rik-ew-*lah*-ta. Auricled. Floating Fern. C and S America.
rotundifolia ro-tund-i-*fo*-lee-a. Round-leaved. Tropical America.

Sambucus săm-*bew*-kus *Caprifoliaceae*. The L. name. Deciduous shrubs. Elder.
canadensis kăn-a-*den*-sis. Of E N America. American Elder.
nigra *nig*-ra. Black (the fruits). Common Elder. Europe, N Africa, W Asia.
racemosa ră-kay-*mō*-sa. In racemes, the elongated inflorescence. Red-berried Elder. Europe, W Asia, Siberia.

Sand Myrtle see *Leiophyllum buxifolium*

Sandersonia sahn-der-*son*-ee-a *Liliaceae*. After John Sanderson (c 1820–1881), who discovered the following in 1851. Tender perennial herb.
aurantiaca ow-răn-tee-*ah*-ka. Orange (the leaves). Chinese Lanterns. S Africa.

Sandwort see *Arenaria*

Sanguinaria sang-gwi-*nah*-ree-a *Papaveraceae*. From L. *sanguis* (blood) referring to the red sap. Perennial herb.
canadensis kăn-a-*den*-sis. Of EN America. Bloodroot.

Sanguisorba sang-gwi-*sor*-ba *Rosaceae*. From L. *sanguis* (blood) and *sorbeo* (to absorb), it was said to stop bleeding. Perennial herbs. Burnet.
canadensis kăn-a-*den*-sis. Of Canada. N America.
obtusa ob-*tew*-sa. Blunt (the leaflets). Japan.
tenuifolia ten-ew-i-*fo*-lee-a. With slender leaflets. N Asia.

Sansevieria săn-sev-ee-*e*-ree-a *Agavaceae*. After Raimond de Sangro, Prince of

Sansevieria (continued)
Sanseviero, 18th-century patron of
horticulture. Tender, evergreen herbs.
  trifasciata tri-fås-ee-*ah*-ta. In three bundles,
  the flower clusters are in groups of 1 to 3.
  Mother-in-law's Tongue. E South Africa.
    'Hahnii' *hahn*-ee-ee. After Hahn.
    'Laurentii' lo-*rent*-ee-ee. After Emile
    Laurent (1861–1904), who discovered
    it.

Santolina sån-to-*leen*-a *Compositae*. From
sanctum linum (holy flax) the L. name for *S.
rosmarinifolia*. Evergreen shrubs.
  chamaecyparissus kå-mie-kew-pa-*ris*-us.
  Dwarf cypress. Lavender Cotton.
  Mediterranean region.
  pinnata pin-*ah*-ta. Pinnate (the leaves).
  Italy.
    neopolitana nee-ah-po-li-*tah*-na. (= *S.
    neapolitana*). Of Naples.
    rosmarinifolia rōs-ma-reen-i-*fo*-lee-a. (= *S.
    virens*). *Rosmarinus*-leaved. Holy Flax.
  Mediterranean region.

Sanvitalia sån-vi-*tah*-lee-a *Compositae*. After
Frederico Sanvitali (1704–61). Annual herb.
  procumbens prō-*kum*-benz. Prostrate.
  Creeping Zinnia. C America.

Saponaria så-pō-*nah*-ree-a *Caryophyllaceae*.
From L. *sapo* (soap), a soap can be made
from *S. officinalis*. Annual and perennial
herbs.
  caespitosa kie-spi-*tō*-sa. Tufted. Pyrenees.
  calabrica ka-*lå*-bri-ka. Of Calabria, Italy.
  E Mediterranean region.
  ocymoides ō-kim-*oi*-dees. Like *Ocimum*.
  Rock Soapwort. S Europe.
  officinalis o-fi-ki-*nah*-lis. Sold as a herb.
  Bouncing Bet, Soapwort. Europe.
  vaccaria see *Vaccaria pyramidata*

Sapphire Flower see *Browallia speciosa*

Sarcococca sar-kō-*ko*-ka *Buxaceae*. From
Gk. *sarcos* (flesh) and *kokkos* (a berry)
referring to the fleshy fruits. Christmas Box,
Sweet Box.
  confusa kon-*few*-sa. Confused (with other
  species). ? China.
  hookeriana hu-ka-ree-*ah*-na. After Sir
  Joseph Hooker. Himalaya.
    digyna di-gi-na. With two styles.
    W China.
  humilis hu-mi-lis. Low-growing. W China.
  ruscifolia rus-ki-*fo*-lee-a. *Ruscus*-leaved.
  China.

Sarracenia så-ra-*sen*-ee-a *Sarraceniaceae*.
After Michael Sarrasin (1659–1734), French
botanist and physician. Carnivorous perennial
herbs. Pitcher Plant.
  flava flah-va. Yellow (the flowers). Yellow
  Pitcher Plant. SE United States.
  leucophylla loo-kō-*fil*-la. (= *S. drummondii*).
  White-leaved. Fiddler's Trumpets, Lace
  Trumpets. SE United States.
  purpurea pur-*pewr*-ree-a. Purple (the
  pitchers). Northern Pitcher Plant. E N
  America.

Sasa sah-sa *Gramineae*. The Japanese name.
Bamboos.
  palmata pahl-*mah*-ta. Hand-like, the
  spreading, clustered leaves. Japan.
  tessellata te-se-*lah*-ta. Tesselated i.e. with a
  minute, squared venation (the leaves).
  China.
  veitchii veech-ee-ee. After Messrs Veitch for
  whom Charles Maries introduced it in
  c 1880. Japan.

Sassafras sås-a-fràs *Lauraceae*. Probably
adapted by French settlers from an
American Indian name. Deciduous tree.
  albidum ål-bi-dum. Whitish (the undersides
  of the leaves). Sassafras. E N America.

Satin Flower see *Clarkia amoena*
Satsuma see *Citrus reticulata*

Satureja såt-ew-*ray*-a *Labiatae*. The L.
name. Sub-shrub and annual herb.
  hortensis hor-*ten*-sis. Of gardens. Summer
  Savoury. Mediterranean region.
  montana mon-*tah*-na. Of mountains.
  Winter Savoury. S Europe.

Saucer Plant see *Aeonium undulatum*

Sauromatum sow-ro-ma-tum *Araceae*. From
Gk. *sauros* (a lizard) referring to the spotted
spathe and its long, tail-like appendage.
Tender perennial herb.
  guttatum gu-*tah*-tum. (= *Arum cornutum*).
  Spotted (the spathe). Monarch of the East.
  Himalaya, SE Tibet.
    venosum vay-*nō*-sum. Conspicuously
    veined.

Savin see *Juniperus sabina*
Savoury, Summer see *Satureja hortensis*
  Winter see *S. montana*

Saxegothaea săks-ee-goth-*ee*-a
Podocarpaceae. After Prince Albert of Saxe-Coburg-Gotha. Evergreen conifer.
conspicua con-*spik*-ew-a. Conspicuous,
distinguished. Prince Albert's Yew. Chile.

Saxifraga săks-*if*-ra-ga *Saxifragaceae*. From
L. *saxum* (a rock) and *frango* (to break), by
growing in rock crevices they appear to break
rocks. Perennial herbs. Saxifrage.
  aizoides ie-zō-i-deez. Like *Aizoon*. Europe.
  aizoon see *S. paniculata*
  × *apiculata* a-pik-ew-*lah*-ta. *S. juniperifolia
  sancta* × *S. marginata rocheliana*. With a
  short, abrupt point (the leaves).
  brunoniana see *S. brunonis*
  brunonis broo-nō-nis. (= *S. brunoniana*)
  After Robert Brown. Himalaya, S Tibet.
  burseriana bur-sa-ree-*ah*-na. After Joachim
  Burser (1583–1649), German physician
  and botanist. E Alps.
  callosa ka-*lō*-sa. Calloused, the lime-
  encrusted leaves. W Mediterranean
  region.
  cochlearis kok-lee-*ah*-ris. Spoon-shaped (the
  leaves). Alps.
  cortusifolia kor-tew-si-*fo*-lee-a. *Cortusa*-
  leaved. E Asia.
    fortunei for-*tewn*-ee-ee. (= *S. fortunei*).
    After Robert Fortune, see *Fortunella*.
    Japan.
  cotyledon ko-ti-*lay*-don. Like *Cotyledon*.
  Alps, Pyrenees, NW Europe.
  cuneifolia kew-nee-i-*fo*-lee-a. With the
  leaves narrowed to the base. Pyrenees,
  Alps, Carpathians.
  diapensioides dee-a-pen-see-*oi*-deez. Like
  Diapensia lapponica. SW Alps.
  fortunei see *S. cortusifolia fortunei*
  grisebachii gree-za-*bahk*-ee-ee. After
  Professor August Heinrich Rudolph
  Grisebach (1814–79), German botanist.
  Balkan Peninsula.
  longifolia long-gi-*fo*-lee-a. Long-leaved.
  Pyrenees.
  moschata mos-*kah*-ta. Musky. C and S
  Europe.
  oppositifolia o-po-si-ti-*fo*-lee-a. With
  opposite leaves. Europe.
  paniculata pa-nik-ew-*lah*-ta. (= *S. aizoon*).
  With flowers in panicles. C and S Europe,
  Norway.
  sarmentosa see *S. stolonifera*
  stolonifera sto-lō-*ni*-fe-ra. (= *S. sarmentosa*).
  Bearing stolons. Mother of Thousands.
  E Asia.
  umbrosa hort. see *S.* × *urbium*
  × *urbium* ur-bee-um. (= *S. umbrosa* hort.)

Saxifraga (continued)
  S. umbrosa × *S. spathularis*. Of towns.
  London Pride.

Scabiosa skăb-ee-ō-sa *Dipsacaceae*. From L.
scabies (itch, scurf), the rough leaves were
said to cure it. Annual and perennial herbs.
Scabious.
  alpina see *Cephalaria alpina*
  atropurpurea aht-rō-pur-*pewr*-ree-a. Deep
  purple. Sweet Scabious. S Europe.
  caucasica kaw-*kă*-si-ka. Of the Caucasus.
  columbaria ko-lum-*bah*-ree-a. Dove-like or
  of doves. Europe, N Africa, Asia.
  graminifolia grah-mi-ni-*fo*-lee-a. With
  grass-like leaves. S Europe.
  lucida loo-ki-da. Shining. C Europe.
  ochroleuca ok-rō-*loo*-ka. Yellowish-white.
  Europe, W Asia.

Scarborough Lily see *Vallota speciosa*
Scarlet Leadwort see *Plumbago indica*
Scarlet Trompetilla see *Bouvardia longiflora*

Schefflera shef-*le*-ra *Araliaceae*. After J. C.
Scheffler. Tender, evergreen shrubs and
trees.
  actinophylla ăk-tin-ō-*fil*-la (= *Brassaia
  actinophylla*). With rayed leaves, referring to
  the radiating leaflets. Queensland Umbrella
  Tree. Queensland.
  arboricola ar-bo-ri-ko-la. (= *Heptapleurum
  arboricolum*). Growing on trees. Taiwan.
  digitata di-gi-*tah*-ta. Lobed like a hand (the
  leaves). Seven Fingers. New Zealand.

Schisandra skis-ăn-dra *Schisandraceae*.
From Gk. *schizo* (to divide) and *aner* (a man)
referring to the well-separated anther cells.
Deciduous and evergreen climbers.
  glaucescens glow-*kes*-enz. Somewhat
  glaucous (the lower leaf surface). C China.
  grandiflora grănd-i-*flō*-ra. Large-flowered.
  Himalaya.
    rubriflora rub-ri-*flō*-ra. Red-flowered.
    China, NE India.
  propinqua prō-*ping*-kwa. Related to
  (another species). Himalaya.
    chinensis chin-en-sis. Of China.

Schizanthus skiz-*ănth*-us *Solanaceae*. From
Gk. *schizo* (to divide) and *anthos* (a flower)
referring to the deeply divided corolla.
Annual herbs. Butterfly Flower. Poor Man's
Orchid.
  pinnatus pi-*nah*-tus. Pinnate (the leaves).
  Chile.
  × *wisetonensis* wiez-ton-*en*-sis. Of Wiseton.

Schizocentron see *Heterocentron*

Schizophragma ski-zō-*frăg*-ma
Hydrangeaceae. From Gk. *schizo* (to divide)
and *phragma* (a wall), parts of the fruit fall
away leaving it skeletonised. Deciduous
climbers.
   hydrangeoides hi-drang-gee-*oi*-deez. Like
   Hydrangea. Japan.
   integrifolia in-teg-ri-*fo*-lee-a. Entire-leaved.
   C China.

Schizostylis ski-zō-*sti*-lis Iridaceae. From
Gk. *schizo* (to divide) and *stylis* (a style), the
style is deeply divided. Perennial herb.
   coccinea kok-*kin*-ee-a. Scarlet (the flowers).
   Kaffir Lily. S Africa.

Schlumbergera shlum-*ber*-ga-ra *Cactaceae*.
After Frederick Schlumberger (1804–65),
Belgian horticulturist, explorer and plant
collector.
   bridgesii bri-*jez*-ee-ee. After Thomas
   Bridges who collected in S America in the
   mid 19th century. Christmas Cactus.
   Brazil.
   × *buckleyi* bŭk-lee-ee. *S. russelliana* × *S.
   truncata*. After W. Buckley, a cactus
   grower.
   gaertneri see *Rhipsalidopsis gaertneri*
   truncata trung-*kah*-ta. (= *Zygocactus
   truncatus*). Abruptly cut off (the ends of
   the shoots). Claw Cactus, Crab Cactus.
   Brazil.

Sciadopitys skee-a-*do*-pi-tis Taxodiaceae.
From Gk. *skiados* (an umbel) and *pitys* (a fir
tree), the leaves appear in whorls like the ribs
of an umbrella.
   verticillata ver-ti-ki-*lah*-ta. Whorled.
   Umbrella Pine. Japan.

Scilla skil-la Liliaceae. The Gk. name for the
sea squill, *Urginea maritima*. Bulbous
perennial herbs.
   bifolia bi-*fo*-lee-a. Two-leaved. S and C
   Europe, Caucasus, Turkey.
   hispanica see *Hyacinthoides hispanica*
   miczenkoana mi-cheng-kō-*ah*-na. (= *S.
   tubergeniana*). After Miczenko.
   Transcaucasus, NW Iran.
   monophyllos mo-nō-*fil*-los. One-leaved. SW
   Europe, N Africa.
   non-scripta see *Hyacinthoides non-scripta*
   ovalifolia see *Ledebouria ovalifolia*
   peruviana pe-roo-vee-*ah*-na. Originally
   thought to be from Peru.

Scilla (continued)
   W Mediterranean region, Portugal.
   sibirica si-*bi*-ri-ka. Siberian. Siberian
   Squill. S Russia, Caucasus, W Asia.
   violacea see *Ledebouria socialis*

Scindapsus skin-*dăp*-sus Araceae. Gk. name
for an ivy-like plant. Tender, evergreen
climber.
   aureus see *Epipremnum aureum*
   pictus pik-tus. Painted (the leaves). Silver
   Vine. SE Asia.
      'Argyraeus' ar-gi-*ree*-us. Silvery (the
      leaves).

Scirpus skir-pus Cyperaceae. The L. name.
Perennial herbs.
   tabernaemontani tă-ber-nie-mon-*tah*-nee.
   After Tabernaemontanus, see *Amsonia
   tabernaemontani*. Europe.
      'Zebrinus' ze-*bree*-nus. Striped.

Scorpion senna see *Coronilla emerus*

Scorzonera skor-zo-*ne*-ra Compositae. From
French *scorzon* (a viper), the root was said to
cure snake bites. Biennial herb.
   hispanica his-*pah*-ni-ka. Spanish.
   Scorzonera. C and S Europe, S Russia.

Scotch Thistle see *Onopordum acanthium*
Screw Pine see *Pandanus*

Scrophularia skrō-few-*lah*-ree-a
Scrophulariaceae. From L. *scrofulae* (a
swelling of the neck glands) referring to
medicinal properties. Perennial herb.
   auriculata ow-rik-ew-*lah*-ta. (= *S.
   aquatica*). Auricled (the leaves). Water
   Figwort. W Europe, N Africa.

Scutellaria sku-te-*lah*-ree-a Labiatae. From
L. *scutella* (a small dish) referring to the
appearance of the calyx in fruit. Tender and
hardy perennial herbs.
   alpina ăl-*peen*-a. Alpine. S and E Europe.
   baicalensis bie-ka-*len*-sis. Of Lake Baikal,
   E Siberia. E Asia.
   costaricana kos-ta-ree-*kah*-na. Of Costa
   Rica.
   indica in-di-ka. Indian. China, Japan.
   scordiifolia skor-dee-i-*fo*-lee-a. With leaves
   like *Scordium* (= *Teucrium*). Asia.

Sea Buckthorn see *Hippophae rhamnoides*
Sea Campion see *Silene vulgaris maritima*
Sea Daffodil see *Pancratium maritimum*
Sea Holly see *Eryngium maritimum*

Sea Kale see *Crambe maritima*
Sea Lily see *Pancratium maritimum*
Sea Squill see *Urginea maritima*
Sea Urchin see *Hakea laurina*
Sedge see *Carex*
  Great Pond see *C. riparia*

Sedum *say*-dum *Crassulaceae*. Classical name for several succulent plants from L. *sedo* (to sit). Tender and hardy succulents.
  acre ahk-ree. Sharp-tasting. Biting Stonecrop. Europe, N Africa, W and N Asia.
  adolphi a-*dolf*-ee. After Adolf Engler. Golden Stonecrop. Mexico.
  album ăl-bum. White (the flowers). Europe, N Africa, W Asia.
  allantoides a-lăn-tō-i-deez. Sausage-like (the leaves). S Mexico.
  bellum bel-um. Beautiful. Mexico.
  brevifolium bre-vi-*fo*-lee-um. Short-leaved. S Europe, N Africa.
  cauticolum kaw-*ti*-ko-lum. Growing on cliffs. N Japan.
  dasyphyllum dăs-i-*fil*-lum. With hairy leaves. S Europe, N Africa.
  dendroideum den-*droi*-dee-um. Tree-like. Mexico.
    praealtum prie-ăl-tum. (= *S. praealtum*). Very tall.
  ewersii ay-*verz*-ee-ee. After Joseph Ewers (1781–1830), a German statesman. Himalaya, N China.
  fastigiatum see *Rhodiola fastigiata*
  floriferum flō-*ri*-fe-rum. Floriferous. China.
  heterodontum see *Rhodiola heterodonta*
  kamtschaticum kămt-*shă*-ti-kum. Of Kamtchatka. NE Asia.
  lineare li-nee-*ah*-ree. Linear (the leaves). E Asia.
  lydium li-dee-um. Of Lydia, W Asia.
  maximum mahks-i-mum. Larger. Europe, Caucasus.
  middendorfianum mi-dan-dorf-ee-*ah*-num. After Middendorf see *Hemerocallis middendorfiana*. C Asia.
  morganianum mor-găn-ee-*ah*-num. After Dr Meredith Morgan who grew it soon after its discovery. Donkey's Tail. Mexico.
  oaxacanum wah-hah-*kah*-num. Of Oaxaca. Mexico.
  oreganum o-ree-*gah*-num. Of Oregon, it was discovered at the mouth of the Oregon River. W N America.
  pachyphyllum pă-ki-*fil*-lum. Thick-leaved. Jelly Beans. S Mexico.
  populifolium pō-pew-li-*fo*-lee-um. *Populus-*

Sedum (continued)
  leaved. Siberia.
  praealtum see *S. dendroideum praealtum*
  reflexum re-*fleks*-um. Reflexed (the leaves on flowering stems). Europe.
  rhodiola see *Rhodiola rosea*
  rosea see *Rhodiola rosea*
  × *rubrotinctum* rub-rō-*tink*-tum. Red-tinged (the leaves). Christmas Cheer.
  rupestre roo-*pes*-tree. Growing on rocks. W Europe.
  sieboldii see-*bōld*-ee-ee. After Siebold who introduced it, see *Acanthopanax sieboldii*. Japan.
  spathulifolium spăth-ew-li-*fo*-lee-um. With spatula-shaped leaves. W N America.
  spectabile spek-*tah*-bi-lee. Spectacular. China, Korea.
  spurium spew-ree-um. False. Caucasus.
  telephium te-*le*-fee-um. Gk. name of a plant. Europe to Japan.

Seersucker Plant see *Geogenanthus undatus*

Selaginella se-lah-gi-*nel*-a *Selaginellaceae*. Diminutive of *selago* (*Lycopodium selago*). Tender, moss-like plants.
  apoda a-*pod*-a. Stalkless (the strobili). E N America.
  involvens in-*vol*-venz. Rolled up. E Asia.
  kraussiana krows-ee-*ah*-na. After C. Ferdinand F. Krauss (1812–90), who introduced it. Spreading Clubmoss. S Africa.
  lepidophylla le-pi-dō-*fil*-la. With scale-like leaves. Resurrection Plant. SW United States, C America.
  pallescens pa-*les*-enz. (= *S. emmeliana*). Rather pale. Moss fern. N and S America.

Selenicereus se-lay-nee-*kay*-ree-us *Cactaceae*. From Gk. *selene* (the moon) and *Cereus* q.v., they flower at night.
  coniflorus kō-ni-*flō*-rus. (= *S. pringlei*). With conical flowers. Mexico.
  grandiflorus grănd-i-*flō*-rus. Large-flowered. Queen of the Night. Jamaica, Cuba.
  megalanthus me-ga-*lănth*-us. Large-flowered. Peru.
  pteranthus te-*rănth*-us. With winged flowers. Princess of the Night. Mexico.
  setaceus say-*tah*-kee-us. Bristly. S America.
  wercklei vair-klee-ee. After Karl Werckle (1866–1924), who studied the flora of Costa Rica. Costa Rica.

Self Heal see *Prunella*

Selinum se-*leen*-um *Umbelliferae*. From Gk.
selinon (celery). Perennial herb.
   tenuifolium ten-ew-i-*fo*-lee-um. With finely
   divided leaves. Himalaya, Assam, Tibet.

Sempervivum sem-per-*veev*-um
Crassulaceae. From L. *semper* (always) and
vivus (alive). Succulent perennials.
Houseleek.
   arachnoideum ă-răk-*noi*-dee-um. With hairs
   like a spiders-web. Cobweb Houseleek.
   Alps, Pyrenees.
   dolomiticum do-lo-*mi*-ti-kum. Of the
   Dolomites.
   grandiflorum gránd-i-*flō*-rum. Large-
   flowered. Alps.
   marmoreum mar-*mo*-ree-um. Mottled.
   Balkans.
      rubrifolium ru-bri-*fo*-lee-um. Red-leaved.
   montanum mon-*tah*-num. Of mountains.
   Europe.
   soboliferum see *Jovibarba sobolifera*
   tectorum tek-*to*-rum. Growing on roofs.
   Common Houseleek, Hen and Chickens.
   Pyrenees to SE Alps.

Senecio se-*ne*-kee-ō *Compositae*. From L.
senex (an old man), referring to the fluffy,
white seed heads. Herbs, shrubs and tender
succulents.
   articulatus ar-tik-ew-*lah*-tus. (= *Kleinia
   articulata*). Jointed. Candle Plant. S Africa.
   bicolor bi-ko-lor. Two-coloured (the leaves
   are greenish above, white beneath). C and
   E Mediterranean region.
      cineraria ki-ne-*rah*-ee-a. (= *S. cineraria.
      S. maritimus. Cineraria maritima*). Ash-
      coloured, the leaves. W and C
      Mediterranean region.
   cineraria see *S. bicolor cineraria*
   clivorum see *Ligularia dentata*
   compactus com-*păk*-tus. Compact. New
   Zealand.
   cruentus kroo-*en*-tus. (= *Cineraria cruenta*).
   Blood-red (the flowers). Canary Islands.
   doronicum do-*ro*-ni-kum. From *Doronicum*
   q.v. Europe, N Africa.
   fulgens ful-genz. (= *Kleinia fulgens*).
   Shining. S Africa.
   galpinii gál-*pin*-ee-ee. (= *Kleinia galpinii*).
   After Galpin, see *Kniphofia galpinii*.
   S Africa.
   greyi gray-ee. After Sir George Grey
   (1812–98), Prime Minister of New
   Zealand. The true plant is rarely grown.
   New Zealand.
   haworthii hay-*werth*-ee-ee. (= *Kleinia*

Senecio (continued)
   tomentosa). After Haworth, see *Haworthia*.
   S Africa.
   herreianus he-ree-*ah*-nus. After H. Herre of
   Stellenbosch. S Africa.
   × *hybridus* hib-ri-dus. (*Cineraria ×
   hybrida*). Hybrid. Cineraria.
   kleinia klie-nee-a. (= *Kleinia neriifolia*). The
   genus in which it is sometimes placed.
   S Africa.
   laxifolius làks-i-*fo*-lee-us. Loose-leaved.
   New Zealand.
   macroglossus măk-rō-*glos*-us. Large-
   tongued, referring to the long ray florets.
   Cape Ivy. S Africa.
   mikanioides mi-kah-nee-*oi*-deez. Like
   Mikania, a related genus. German Ivy.
   S Africa.
   monroi mon-*rō*-ee. After Sir David Monro
   (1813–77), speaker in the NZ Parliament
   who sent plants to Kew. A shepherd
   collected the type specimen for him. New
   Zealand.
   pendulus pen-dew-lus. (= *Kleinia pendula*).
   Pendulous. N Africa, Arabia.
   petasitis pe-ta-*see*-tis. Hat-like (the leaves).
   California Geranium. Mexico.
   przewalskii see *Ligularia przewalskii*
   rowleyanus rō-lee-*ah*-nus. After Gordon D.
   Rowley (born 1921), a succulent
   enthusiast. String of Beads. SW Africa.
   scandens skăn-denz. Climbing. E Asia.
   serpens ser-penz. (= *Kleinia repens*). Snake-
   like. S Africa.
   tanguticus tang-*gew*-ti-kus. (= *Ligularia
   tangutica*). Of Gansu (Kansu). China.

Senna see *Cassia*
   American see *C. marylandica*
Sensitive Plant see *Mimosa pudica*

Sequoia se-*kwoy*-a *Taxodiaceae*. After
Sequoiah (1770–1843), son of a British
merchant by a Cherokee Indian woman.
Evergreen conifer.
   sempervirens sem-per-*vi*-renz. Evergreen.
   Coast Redwood. California, Oregon.

Sequoiadendron se-kwoy-a-*den*-dron
Taxodiaceae. From *Sequoia* q.v. and Gk.
dendron (a tree). Evergreen conifer.
   giganteum gi-*găn*-tee-um. Very large. Big
   Tree, Giant Redwood, Wellingtonia.
   California.

Service Tree see *Sorbus domestica*
   Wild see *S. torminalis*
Service Tree of Fontainbleau see *S. latifolia*

Serviceberry see *Amelanchier*

Setcreasia set-*krees*-ee-a *Commelinaceae*.
Derivation obscure. Tender perennial herb.
   pallida pă-li-da. (= *S. purpurea*). Pale (the
   flowers of some forms). Purple Heart.
   Mexico.
   striata hort. see *Callisia elegans*

Setiechinopsis say-tee-e-kee-*nop*-sis
Cactaceae. From L. *seta* (a bristle) and
Echinopsis q.v.
   mirabilis mee-*rah*-bi-lis. Wonderful.
   Argentina.

Seven Fingers see *Schefflera digitata*
Shaddock see *Citrus maxima*
Shallon see *Gaultheria shallon*
Shallot see *Allium cepa* Aggregatum
Shamrock Pea see *Parochetos communis*
Shasta Daisy see *Chrysanthemum* × *superbum*
Sheep Laurel see *Kalmia angustifolia*
Sheep's Bit see *Jasione*
Shell Flower see *Moluccella laevis*

Shibataea shi-ba-*tie*-a *Gramineae*. After
Keita Shibata (1877–1949), Japanese
botanist. Bamboo.
   kumasasa kew-ma-*sah*-sa. Japanese name of
   a bamboo. Japan.

Shooting Star see *Dodecatheon*

Shortia *short*-ee-a *Diapensiaceae*. After
Charles W. Short (1794–1863), a Kentucky
botanist. Evergreen, perennial herbs.
   galacifolia ga-lăk-i-*fo*-lee-a. With leaves like
   Galax. Oconee Bells. E N America.
   soldanelloides sol-da-nel-*oi*-deez. Like
   Soldanella. Japan.
   uniflora ew-ni-*flo*-ra. With solitary flowers.
   Japan.

Shrimp Plant see *Justicea brandegeana*
Siberian Squill see *Scilla sibirica*

Sidalcea see-*dăl*-kee-a *Malvaceae*. From
Sida and *Alcea*, related genera. Perennial
herbs. SW United States.
   candida kăn-di-da. White (the flowers).
   malviflora măl-vi-*flo*-ra. *Malva*-flowered.

Silene si-*lay*-nee *Caryophyllaceae*. Gk. name
for a related plant. Campion.
   acaulis a-*kaw*-lis. Stemless. Moss Campion.
   Europe.

Silene (continued)
   alpestris ăl-*pes*-tris. Of lower mountains.
   E Alps.
   armeria ar-*me*-ree-a. See *Armeria*. C, S and
   E Europe.
   coeli-rosea koy-lee-ro-see-a. (= *Lychnis coeli-
   rosea*). As the common name, Rose of
   Heaven. SW Europe.
   compacta com-*păk*-ta. Compact. SE
   Europe.
   pendula pen-dew-la. Pendulous (the
   flowers). Nodding Catchfly.
   Mediterranean region, W Asia.
   schafta shăf-ta. The native name. Caucasus.
   vulgaris vul-*gah*-ris. Common. Europe,
   N Africa, W Asia.
      maritima ma-*ri*-ti-ma. (= *S. maritima*).
      Growing near the sea. Sea Campion.
      W Europe (coasts).

Silk Tree see *Albizia julibrissin*
Silky Oak see *Grevillea robusta*
Silver Bell see *Halesia*
Silver Berry see *Elaeagnus commutata*
Silver Crown see *Cotyledon undulata*
Silver Gum see *Eucalyptus cordata*
Silver Squill see *Ledebouria socialis*
Silver Torch see *Cleistocactus straussii*
Silver Tree see *Leucadendron argenteum*
Silver Vine see *Actinidia polygama*,
Scindapsus pictus

Silybum si-li-bum *Compositae*. From *silybon*
the Gk. name for a similar plant. Annual or
biennial herb.
   marianum mă-ree-*ah*-num. Of the Virgin
   Mary who is said to have caused the white
   mottling of the leaves by dropping milk on
   them. Our Lady's Milk Thistle.
   Mediterranean region.

Sinningia si-*ning*-gee-a *Gesneriaceae*. After
William Sinning (1794–1874). Tender,
perennial herbs. Brazil.
   cardinalis kar-di-*nah*-lis. (= *Rechsteineria
   cardinalis*). Scarlet. Cardinal Flower.
   eumorpha ew-*morf*-a. Of good shape.
   leucotricha loo-*ko*-tri-ka. (= *Rechsteineria
   leucotricha*). With white hairs. Brazilian
   Edelweiss.
   macropoda ma-*kro*-po-da. (= *Rechsteineria
   cyclophylla*). With a large stalk.
   pusilla pu-*sil*-la. Dwarf.
   regina ray-*geen*-a. Queen. Cinderella
   Slippers.
   speciosa spe-kee-*ō*-sa. Showy. Gloxinia.

Sisyrinchium si-si-*ring*-kee-um *Iridaceae*.
The Gk. name of a plant. Perennial herbs.
  angustifolium ang-gus-ti-*fo*-lee-um. Narrow-
leaved. E N America.
  bellum be-lum. Beautiful. California.
  bermudiana ber-mew-dee-*ah*-na. An early
name for the plant, from Bermuda.
  brachypus bră-ki-pus. Short-stalked (the
flowers). W N America.
  californicum kă-li-*forn*-i-kum. Of California.
W N America.
  douglasii dŭg-*lăs*-ee-ee. After Douglas, see
Iris douglasii. W N America.
  macounii ma-*koon*-ee-ee. After Professor
John Macoun who first collected it. British
Columbia.
  striatum stree-*ah*-tum. Striped (the flowers).
Argentina, Chile.

Skimmia skim-ee-a *Rutaceae*. From the
Japanese name, *Miyami-Shikimi*. Evergreen
shrubs.
  anquetilia ang-kwe-*ti*-lee-a. (= *S. laureola*
hort. in part). An old generic name, a
synonym of *Skimmia* after Anquetil-
Duperron, friend of Decaisne who first
described it. W Himalaya.
  × *foremanii* for-*măn*-ee-ee. (= *S. japonica*
'Foremanii'). *S. japonica* × *S. reevesiana*.
After Foreman, a Scottish nurseryman.
  japonica ja-*pon*-i-ka. Of Japan. Japan,
Philippines.
    'Foremannii' see *S.* × *foremanii*
    'Rubella' ru-*bel*-la. Reddish (the flower
buds).
  laureola low-*ree*-o-la. Like a small laurel.
Himalaya, W China.
  reevesiana reevz-ee-*ah*-na. After John
Reeves (1774–1856), a tea inspector who
introduced Chinese plants. China.

Skull-cap see *Scutellaria*
Skunk Cabbage see *Lysichiton americanum*,
Symplocarpus foetidus
Slipper Flower see *Pedilanthes tithymaloides*
Slipperwort see *Calceolaria*
Sloe see *Prunus spinosa*
Small-leaved Gum see *Eucalyptus parvifolia*

Smilacina smee-la-*keen*-a *Liliaceae*.
Diminutive of *Smilax* q.v. Perennial herbs.
N America.
  racemosa ră-kay-*mō*-sa. With flowers in
racemes. False Spikenard.
  stellata ste-*lah*-ta. Star-like (the flowers).
Star-flowered Lily of the Valley.

Smilax smee-lăks *Liliaceae*. The Gk. name.
Evergreen and deciduous climbers.
  aspera a-*spe*-ra. Rough (the stems). Rough
Bindweed. Mediterranean region to
W Asia.
  excelsa eks-*kel*-sa. Tall. E Europe, W Asia.
  rotundifolia ro-tun-di-*fo*-lee-a. Round-
leaved. Horse Briar. E N America.

Smilax (of florists) see *Asparagus asparagoides*

Smithiantha smith-ee-*ănth*-a *Gesneriaceae*.
After Matilda Smith (1854–1926), who drew
for the Botanical Magazine. Tender perennial
herbs. Temple Bells. Mexico.
  cinnabarina ki-na-ba-*reen*-a. Scarlet.
  fulgida ful-gi-da. Shining.
  multiflora mul-ti-*flō*-ra. Many-flowered.
  zebrina ze-*breen*-a. Striped (the leaves).

Smoke Tree see *Cotinus coggygria*
Snake Vine see *Hibbertia scandens*
Snake's-head Iris see *Hermodactylus tuberosus*
Snapdragon see *Antirrhinum majus*
Sneezeweed see *Helenium autumnale*
Sneezewort see *Achillea ptarmica*
Snow Gum see *Eucalyptus niphophila*
  Tasmanian see *E. coccifera*
Snow-in-Summer see *Cerastium tomentosum*
Snow on the Mountain see *Euphorbia
marginata*
Snowberry see *Symphoricarpos albus*
Snowdrop see *Galanthus*
  Common see *G. nivalis*
Snowdrop Tree see *Halesia*
  Mountain see *H. monticola*
Snowflake see *Leucojum*
  Spring see *L. vernum*
  Summer see *L. aestivum*
Snowy Mespilus see *Amelanchier ovalis*
Soapwort see *Saponaria officinalis*
  Rock see *S. ocymoides*

Soehrensia bruchii see *Lobivia bruchii*

Solanum so-*lah*-num *Solanaceae*. L. name of
a plant. Annual and perennial herbs, shrubs
and climbers.
  aviculare a-vik-ew-*lah*-ree. Of small birds.
Kangaroo Apple. New Zealand, Australia.
  capsicastrum kăp-si-*kăs*-trum. Pepper-like
(the fruit). False Jerusalem Cherry,
Winter Cherry. Brazil.
  crispum kris-pum. Wavy-edged (the leaves).
Chile, Peru.
  jasminoides yas-min-*oi*-dees. Jasmine-like.
S America.

Solanum (continued)
  melongena me-lon-*zhee*-na. From *melongene*, an old French name. Aubergine, Egg Plant. Africa, Asia.
  pseudocapsicum soo-dō-*kăp*-si-kum. False *Capsicum*. Jerusalem Cherry, Christmas Cherry. Old World tropics.
  tuberosum tew-be-*rō*-sum. Tuberous. Potato. Andes.

Soldanella sol-da-*nel*-la *Primulaceae*. Diminutive of Italian *soldo* (a small coin), referring to the rounded leaves. Perennial herbs.
  alpina ăl-*peen*-a. Alpine. Europe.
  minima *mi*-ni-ma. Smaller. E Alps.
  montana mon-*tah*-na. Of mountains. C and E Europe.
  pindicola pin-*di*-ko-la. Growing in the Pindus Mountains. NW Greece.
  pusilla pu-*sil*-la. Dwarf. Alps.
  villosa vi-*lō*-sa. Softly hairy. W Pyrenees.

Soleirolia so-lay-*rol*-ee-a *Urticaceae*. After Joseph Francois Soleirol (1796–1863), who collected in Corsica.
  soleirolii so-lay-*rol*-ee-ee. (= *Helxine soleirolii*). As above. Baby's Tears, Mind your own Business. W Mediterranean region.

Solidago so-li-*dah*-gō *Compositae*. From L. *solido* (to make whole or strengthen), referring to medicinal properties. Perennial herbs. Golden Rod.
  canadensis kăn-a-*den*-sis. Of Canada. N America.
  virgaurea virg-*ow*-ree-a. (= *S. brachystachys*). A golden rod. Europe, N Africa, W Asia.

× **Solidaster** so-li-*dăs*-ter *Compositae*. Intergeneric hybrid, from the names of the parents. *Aster* × *Solidago*. Perennial herb.
  luteus loo-tee-us. *Aster ptarmicoides* × *Solidago* sp. Yellow (the flowers).

Sollya so-lee-a *Pittosporaceae*. After Richard Horsman Solly (1778–1858), English botanist. Tender, evergreen climbers. Australia.
  heterophylla he-te-rō-*fil*-la. (= *S. fusiformis*). With variable leaves. Australian Bluebell Creeper.
  parviflora par-vi-*flō*-ra. Small-flowered.

Solomon's Seal see *Polygonatum*

Sonerila so-*ne*-ri-la *Melastomataceae*. From *soneri-ila* the Malabar name. Tender herbs.
  margaritacea mar-ga-ri-*tah*-kee-a. Pearly, the silvery-white spotted leaves. SE Asia.
  orientalis o-ree-en-*tah*-lis. Eastern. Burma.

Sophora so-*fo*-ra *Leguminosae*. From the Arabic name. Deciduous and evergreen trees and shrubs.
  davidii dă-*vid*-ee-ee. (= *S. viciifolia*). After David, see *Davidia*. China.
  japonica ja-*pon*-i-ka. Of Japan. Japanese Pagoda Tree. China.
  microphylla mik-rō-*fil*-la. Small-leaved. New Zealand.
  tetraptera tet-*răp*-te-ra. Four-winged (the pod). Kowhai. New Zealand.

× **Sophrocattleya** sō-frō-*kăt*-lee-a *Orchidaceae*. Intergeneric hybrids, from the names of the parents. *Cattleya* × *Sophronitis*. Greenhouse orchids.

× **Sophrolaelia** sō-frō-*lie*-lee-a *Orchidaceae*. Intergeneric hybrids, from the names of the parents. *Laelia* × *Sophronitis*. Greenhouse orchids.

× **Sophrolaeliocattleya** sō-frō-lie-lee-ō-*kăt*-lee-a *Orchidaceae*. Intergeneric hybrids, from the names of the parents. *Cattleya* × *Laelia* × *Sophronitis*. Greenhouse orchids.

Sophronitis sō-*frō*-ni-tis *Orchidaceae*. From Gk. *sophron* (modest), the flowers are small. Greenhouse orchids. Brazil.
  cernua *kern*-ew-a. Nodding.
  coccinea kok-*kin*-ee-a. (= *S. grandiflora. S. rosea*). Scarlet.
  violacea vee-o-*lah*-kee-a. Violet.

Sorbaria sor-*bah*-ree-a *Rosaceae*. From *Sorbus* q.v. a related genus. Deciduous shrubs.
  aitchisonii aych-i-*son*-ee-ee. After Dr John Aitchison (1836–98), a British physician and botanist who discovered it. Afghanistan to Kashmir.
  arborea ar-*bo*-ree-a. Tree-like. China.
  sorbifolia sor-bi-*fo*-lee-a. N Asia.
  tomentosa tō-men-*tō*-sa. Hairy. Himalaya.

Sorbus *sor*-bus *Rosaceae*. L. name for the service tree. (= *S. domestica*). Deciduous trees and shrubs.
  alnifolia ăl-ni-*fo*-lee-a. *Alnus*-leaved. E Asia.

Sorbus (continued)
aria ah-ree-a. An old name for this tree.
Whitebeam. Europe.
  'Chrysophylla' kris-ō-*fil*-la. Golden-
  leaved.
  'Lutescens' loo-*tes*-enz. Yellowish, the
  creamy white young leaves.
  'Majestica' mah-*yes*-ti-ka. (=
  'Decaisneana'). Majestic.
aucuparia ow-kew-*pah*-ree-a. From L. *avis*
(a bird) and *capere* (to catch), the fruits
attract birds. Mountain Ash, Rowan.
  'Asplenifolia' a-splay-ni-*fo*-lee-a.
  Asplenium-leaved.
  'Edulis' ed-*ew*-lis. Edible (the fruit).
cashmiriana kash-mi-ree-*ah*-na. Of
Kashmir. W Himalaya.
commixta kom-*miks*-ta. Mixed together.
Japan, Korea.
cuspidata kus-pi-*dah*-ta. Abruptly sharp-
pointed (the leaves). Himalaya.
domestica do-*mes*-ti-ka. Cultivated. Service
Tree. S and E Europe, Caucasus,
N Africa.
esserteauiana e-ser-tō-ee-*ah*-na. After Dr
Esserteau, a bacteriologist who helped
Ernest Wilson in China. China.
hupehensis hew-pee-*hen*-sis. Of Hubei
(Hupeh). China.
hybrida hib-ri-da. Originally thought to be
a hybrid. Scandinavia.
intermedia in-ter-*me*-dee-a. Intermediate
(between the wild service tree and a
mountain ash). Swedish Whitebeam.
Scandinavia.
× *kewensis* kew-*en*-sis. *S. aucuparia* × *S.
pohuashanensis*. Of Kew.
koehneana kur-nee-*ah*-na. After Bernard
Adalbert Emil Koehne (1848–1918).
W China.
latifolia lah-tee-*fo*-lee-a. Broad-leaved.
Service Tree of Fontainbleau. France.
poteriifolia po-te-ree-i-*fo*-lee-a. (= *S.
pygmaea*). With leaves like *Poterium*.
China, N Burma.
pygmaea see *S. poteriifolia*
reducta re-*duk*-ta. Dwarf. China.
sargentiana sar-jent-ee-*ah*-na. After
Sargent, see *Prunus sargentii*. China.
scalaris ska-*lah*-ris. Ladder-like (the
leaves). China.
× *thuringiaca* thu-ring-gee-*ah*-ka. *S. aria* ×
S. aucuparia. Of Thuringia, Germany.
Wild with the parents.
torminalis tor-mi-*nah*-lis. Effective against
colic. Wild Service Tree. Europe,
N Africa, SW Asia.
vilmorinii vil-mo-*rin*-ee-ee. After Maurice

Sorbus (continued)
Vilmorin, who received seeds from Delavay
in 1889. China.

Sorrel, Common see *Rumex acetosa*
  French see *R. scutatus*
Sorrel Tree see *Oxydendrum arboreum*
Southern Beech see *Nothofagus*
Southernwood see *Artemisia abrotanum*
Sowbread see *Cyclamen*
Spanish Bayonet see *Yucca aloifolia*
Spanish Shawl see *Heterocentron elegans*
Spanish Moss see *Tillandsia usneoides*

Sparaxis spa-*răks*-is *Iridaceae*. From Gk.
sparasso (to tear) referring to the lacerated
spathe bracts. Cormous perennial herbs.
Wandflower. S Africa.
  grandiflora grănd-i-*flō*-ra. Large-flowered.
  tricolor *tri*-ko-lor. Three-coloured.
  Harlequin Flower.

Sparmannia spar-*măn*-ee-a *Tiliaceae*. After
Dr Andreas Sparrman (1748–1820), Swedish
botanist. Tender shrub.
  africana ăf-ri-*kah*-na. African. African
  Hemp. S Africa.

Spartina spar-*teen*-a *Gramineae*. From Gk.
spartion (esparto grass). Perennial grass.
  pectinata pek-ti-*nah*-ta. Comb-like (the one-
  sided spikes). N America.

Spartium *spar*-tee-um *Leguminosae*. From
Gk. *spartion* (esparto grass), they were both
used for cordage. Nearly leafless shrub.
  junceum *yung*-kee-um. Rush-like.
  S Europe, N Africa, W Asia.

Spartocytisus nubigenus see *Cytisus
supranubias*

Spathiphyllum spă-thi-*fil*-lum *Araceae*. From
Gk. *spathe* and *phyllon* (a leaf) referring to
the leaf-like spathe. Tender herb.
  wallisii wo-*lis*-ee-ee. After Gustave Wallis
  who introduced it from Colombia in 1824.

Spatterdock see *Nuphar advena*
Spearmint see *Mentha* × *spicata*

Specularia speculum-veneris see *Legousia
speculum-veneris*

Speedwell see *Veronica*
Speedy Jenny see *Tradescantia fluminensis*

Sphaeralcea sfie-*răl*-kee-a *Malvaceae*. From
Gk. *sphaira* (a globe) and *Alcea* a related

Sphaeralcea (continued)
genus, referring to the spherical fruits. Subshrubby perennials.

coccinea kok-*kin*-ee-a. (= *Malvastrum coccineum*). Scarlet. W United States.
munroana mūn-rō-*ah*-na. After Munro. W United States.
purpurata pur-pew-*rah*-ta. (= *Malvastrum campanulatum*). Purplish (the flowers). Chile.

Spice Bush see *Lindera benzoin*
Spider Flower see *Cleome hassleriana*
Spider Lily see *Hymenocallis*
  Golden see *Lycoris africana*
  Red see *Lycoris radiata*
Spider Plant see *Chlorophytum comosum*
Spike Heath see *Bruckenthalia spiculifolia*
Spinach see *Spinacia oleracea*
Spinach Beet see *Beta vulgaris*

Spinacia spee-*nah*-kee-a *Chenopodiaceae*. A medieval L. name probably originally from L. *spina* (a spine) referring to the spiny seeds. Annual herb.

oleracea o-le-*rah*-kee-a. Vegetable-like. Spinach. SW Asia.

Spindle Tree see *Euonymus europaeus*
Spinning Gum see *Eucalyptus perriniana*

Spiraea spee-*rie*-a *Rosaceae*. From Gk. *speiraira*, a plant used in garlands. Deciduous shrubs.

albiflora see *S. japonica* 'Albiflora'
'Arguta' ar-*gew*-ta. Sharply toothed (the leaves). Bridal Wreath.
× *billiardii* bi-lee-*ard*-ee-ee. After Billiard, the raiser.
× *bumalda* see *S. japonica* 'Bumalda'
japonica ja-*pon*-i-ka. Of Japan.
  'Albiflora' ăl-bi-*flō*-ra. (= *S. albiflora*). White-flowered.
  'Bullata' bu-*lah*-ta. With puckered leaves.
  'Bumalda' bew-*mahl*-da. (= *S.× bumalda*). After Bumaldus.
nipponica ni-*pon*-i-ka. Of Japan.
prunifolia proo-ni-*fo*-lee-a. *Prunus*-leaved. China.
salicifolia să-li-ki-*fo*-lee-a. *Salix*-leaved. Bridewort. Europe to Japan.
thunbergii thun-*berg*-ee-ee. After Thunberg, see *Thunbergia*. China.
trilobata tri-lo-*bah*-ta. Three-lobed (the leaves). N Asia.
× *vanhouttei* văn-*hoot*-ee-ee. *S. cantoniensis* × *S. trilobata*. After L. B. van Houtte

Spiraea (continued)
(1810–76), Belgian nurseryman.

Spleenwort see *Asplenium*
  Black see *A. adiantum-nigrum*
  Ebony see *A. platyneuron*
  Green see *A. viride*
  Hanging see *A. flaccidum*
  Maidenhair see *A. trichomanes*
  Mother see *A. bulbiferum*
  Sea see *A. marinum*
Spotted Evergreen see *Aglaonema costatum*
Spreading Clubmoss see *Selaginella kraussiana*
Spring Meadow Saffron see *Bulbocodium vernum*
Spruce see *Picea*
  Black see *P. mariana*
  Brewer see *P. breweriana*
  Colorado see *P. pungens*
  Dragon see *P. asperata*
  Hondo see *P. jezoensis hondoensis*
  Norway see *P. abies*
  Serbian see *P. omorika*
  Sitka see *P. sitchensis*
  White see *P. glauca*
  Yezo see *P. yezoensis*
Spurge Laurel see *Daphne laureola*
Squirrel Tail Grass see *Hordeum jubatum*

Stachys stă-kis *Labiatae*. From Gk. *stachys* (a spike) referring to the inflorescence. Perennial herbs.

affinis a-*fee*-nis. Related to. Chinese Artichoke. China.
byzantina bi-zan-*teen*-a. (= *S. lanata*. *S. olympica*). Of Istanbul (Byzantium). Lamb's Tongue.
corsica kor-si-ka. Of Corsica. Corsica, Sardinia.
macrantha ma-*krănth*-a. Large-flowered. Bishop's Wort. Caucasus.
monieri mo-nee-*e*-ree. (= *S. densiflora*). After Monier. Alps, Pyrenees.
nivea ni-vee-a. Snow-white (the flowers). Syria.

Stachyurus stă-kee-*ew*-rus *Stachyuraceae*. From Gk. *stachys* (a spike) and *oura* (a tail) referring to the slender racemes. Deciduous shrubs.

chinensis chin-*en*-sis. Of China.
praecox prie-koks. Early (flowering). Japan.

Standing Cypress see *Ipomopsis rubra*

Stanhopea stăn-*hō*-pee-a *Orchidaceae*. After Philip Henry, 4th Earl of Stanhope

Stanhopea (continued)
(1781–1855). Greenhouse orchids.
costaricensis kos-ta-ree-*ken*-sis. Of Costa Rica.
grandiflora grănd-i-*flō*-ra. (= *S. eburnea*). Large-flowered. NS America.
oculata ok-ew-*lah*-ta. With an eye. C America.
tigrina ti-*green*-a. Striped like a tiger. Mexico.
wardii word-ee-ee. After Mr Ward who sent it to England. C America.

Stapelia sta-*pel*-ee-a *Asclepiadaceae*. After Johannes Bodaeus von Stapel (died 1631). Tender succulents. S Africa.
gigantea gi-*găn*-tee-a. Very large.
grandiflora grănd-i-*flō*-ra. Large-flowered.
hirsuta hir-*soo*-ta. Hairy (the corolla).
pillansii pi-*lănz*-ee-ee. After Dr Neville Stewart Pillans (1884–1964) of Pretoria.
variegata vă-ree-a-*gah*-ta. Variegated (the corolla). Starfish Plant, Toad Plant.

Staphylea sta-*fi*-lee-a *Staphyleaceae*. From Gk. *staphyle* (a cluster) referring to the inflorescence. Deciduous shrubs. Bladder-nut.
colchica kol-ki-ka. Of Colchis, W Asia. Caucasus.
holocarpa ho-lo-*kar*-pa. With an unlobed fruit.
pinnata pi-*nah*-ta. Pinnate (the leaves). SE Europe, W Asia.

Star-flowered Lily of the Valley see *Smilacina stellata*
Star of Bethlehem see *Ornithogalum umbellatum*
Star of the Veldt see *Dimorphotheca sinuata*
Starfish Plant see *Cryptanthus acaulis, Stapelia variegata*
Stars of Persia see *Allium christophii*

Stauntonia stawn-*ton*-ee-a *Lardizabalaceae*. After Sir George Leonard Staunton (1737–1801). Evergreen climber.
hexaphylla heks-a-*fil*-la. With six leaves (leaflets). Japan, China.

Stenocactus sten-ō-*kăk*-tus *Cactaceae*. From Gk. *stenos* (narrow) and *Cactus* q.v. The following are often listed under *Echinofossulocactus*. Mexico.
coptonogonus kop-ton-ō-*gō*-nus. With notched ribs.
hastatus hăs-*tah*-tus. Spear-shaped (some of the spines).

Stenocactus (continued)
lancifer lăn-ki-fer. Lance-bearing.
multicostatus mul-ti-kos-*tah*-tus. Many-ribbed.
violaciflorus vee-o-lah-ki-*flō*-rus. With violet flowers.
zacatecasensis ză-ka-te-ka-*sen*-sis. Of Zacatecas.

Stenocarpus sten-ō-*kar*-pus *Proteaceae*. From Gk. *stenos* (narrow) and *karpos* (a fruit). Tender shrub.
sinuatus sin-ew-*ah*-tus. Wavy-edged (the leaves). Firewheel Tree. E Australia.

Stephanandra ste-fa-*năn*-dra *Rosaceae*. From Gk. *stephanos* (a crown) and *andros* (a man), the stamens form a wreath around the capsule. Deciduous shrubs.
incisa in-*kee*-sa. Deeply cut (the leaves). Japan, Korea.
tanakae ta-*nah*-kie. After Tanaka, a Japanese botanist. Japan.

Stephanotis ste-fa-*nō*-tis *Asclepiadaceae*. Gk. name for myrtle which was used to make crowns, from *stephanos* (a crown) and *otos* (an ear) referring in this case to auricles in the staminal crown. Tender, evergreen climber.
floribunda flō-ri-*bun*-da. Profusely flowering. Wax Flower, Madagascar Jasmine. Madagascar.

Sternbergia stern-*berg*-ee-a *Amaryllidaceae*. After Count Kaspar von Sternberg (1761–1838), Austrian botanist. Bulbous perennial herbs.
candida kăn-di-da. White. SW Turkey.
clusiana klooz-ee-*ah*-na. After Clusius, see *Gentiana clusii*. W Asia.
colchiciflora kol-ki-ki-*flō*-ra. *Colchicum*-flowered. S Europe, W Asia.
fischeriana fi-sha-ree-*ah*-na. After Fischer who collected the type specimen. Caucasus, W Asia.
lutea loo-tee-a. Yellow. Mediterranean region to Iran, Caucasus.

Stetsonia stet-*son*-ee-a *Cactaceae*. After Francis Lynde Stetson of New York.
coryne ko-ri-nay. A club. Toothpick Cactus. Argentina.

Stewartia see *Stuartia*
Stinking Benjamin see *Trillium erectum*

Stipa *stee*-pa *Gramineae*. From Gk. *tuppe* (tow, fibre), *S. tenacissima* is esparto grass

Stipa (continued)
from which cordage is made. Perennial grasses.
 barbata bar-*bah*-ta. Bearded. SW Europe.
 calamagrostis see *Achnatherum calamagrostis*
 gigantea gi-*găn*-tee-a. Very large. Spain, Portugal, Morocco.
 pennata pe-*nah*-ta. Feathery (the inflorescence). S and C Europe, Asia.

Stock see *Matthiola*
 Brompton see *M. incana*
 Night-scented see *M. longipetala bicornis*
 Ten Weeks see *M. incana* Annua
 Virginia see *Malcolmia maritima*

Stokesia stōks-ee-a *Compositae*. After Dr Jonathan Stokes (1755–1831), Edinburgh physician and friend of Linnaeus the younger. Perennial herb.
 laevis lie-vis. Smooth. Stokes Aster. SE United States.

Stone Cress see *Aethionema*
Stonecrop see *Sedum*
 Biting see *S. acre*
 Golden see *S. adolphi*
Stranvaesia davidiana see *Photinia davidiana*
Strap Flower see *Anthurium crystallinum*

Stratiotes strǎ-tee-ō-teez *Hydrocharitaceae*. Gk. name for *Pistia stratiotes*. Aquatic herb.
 aloides a-lō-i-deez. Like *Aloe* (the leaves). Water Soldier. Europe.

Strawberry see *Fragaria*
 Alpine see *F. vesca*
 Garden see *F. × ananassa*
 Hautbois see *F. moschata*
 Mock see *Duchesnea indica*
Strawberry Tree see *Arbutus unedo*

Strelitzia stre-*lits*-ee-a *Strelitziaceae*. After Charlotte of Mecklenberg-Strelitz (1744–1818), Queen to George III. Tender herbaceous perennials. S Africa.
 alba ǎl-ba. (= *S. augusta*). White (the flowers).
 reginae ray-*geen*-ie. (= *S. parvifolia*). Of the Queen (see above). Bird of Paradise Flower, Crane Lily.

Streptocarpus strep-to-*kar*-pus *Gesneriaceae*. From Gk. *streptos* (twisted) and *karpos* (a fruit), the fruits are spirally twisted. Tender perennial herbs. Cape Primrose.

Streptocarpus (continued)
 dunnii dūn-ee-ee. After its discoverer, Edward John Dunn (1844–1937). Red Nodding Bells. S Africa.
 holstii holst-ee-ee. After C. H. E. W. Holst (1865–94), a German gardener who travelled in E Africa. E Africa.
 × *hybridus* hib-ri-dus. Hybrid.
 polyanthus po-lee-*ănth*-us. Many-flowered. S Africa.
 rexii reks-ee-ee. After George Rex, son of George III, on whose property it was discovered. S Africa.
 saxorum săks-ō-rum. Growing on rocks. False African Violet. E Africa.
 wendlandii vend-*lănd*-ee-ee. After Hermann Wendland (1825–1903), a German botanist. Royal Nodding Bells. S Africa.

Streptosolen strep-to-*sō*-len *Solanaceae*. From Gk. *streptos* (twisted) and *solen* (a tube) the corolla tube is spirally twisted. Tender, evergreen shrub.
 jamesonii jaym-*son*-ee-ee. After Dr William Jameson, professor of botany at Quito. Marmalade Bush. N Andes.

String of Beads see *Senecio rowleyanus*

Strobilanthes stro-bi-*lănth*-eez *Acanthaceae*. From Gk. *strobilos* (a cone) and *anthos* (a flower) referring to the dense inflorescence. Tender shrub.
 dyerianus die-a-ree-*ah*-nus. After Sir William Thistleton-Dyer (1843–1928) who introduced it to Kew. Persian Shield. Burma.

Stromanthe strō-*mănth*-ee *Marantaceae*. From Gk. *stroma* (a bed) and *anthos* (a flower) referring to the inflorescence. Tender perennial herbs. Brazil.
 amabilis a-*mah*-bi-lis. Beautiful.
 sanguinea sang-*gwin*-ee-a. Blood-red (the bracts).

Stuartia stew-*art*-ee-a *Theaceae*. (*Stewartia*). After John Stuart (1713–92), 3rd Earl of Bute. Deciduous shrubs and trees.
 malacodendron mǎ-la-kō-*den*-dron. Mallow tree. SE United States.
 ovata ō-*vah*-ta. Ovate (the leaves). SE United States.
 pseudocamellia soo-dō-ka-*mel*-ee-a. False Camellia. Japan.
  koreana ko-ree-*ah*-na. (= *S. koreana*). Of Korea.

Stuartia (continued)
  serrata se-*rah*-ta. Saw-toothed (the leaves).
  Japan.
  sinensis si-*nen*-sis. Of China.

Sturt's Desert Pea see *Clianthus formosus*

Stylophorum sti-*lo*-fo-rum *Papaveraceae*.
From Gk. *stylos* (a style) and *phoros* (bearing)
referring to the long style, a distinctive
character of the genus. Perennial herb.
  diphyllum di-*fil*-lum. Two-leaved.
  Celandine Poppy. E United States.

Styrax sti-räks *Styracaceae*. The Gk. name
for *S. officinalis*. Deciduous trees and
shrubs.
  hemsleyana hemz-lee-*ah*-na. After William
  Botting Hemsley (1843–1924). China.
  japonica ja-*pon*-i-ka. Of Japan. Japan,
  China, Korea.
  obassia ō-*bă*-see-a. From the Japanese
  name. Japan, China, Korea.
  wilsonii wil-*son*-ee-ee. After Wilson who
  introduced it in 1908, see *Magnolia
  wilsonii*. W Sichuan.

Sugar Beet see *Beta vulgaris*
Sugarberry see *Celtis laevigata*
Sumach see *Rhus*
  Smooth see *R. glabra*
  Stag's Horn see *R. typhina*
Summer Cypress see *Kochia scoparia*
Summer Hyacinth see *Galtonia candicans*
Summer Torch see *Billbergia pyramidalis*
Sun Plant see *Portulaca grandiflora*
Sun Rose see *Helianthemum*
Sundew see *Drosera*
Sundrops see *Oenothera fruticosa*
Sunflower see *Helianthus annuus*
Swamp Cypress see *Taxodium distichum*
Swan River Daisy see *Brachycome iberidifolia*
Swede see *Brassica napus* Napobrassica
Swedish Ivy see *Plectranthus australis*
Sweet Alyssum see *Lobularia maritima*
Sweet Bay see *Magnolia virginiana*
Sweet Bergamot see *Monarda didyma*
Sweet Box see *Sarcococca*
Sweet Briar see *Rosa eglanteria*
Sweet Cicely see *Myrrhis odorata*
Sweet Fern see *Comptonia peregrina*
Sweet Flag see *Acorus calamus*
Sweet Four o'Clock Plant see *Mirabilis
longiflora*
Sweet Gale see *Myrica gale*
Sweet Gum see *Liquidambar styraciflua*
Sweet Pea see *Lathyrus odoratus*
Sweet Pepper Bush see *Clethra alnifolia*

Sweet Potato Vine see *Ipomoea batatas*
Sweet Rocket see *Hesperis matronalis*
Sweet Sop see *Annona squamosa*
Sweet Sultan see *Centaurea moschata*
Sweet William see *Dianthus barbatus*
Sweet Woodruff see *Galium odoratum*
Swiss Chard see *Beta vulgaris*
Swiss Cheese Plant see *Monstera deliciosa*
Sycamore see *Acer pseudoplatanus*

Sycopsis si-*kop*-sis *Hamamelidaceae*. From
Gk. *sykon* (a fig) and -*opsis* indicating
resemblance. Evergreen shrub or tree.
  sinensis si-*nen*-sis. Of China.

Symphoricarpos sim-fo-ree-*kar*-pos
Caprifoliaceae. From Gk. *symphorein* (bear
together) and *karpos* (a fruit) referring to the
clustered fruits. Deciduous shrubs.
  albus äl-bus. White (the fruit). Snowberry.
  N America.
    laevigatus lie-vi-*gah*-tus. (= *S. rivularis*).
    Smooth (the leaves and shoots).
    W N America.
  × *chenaultii* she-*nolt*-ee-ee. *S. orbiculatus* ×
  S. microphyllus. After Chenault, French
  nurseryman who raised it.
  orbiculatus or-bik-ew-*lah*-tus. Orbicular
  (the fruit). Indian Currant. United States,
  Mexico.

Symphytum sim-fi-tum *Boraginaceae*. The
Gk. name from *symphysis* (growing together
of bones) and *phyton* (a plant), it was reputed
to heal broken bones. Perennial herbs.
Comfrey.
  caucasicum kaw-*kă*-si-kum. Of the
  Caucasus.
  grandiflorum gränd-i-*flō*-rum. Large-
  flowered. Caucasus.
  orientale o-ree-en-*tah*-lee. Eastern. W Asia.
  × *uplandicum* up-*lănd*-i-kum. *S. asperum* ×
  S. officinale. Of Uppland, Sweden.
  Russian Comfrey. Caucasus.

Symplocarpus sim-plo-*kar*-pus *Araceae*.
From Gk. *symploke* (a connection) and *karpos*
(a fruit), the ovaries combine to make a single
fruit. Perennial bog-garden herb.
  foetidus foy-ti-dus. Foetid. Skunk Cabbage.
  N America, NE Asia, Japan.

Symplocos sim-plo-kos *Symplocaceae*. From
Gk. *symploke* (a connection), the stamens are
united. Deciduous shrub.
  paniculata pa-nik-ew-*lah*-ta. With flowers
  in panicles. Himalaya, China, Japan.

Syngonium sin-*gon*-ee-um *Araceae*. From Gk. *syn* (together) and *gone* (womb) referring to the fused ovaries. Tender climbers.

angustatum ang-gus-*tah*-tum. Narrowed. Arrowhead Vine. C America.

auritum ow-*ree*-tum. Eared (the outer leaf segments). Five Fingers. Caribbean.

podophyllum po-dō-*fil*-lum. With stoutly-stalked leaves. Arrowhead Vine. C America.

vellozianum ve-lō-zee-*ah*-num. After J. de la Concepcion Velloss (1742–1811), Brazilian monk and botanist. Brazil.

Synnema triflorum see *Hygrophila difformis*

Syringa si-*ring*-ga *Oleaceae*. From Gk. *syrinx* (a pipe) referring to the hollow stems. For the same reason, *Philadelphus* has long been known by this name. Deciduous shrubs. Lilac.

× *chinensis* chin-*en*-sis. *S. laciniata* × *S. vulgaris*. Originally thought to be a native of China. Rouen Lilac.

× *hyacinthiflora* hee-a-kinth-i-*flō*-ra. *S. oblata* × *S. vulgaris*. With hyacinth-coloured flowers.

josikaea jo-si-*kie*-a. After Baroness von Josika. E Europe.

julianae yoo-lee-*ah*-nie. Named by Schneider after his wife Juliana. W China.

laciniata la-kin-ee-*ah*-ta. Deeply cut (the leaves). W China.

meyeri *may*-a-ree. After F. N. Meyer who introduced it to America in 1908. N China (Cult.).

'Palibin' *pă*-li-bin. (= *S. palibiniana* hort. *S. velutina* hort.). From *S. palibiniana* which it was originally thought to be, after Ivan Vladimirovich Palibin (1872–1949), a Russian botanist.

microphylla mik-rō-*fil*-la. Small-leaved. N and W China.

palibiniana hort. see *S. meyeri* 'Palibin'.

× *persica* *per*-si-ka. *S. afghanica* × *S. laciniata*. Of Iran (Persia) where it has long been cultivated. Persian Lilac.

× *prestoniae* pres-*ton*-ee-ie. *S. reflexa* × *S. villosa*. After Dr Isabella Preston (1881–1965) of the Central Experimental Farm, Ottawa, who raised it in 1920.

reflexa re-*fleks*-a. Reflexed (the corolla lobes). C China.

velutina hort. see *S. meyeri* 'Palibin'.

vulgaris vul-*gah*-ris. Common. Common Lilac. E Europe.

T

Tacca *tă*-ka *Taccaceae*. From *taka* the Indonesian name. Tender perennial herbs.

chantrieri shon-tree-*e*-ree. After Chantrier Frères, French nurserymen. Cat's Whiskers, Devil Flower. SE Asia.

integrifolia in-teg-ri-fo-lee-a. (= *T. aspera*). With entire leaves. Bat Plant. SE Asia.

leontopetaloides lee-on-to-pe-ta-*loi*-deez. Like *Leontopetalon* (*Leontice*). Old World tropics.

Tagetes ta-*gay*-teez *Compositae*. From *Tages* an Etruscan deity, the grandson of Jupiter, who sprang from the ploughed earth. Annual herbs. Marigold. Mexico, C America.

erecta e-*rek*-ta. Erect. African Marigold.

lucida loo-ki-da. Bright.

patula *păt*-ew-la. Spreading. French Marigold.

tenuifolia ten-ew-i-*fo*-lee-a. (= *T. signata*). With finely divided leaves. Signet Marigold.

Tail Flower see *Anthurium crystallinum*
Tailor's Patch see *Crassula lactea*

Taiwania tie-*wahn*-ee-a *Taxodiaceae*. From Taiwan. Semi-hardy, evergreen conifer.

cryptomerioides krip-to-me-ree-*oi*-deez. Like *Cryptomeria*. Taiwan.

Talinum ta-*leen*-um *Portulacaceae*. Derivation obscure, possibly an African name. Tender succulents. Fameflower.

caffrum *kăf*-rum. Of S Africa. Tropical and S Africa.

guadalupense gwah-da-loop-*en*-see. Of Guadalupe.

portulacifolium por-tew-lah-ki-*fo*-lee-um. *Portulaca*-leaved. India, Arabia, Africa.

Tamarix *tă*-ma-riks *Tamaricaceae*. The L. name. Deciduous shrubs. Tamarisk.

gallica *gă*-li-ka. Of France. W Europe, N Africa.

germanica see *Myricaria germanica*

parviflora par-vi-*flō*-ra. Small-flowered. SE Europe.

ramosissima rah-mō-*si*-si-ma. (= *T. pentandra*). Much branched. S Russia to China.

tetrandra tet-*răn*-dra. With four stamens. E Europe, W Asia.

Tanacetum tăn-a-*set*-um *Compositae*. From the medieval L. name of a herb. Perennial herbs.
  argenteum ar-*gen*-tee-um. Silvery. Turkey.
  corymbosum ko-rim-*bō*-sum. (= *Chrsyanthemum corymbosum*). With flowers in corymbs. Europe.
  densum den-sum. (= *Chrysanthemum densum*). Compact. W Asia.
  haradjanii hă-ra-*dyahn*-ee-ee. (= *Chrysanthemum haradjanii*). After Haradjian. Turkey.
  herderi her-da-ree. After Herder. Turkestan.
  parthenium par-*then*-ee-um. (= *Chrysanthemum parthenium*). From *parthenion*, the Gk. name of a plant. Bachelor's Buttons. Balkan Peninsula.
  praeteritum prie-*te*-ri-tum. Overlooked. Turkey.
  vulgare vul-*gah*-ree. (= *Chrysanthemum vulgare*). Common. Tansy. Europe, Asia.

Tanakaea tăn-a-*kie*-a *Saxifragaceae*. After Yoshio Tanaka (1836–1916), Japanese botanist. Perennial herb.
  radicans rah-di-kănz. With rooting stems. Japanese Foam Flower. China, Japan.

Tangerine see *Citrus reticulata*
Tansy see *Tanacetum vulgare*
Tarragon, French see *Artemisia dracunculus*
  Russian see *A. dracunculus inodora*
Tasmanian Blue Gum see *Eucalyptus globulus*
Tassel Flower see *Emilia javanica*
Tassel Hyacinth see *Muscari comosum*

Taxodium tăks-o-dee-um *Taxodiaceae*. From *Taxus* q.v. and Gk. *eidos* (resemblance). Deciduous conifers. SE United States.
  ascendens a-sen-denz. Ascending (the ultimate shoots). Pond Cypress.
    Nutans *new*-tănz. Nodding (the ultimate shoots).
  distichum dis-ti-kum. In two ranks (the leaves). Swamp Cypress.

Taxus tăks-us *Taxaceae*. The L. name. Evergreen trees. Yew.
  baccata bah-*kah*-ta. Berry-bearing. Common Yew. Europe, N Africa, W Asia.
    'Dovastoniana' dŭv-a-ston-ee-*ah*-na. After John Dovaston of West Felton where the original tree grows. West Felton Yew.
    'Fastigiata' fa-stig-ee-*ah*-ta. With upright branches. Irish Yew.

Taxus (continued)
    'Standishii' stăn-*dish*-ee-ee. After the Standish nursery, Ascot who distributed it.
  × *media* me-dee-a. *T. baccata* × *T. cuspidata*. Intermediate (between the parents).
    'Hatfieldii' hăt-*feeld*-ee-ee. After T. D. Hatfield of the Hunnewell Arboretum, who raised it.
    'Hicksii' *hiks*-ee-ee. After Henry Hicks who introduced it to cultivation.

Tea Plant see *Camellia sinensis*
Tea Tree see *Leptospermum scoparium*
Teasel see *Dipsacus fullonum*

Tecoma te-*kō*-ma *Bignoniaceae*. From the Mexican name. Tender shrub.
  stans stănz. Erect. Yellow Elder. SE United States, C and S America.

Tecomaria te-kō-*mah*-ree-a *Bignoniaceae*. From *Tecoma* q.v. a related genus. Tender, scrambling shrub.
  capensis ka-*pen*-sis. Of the Cape of Good Hope. Cape Honeysuckle. S Africa.

Tecophilaea te-ko-fi-*lie*-a *Tecophilaeaceae*. After Tecophila Billoti, 18th-century Italian botanical artist. Semi-hardy, cormous perennial.
  cyanocrocus see-ăn-ō-*krō*-kus. Blue Crocus. Chilean Crocus. Chile.

Teddy Bear Plant see *Cyanotis kewensis*
Telegraph Plant see *Desmodium gyrans*

Telekia te-*le*-kee-a *Compositae*. After Samuel Teleki di Szék, a Hungarian nobleman and botanical patron. Perennial herbs.
  speciosa spe-kee-ō-sa. (= *Buphthalmum speciosum*). Showy. SE Europe, Caucasus, W Asia.
  speciosissima spe-kee-ō-*si*-si-ma. (= *Buphthalmum speciosissimum*). Very showy. N Italy.

Tellima te-li-ma *Saxifragaceae*. Anagram of *Mitella*, a related genus. Perennial herb.
  grandiflora grănd-i-*flō*-ra. Large-flowered. W N America.

Telopea tay-*lō*-pee-a *Proteaceae*. From Gk. *telopos* (seen from afar) referring to the showy flowers. Evergreen, tender and semi-hardy shrubs.
  speciosissima spe-kee-ō-*si*-si-ma. Very

Telopea (continued)
 showy. Waratah. New South Wales.
 truncata trung-*kah*-ta. Abruptly cut off (the
 leaves). Tasmanian Waratah. Tasmania.

Temple Bells see *Smithiantha*
Tendergreen see *Brassica rapa* Perviridis

Tephrosia te-*fros*-ee-a *Leguminosae*. From
Gk. *tephros* (ash-coloured) referring to the
leaves. Tender shrub.
 grandiflora gránd-i-*flō*-ra. Large-flowered.
 S Africa.

Tetracentron tet-ra-*ken*-tron *Tetracentraceae*.
From Gk. *tetra* (four) and *kentron* (a spur)
referring to the four-spurred fruit. Deciduous
tree.
 sinense si-*nen*-see. Of China. China,
 Himalaya, N Burma.

Tetrastigma tet-ra-*stig*-ma *Vitaceae*. From
Gk. *tetra* (four) and *stigma*, the stigma is four-
lobed. Tender, evergreen climber.
 voinierianum vwū-nee-e-ree-*ah*-num. After
 M. Voinier, chief veterinary surgeon with
 the French army in Hanoi. Chestnut Vine.
 Laos.

Teucrium *toyk*-ree-um *Labiatae*. From
teukrion, the Gk. name. Herbs and shrubs.
Germander.
 aroanium ă-rō-*ah*-nee-um. Of Aroania.
 S Greece.
 chamaedrys ka-*mie*-dris. From *chamaidrys*
 (dwarf oak), the Gk. name. Wall
 Germander. Europe, SW Asia, N Africa.
 creticum kray-ti-kum. (= *T.*
 rosmarinifolium). Of Crete. E
 Mediterranean region.
 fruticans froo-ti-kănz. Shrubby. Shrubby
 Germander. W Mediterranean region.
 marum mah-rum. The Gk. name. Cat
 Thyme. W Mediterranean islands.
 polium po-lee-um. (= *T. aureum*). The Gk.
 name. S Europe.
 pyrenaicum pi-ray-*nah*-i-kum. Of the
 Pyrenees.
 subspinosum sub-spee-*nō*-sum. Somewhat
 spiny. Mallorca.

Thalictrum tha-*lik*-trum *Ranunculaceae*. Gk.
name for a plant. Perennial herbs. Meadow
Rue.
 alpinum ăl-*peen*-um. Alpine. N hemisphere
 (arctic and alpine regions).
 aquilegiifolium ă-kwi-lee-gee-i-*fo*-lee-um.
 Aquilegia-leaved. Europe, Asia.

Thalictrum (continued)
 delavayi de-la-*vay*-ee. After Delavay, see
 Abies delavayi. W China.
 dipterocarpum dip-te-rō-*kar*-pum. With a
 two-winged fruit. W China.
 flavum flah-vum. Yellow (the flowers).
 Europe, Asia.
  glaucum glow-kum. (= *T.*
  speciosissimum). Glaucous. Spain,
  Portugal.
 kiusianum kee-oo-see-*ah*-num. Of Kyushu.
 Japan.

Thelocactus thay-lō-*kăk*-tus *Cactaceae*.
From Gk. *thele* (a nipple) and *Cactus* q.v.
 bicolor bi-ko-lor. Two-coloured (the
 flowers). Glory of Texas. S Texas,
 N Mexico.
 lophothele lo-*fo*-thay-lay. With crested
 nipples. N Mexico.

Thermopsis ther-*mop*-sis *Leguminosae*. From
Gk. *thermos* (a lupin) and *-opsis* indicating
resemblance. Perennial herbs.
 caroliniana kă-ro-lin-ee-*ah*-na. Of Carolina.
 Carolina Lupin. SE United States.
 lupinoides loo-peen-*oi*-deez. (= *T.*
 lanceolata). Like *Lupinus*. Siberia, Alaska.
 mollis mol-lis. Softly hairy. E United States.
 montana mon-*tah*-na. Of mountains.
 W United States.

Thlaspi *thlăs*-pee *Cruciferae*. Gk. name for a
similar plant. Perennial, alpine herb.
 rotundifolium ro-tund-i-*fo*-lee-um. With
 round (basal) leaves. Alps.

Three Birds Flying see *Linaria triornithophora*
Thrift see *Armeria*
 Common see *A. maritima*
 Jersey see *A. alliacea*
Throatwort see *Campanula trachelium*,
Trachelium caeruleum

Thuja *thoo*-ya *Cupressaceae*. Gk. name of a
juniper. Evergreen conifers.
 koraiensis ko-rie-*en*-sis. Of Korea.
 occidentalis ok-ki-den-*tah*-lis. Western.
 Western White Cedar. E N America.
 orientalis o-ree-en-*tah*-lis. Eastern. China.
 plicata pli-*kah*-ta. Plaited, the appearance
 of the shoots. Western Red Cedar. W N
 America.
 standishii stăn-*dish*-ee-ee. After John
 Standish (1814–75), a nurseryman for
 whom Robert Fortune introduced it.
 Japan.

Thujopsis thoo-*yop*-sis *Cupressaceae*. From *Thuja* q.v. and Gk. *-opsis* indicating resemblance. Evergreen conifer.
  dolabrata do-lah-*brah*-ta. Hatchet-shaped (the leaves). Japan.

Thunbergia thun-*berg*-ee-a *Acanthaceae*. After Carl Peter Thunberg (1743–1828), Dutch physician and botanist who introduced many Japanese plants. Tender climbers.
  alata ah-*lah*-ta. Winged (the petioles). Black-eyed Susan. Tropical Africa.
  coccinea kok-*kin*-ee-a. Scarlet. India.
  fragrans frah-granz. Fragrant. India, Sri Lanka.
  grandiflora gränd-i-*flō*-ra. Large-flowered. Clock Vine, Blue Trumpet Vine. India.
  gregorii gre-*go*-ree-ee. After Dr J. W. Gregory who collected the type specimen. Tropical Africa.
  myosorensis mie-o-sor-*ren*-sis. Of Mysore. Nilgiri Hills, India.
  natalensis nä-ta-*len*-sis. Of Natal. S Africa.

Thunia thun-ee-a *Orchidaceae*. After Franz A. Graf von Thun (1786–1873). Greenhouse orchids.
  alba äl-ba. White (the flowers). N India, Burma.
    marshalliana mar-shäl-ee-*ah*-na. (= *T. marshalliana*). After Mr Marshall who grew it.
  bensoniae ben-*son*-ee-ie. After the wife of Robert Benson (1822–94), who collected orchids in Burma. Burma.

Thymus *tiem*-us *Labiatae*. The Gk. name. Perennial herbs and dwarf shrubs. Thyme.
  azoricus see *T. caespitosus*
  caespitosus kie-spi-*tō*-sus. (= *T. azoricus. T. micans*). Tufted. Portugal, Spain, Azores.
  carnosus kar-*nō*-sus. (= *T. nitidus* hort.). Fleshy. Portugal.
  cilicius ki-*li*-kee-us. Of Cilicia (S Turkey).
  citriodorus kit-ree-o-*dō*-rus. *T. pulegioides* × *T. vulgaris*. Lemon-scented. Lemon Thyme.
  doerfleri durf-la-ree. After J. D. Doerfler who discovered it in 1916. Albania.
  drucei *droos*-ee-ee. After George Claridge Druce (1850–1932), English amateur botanist. Europe.
  herba-barona her-ba-ba-*ron*-a. The Corsican name. Caraway Thyme. Corsica, Sardinia.
  lanuginosus hort. see *T. pseudolanuginosus*
  membranaceus mem-bra-*nah*-kee-us. Membranaceous (the bracts). SE Spain.
  micans see *T. caespitosus*

Thymus (continued)
  nitidus hort. see *T. carnosus*
  praecox prie-koks. Early (flowering). SW and C Europe.
  pseudolanuginosus soo-dō-lah-new-gi-*nō*-sus. False *T. lanuginosus* (woolly). Cult.
  vulgaris vul-*gah*-ris. Common. Common Thyme, French Thyme, Garden Thyme. S Europe.
  serpyllum ser-*pil*-lum. The L. name for thyme. Wild Thyme. Europe.

Tiarella tee-a-*rel*-la *Saxifragaceae*. Diminutive of Gk. *tiara* (a small crown) referring to the fruit. Perennial herbs.
  cordifolia kor-di-*fo*-lee-a. With heart-shaped leaves. Foam Flower. E N America.
  wherryi *we*-ree-ee. (= *T. collina*). After its discoverer, Edgar Theodore Wherry (born 1885), an American botanist. SE United States.

Tibouchina ti-boo-*chee*-na *Melastomataceae*. From the native name. Tender shrub.
  urvilleana ur-vil-ee-*ah*-na. (= *T. semidecandra* hort.). After Jules Sébastian César Dumont d'Urville (1790–1844), a French naval officer. Glory Bush. Brazil.

Tickseed see *Coreopsis*
Tiger Lily see *Lilium lancifolium*
Tiger's Jaws see *Faucaria tigrina*

Tigridia ti-*gri*-dee-a *Iridaceae*. From L. *tigris* (a tiger), the flowers are spotted like the S American tiger (jaguar). Semi-hardy bulbous herbs.
  pavonia pah-*vō*-nee-a. Peacock-like. Peacock Tiger Flower.
  violacea vee-o-*lah*-kee-a. Violet. Mexico.

Tilia *tee*-lee-a *Tiliaceae*. The L. name. Lime, Linden. Deciduous trees.
  cordata kor-*dah*-ta. Heart-shaped (the leaves). Small-leaved Lime. Europe.
  × *euchlora* ew-*klō*-ra. From Gk. *eu* (good) and *chloros* (green) referring to the bright green leaves.
  × *europaea* oy-ro-*pie*-a. *T. cordata* × *T. platyphyllos*. European. Common Lime. Europe.
  oliveri o-*li*-va-ree. After Oliver. C China.
  'Petiolaris' pe-tee-o-*lah*-ris. With a petiole, the long petiole distinguishes it from *T. tomentosa*. Weeping Silver Lime.
  platyphyllos plä-tee-*fil*-los. Broad-leaved. Broad-leaved Lime. Europe, SW Asia.

Tilia (continued)
'Rubra' *rub*-ra. Red (the young shoots).
Red-twigged Lime.
tomentosa tō-men-*tō*-sa. Hairy (the
undersides of the leaves). European White
Lime. SE Europe.

Tillandsia ti-*lǎndz*-ee-a *Bromeliaceae*. After
Elias Til-Landz (died 1693), Swedish
botanist. Tender epiphytic herbs.
cyanea see-*ǎn*-ee-a. Blue (the flowers). Pink
Quill. Ecuador.
lindenii lin-*den*-ee-ee. After J. J. Linden, a
Belgian nurseryman. Blue Flowered
Torch. Peru.
tenuifolia ten-ew-i-*fo*-lee-a. (= *T. pulchella*).
Slender-leaved. S America.
usneoides us-nee-*oi*-deez. Like *Usnea*, a
lichen that hangs from trees in a similar
way. Spanish Moss. SE United States,
C and S America.

Tingiringi Gum see *Eucalyptus glaucescens*

Tithonia tee-*thō*-nee-a *Compositae*. After
Tithonus, a beautiful youth and King of
Troy. He was loved by Aurora who turned
him into a grasshopper. Annual herb.
rotundifolia ro-tun-di-*fo*-lee-a. Round-
leaved. Mexican Sunflower. Mexico,
C America.

Toad Lily see *Tricyrtis hirta*
Toad Plant see *Stapelia variegata*
Toadflax see *Linaria*
  Alpine see *L. alpina*
  Common see *L. vulgaris*
Toadshade see *Trillium sessile*
Tobacco Plant see *Nicotiana tabacum*

Tolmiea tol-*mee*-a *Saxifragaceae*. After Dr
William Fraser Tolmie (1830–86), Scottish
physician and botanist. Perennial herb.
menziesii men-*zeez*-ee-ee. After Menzies,
see *Menziesia*. Piggy-back Plant. W N
America.

Tomato see *Lycopersicum esculentum*
Toothache Tree see *Zanthoxylum
americanum*
Torch Lily see *Kniphofia*

Torenia to-*ren*-ee-a *Scrophulariaceae*. After
the Rev. Olof Toren (1718–53). Annual
herbs.
baillonii bay-*lon*-ee-ee. After Henri Baillon
(1827–95), French botanist. Indochina.
fournieri four-nee-*e*-ree. After Eugène

Torenia (continued)
Pièrre Nicolas Fournier (1834–84), French
botanist. Wishbone Flower. Vietnam.

Torreya to-*ree*-a *Taxaceae*. After John
Torrey (1796–1873), American botanist.
Evergreen trees.
californica kǎ-li-*forn*-i-ka. Of California.
California Nutmeg.
nucifera new-*ki*-fe-ra. Nut-bearing. Japan.

Tortoise Plant see *Dioscorea elephantipes*
Touch-me-not see *Impatiens noli-tangere*

Tovara tō-*vah*-ra *Polygonaceae*. After Simon
a Tovar, 16th-century Spanish physician.
Perennial herb.
virginiana vir-jin-ee-*ah*-na. Of Virginia.
E N America.

Townsendia town-*zend*-ee-a *Compositae*.
After David Townsend (1787–1858),
American botanist. Perennial herbs.
exscapa eks-*skah*-pa. Without a scape. W N
America.
formosa for-*mō*-sa. Beautiful. SW United
States.
parryi pa-ree-ee. After Charles Christopher
Parry (1823–90), American botanical
explorer. W N America.

Trachelium tra-*kay*-lee-um *Campanulaceae*.
From Gk. *trachelos* (a neck) referring to
supposed medicinal properties. Perennial
herbs.
caeruleum kie-*ru*-lee-um. Deep blue (the
flowers). Throatwort. Mediterranean
region.

Trachelospermum tra-kay-lō-*sperm*-um
Apocynaceae. From Gk. *trachelos* (a neck)
and *sperma* (a seed) referring to the narrow
seeds. Evergreen climbers.
asiaticum ah-see-*ah*-ti-kum. Asian. Japan,
Korea.
jasminoides yas-min-*oi*-deez. Jasmine-like
(the fragrant flowers). China, Japan.
  'Japonicum' ja-*pon*-i-kum. (= *T. majus*
  hort.). Of Japan.

Trachycarpus trǎ-kee-*kar*-pus *Palmae*.
From Gk. *trachys* (rough) and *karpos* (a
fruit). The hardiest palm.
fortunei for-*tewn*-ee-ee. (= *Chamaerops
excelsa* hort.). After Robert Fortune who
sent plants to Kew. See *Fortunella*. Chusan
Palm, Fan Palm. C and S China.

Trachymene tră-kee-*may*-nee *Umbelliferae*. From Gk. *trachys* (rough) and *meninx* (a membrane) referring to the fruit. Annual herb.
caerulea kie-*ru*-lee-a. (= *Didiscus caeruleus*). Blue (the flowers). Blue Lace Flower. Australia.

Trachystemon tră-kee-*stay*-mon *Boraginaceae*. From Gk. *trachys* (rough) and *stemon* (a stamen) referring to the rough stamens of one species. Perennial herb.
orientalis o-ree-en-*tah*-lis. Eastern. SE Europe, W Asia.

Tradescantia tră-des-*kănt*-ee-a *Commelinaceae*. After John Tradescant (1608–62). Tender and hardy perennial herbs.
albiflora ăl-bi-*flō*-ra. White-flowered. Inch Plant, Wandering Jew. S America.
× *andersoniana* ăn-der-son-ee-*ah*-na. (= *T. virginiana* hort.). After Anderson.
blossfeldiana blos-feld-ee-*ah*-na. After Robert Blossfeld a Potsdam nurseryman. Flowering Inch Plant. Argentina.
fluminensis floo-min-*en*-sis. Of Rio de Janeiro. Speedy Jenny. Brazil.
navicularis nah-vik-ew-*lah*-ris. Boat-shaped (the leaves). Chain Plant. Mexico, Peru.
sillamontana si-la-mon-*tah*-na. Of Cerro de la Silla, where it was found. White Velvet. Mexico.

Tragopogon tră-go-*pō*-gon *Compositae*. From Gk. *tragos* (a goat) and *pogon* (a beard) referring to the pappus. Biennial herb.
porrifolius po-ri-*fo*-lee-us. With leaves like *Allium porrum*. Salsify. S Europe.

Trailing Arbutus see *Epigaea repens*
Trailing Watermelon Begonia see *Pellionia daveauana*
Transvaal Daisy see *Gerbera jamesonii*

Trapa tră-pa *Trapaceae*. From L. *calcitrappa* (caltrop, a four-pointed weapon) referring to the four-horned fruit of *T. natans*. Aquatic perennials.
bicornis bi-*korn*-is. Two-horned (the fruit).
natans nă-tănz. Floating. Water Chestnut, Water Caltrop. Europe, Asia, Africa.

Traveller's Joy see *Clematis vitalba*
Treasure Flower see *Gazania rigens*
Tree of Heaven see *Ailanthus altissima*

Trichocereus tri-kō-*kay*-ree-us *Cactaceae*. From Gk. *trichos* (hair) and *Cereus* q.v. referring to the hairy areoles.
candicans kăn-di-kănz. White (the flowers). Argentina.
chiloensis ki-lō-*en*-sis. Of Chiloe. Chile.
coquimbanus kō-kim-*bah*-nus. Of Coquimba. Chile.
schickendantzii shi-kan-*dănts*-ee-ee. After Schickendantz. Argentina.
spachianus spăch-ee-*ah*-nus. After Edouard Spach (1801–79). Golden Column, White Torch Cactus. Argentina.

Tricyrtis tri-*kur*-tis *Liliaceae*. From Gk. *tri* (three) and *kyrtos* (humped) referring to the swollen bases of the three outer petals. Perennial herbs.
formosana for-mō-sah-na. (= *T. stolonifera*). Of Taiwan (Formosa).
hirta hir-ta. Hairy. Toad Lily. Japan.
macrantha ma-*krănth*-a. Large-flowered. Japan.
macropoda ma-*kro*-po-da. With a large stalk. Japan.

Trifolium tri-*fo*-lee-um *Leguminosae*. The L. name from *tri*- (three) and *folium* (a leaf), the leaves have three leaflets. Perennial herb.
repens ree-penz. Creeping. White Clover. Europe, Asia, N Africa.
   'Purpurascens Quadrifolium' pur-pew-*răs*-enz-kwod-ri-*fo*-lee-um. Purplish and four-leaved, the leaves have four bronze-purple leaflets.

Trillium tril-lee-um *Liliaceae*. From L. *tri*- (three) they have three leaves and the floral parts are in threes. Herbaceous perennials.
cernuum ker-new-um. Nodding (the flowers). E N America.
erectum e-*rek*-tum. Erect (the flowers). Stinking Benjamin. E N America.
grandiflorum grănd-i-*flō*-rum. Large-flowered. Wake Robin. E N America.
ovatum ō-*vah*-tum. Ovate (the leaves). W N America.
rivale ree-*vah*-lee. Growing by streams. W N America.
sessile se-si-lee. Sessile (the flowers). Toadshade. N America.
undulatum un-dew-*lah*-tum. Wavy-edged (the petals). Painted Wood-lily. E N America.

Triplet Lily see *Brodiaea coronaria*

Triteleia tri-te-*lay*-a *Amaryllidaceae*. From Gk. *tri* (three) and *teleios* (perfect), the floral parts are in threes. Cormous perennial herbs.
bridgesii bri-*jez*-ee-ee. After Thomas Bridges (1807–65), who collected in S America and California. W N America.
crocea krō-kee-a. Saffron yellow. W N America.
grandiflora grănd-i-*flō*-ra. Large-flowered. W N America.
hyacinthina hee-a-kinth-*ee*-na. Hyacinth-coloured. W N America.
ixioides iks-ee-*oi*-deez. Like *Ixia*. C California.
laxa lăks-a. (= *Brodiaea laxa*). Loose (the inflorescence). W N America.
peduncularis pe-dunk-ew-*lah*-ris. With a flower stalk, referring to the long scape. California.
× *tubergeniana* tew-ber-gen-ee-*ah*-na. *T. laxa* × *T. peduncularis*. After the van Tubergen nursery, where it was raised.
uniflora see *Ipheion uniflorum*

Tritonia tri-tō-nee-a *Iridaceae*. From *Triton* (a weather cock), the stamens point in different directions. Semi-hardy or tender, perennial herbs. S Africa.
crocata krō-*kah*-ta. Saffron yellow.
× *crocosmiiflora* see *Crocosmia* × *crocosmiiflora*
flavida flah-vi-da. Yellow.

Trochodendron tro-ko-*den*-dron *Trochodendraceae*. From Gk. *trochos* (a wheel) and *dendron* (a tree) referring to the arrangement of the stamens. Evergreen tree.
aralioides a-rah-lee-*oi*-deez. Like *Aralia*. Japan, Taiwan, Korea.

Trollius tro-lee-us *Ranunculaceae*. From Swiss–German *Trollblume* (Globeflower). Perennial herbs. Globeflower.
acaulis a-*kaw*-lis. Stemless. Himalaya.
× *cultorum* kul-*to*-rum. Cultivated.
europaeus oy-rō-*pie*-us. European. Europe, N America.
pumilus *pew*-mi-lus. Dwarf. Himalaya, W China.

Tropaeolum tro-*pie*-o-lum *Tropaeolaceae*. From Gk. *tropaion* (a trophy), originally used for the trunk of a tree on which were fixed the shields and helmets of a defeated enemy. Linnaeus was reminded of this on seeing a plant growing on a post, the leaves representing the shields and the flowers the helmets. Hardy and semi-hardy, annual and

Tropaeolum (continued)
perennial herbs.
azureum a-*zew*-ree-um. Blue (the flowers). C Chile.
majus mah-yus. Larger. Nasturtium. S America.
peltophorum pel-*to*-fo-rum. Shield bearing, referring to the leaves. Andes.
peregrinum pe-re-*green*-um. (= *T. canariense*). Foreign. Canary Creeper. Andes.
polyphyllum po-lee-*fil*-lum. With many leaves (leaflets). Chile.
speciosum spe-kee-ō-sum. Showy. Flame Creeper. S Chile.
tuberosum tew-be-*rō*-sum. Tuberous. N Andes.

Trout Lily see *Erythronium revolutum*
Trumpet Vine see *Campsis radicans*

Tsuga tsoo-ga *Pinaceae*. From the Japanese name. Evergreen trees. Hemlock.
canadensis kăn-a-*den*-sis. Of E N America. Eastern Hemlock.
heterophylla he-te-rō-*fil*-la. With variable leaves. Western Hemlock. W N America.
mertensiana mer-tenz-ee-*ah*-na. After Karl Heinrich Mertens (1795–1830), a German botanist who discovered it in 1827. Mountain Hemlock. W N America.

Tuberose see *Polianthes tuberosa*
Tulip Tree see *Liriodendron tulipifera*
  Chinese see *L. chinense*

Tulipa tew-li-pa *Liliaceae*. From Turkish *tulband* (a turban), originally used in a descriptive sense but was thought to be a name. Bulbous perennial herbs. Tulip.
acuminata a-kew-mi-*nah*-ta. Long-pointed (the petals). Horned Tulip. Cult.
aucheriana ow-ka-ree-*ah*-na. After P. M. R. Aucher-Eloy (1792–1838). W Iran.
batalinii bah-ta-*lin*-ee-ee. After A. F. Batalin (1847–96), a Russian botanist. C Asia.
biflora bi-*flō*-ra. Two-flowered. SE Europe to C Asia and Siberia.
clusiana klooz-ee-*ah*-na. After Clusius, see *Gentiana clusii*. Lady Tulip. Iran, Himalaya, Tibet.
  stellata ste-*lah*-ta. (= *T. stellata*). Star-like (the flowers).
eichleri iek-la-ree. After Eichler. Transcaucasus, NW Iran.

Tulipa (continued)
fosteriana fos-ta-ree-*ah*-na. After Foster.
C Asia.
greigii greeg-ee-ee. After General Greig,
President of the Imperial Russian
Horticultural Union. C Asia.
humilis hu-mi-lis. (= *T. pulchella*). Low-
growing. E Turkey, N Iraq, NW Iran.
kaufmanniana kowf-măn-ee-*ah*-na. After
General von Kaufmann, governor-general
of the region in which it was found. Water-
lily Tulip. C Asia.
linifolia leen-i-*fo*-lee-a. *Linum*-leaved.
C Asia.
marjolettii see *T. praecox*
orphanidea or-fa-*nid*-ee-a. After Dr
Orphanides, professor of botany at
Athens. SE Europe, W Turkey.
praecox prie-koks. (= *T. marjolettii*). Early
(flowering). ? C Asia.
praestans prie-stahnz. Distinguished.
C Asia.
pulchella see *T. humilis*
stellata see *T. clusiana stellata*
sylvestris sil-*ves*-tris. Of woods. Italy, Sicily,
Sardinia.
tarda tar-da. Late. C Asia.
tubergeniana tew-ber-gen-ee-*ah*-na. After
van Tubergen, the Dutch bulb nursery.
C Asia.
turkestanica tur-kes-*tahn*-i-ka. Of
Turkestan. C Asia, NW China.
urumiensis u-roo-mee-*en*-sis. Of Lake
Urumia. NW Iran.
wilsoniana wil-son-ee-*ah*-na. After Mr G.
F. Wilson. C Asia.

Tunica saxifraga see *Petrorhagia saxifraga*
Tupelo see *Nyssa sylvatica*
Turnip see *Brassica rapa*
Turk's Cap Lily see *Lilium martagon*
Turtle Head see *Chelone*
Tutsan see *Hypericum androsaemum*
Tweedia caerulea see *Oxypetalum caeruleum*
Twin Flower see *Linnaea borealis*
Twinspur see *Diascia barberiae*

Typha tee-fa *Typhaceae*. The Gk. name.
Aquatic or bog-garden perennial herbs.
Reed-mace.
angustifolia ang-gus-ti-*fo*-lee-a. Narrow-
leaved. Europe, Asia, N America.
latifolia lah-tee-*fo*-lee-a. Broad-leaved.
Europe, Asia, N America.
minima mi-ni-ma. Smaller. Europe.

U

Ulex ew-leks *Leguminosae*. The L. name.
Spiny shrubs. Furze, Gorse, Whin.
europaeus oy-rō-*pie*-us. European. Common
Gorse. W and C Europe.
gallii gă-lee-ee. Of France. W Europe.
minor mi-nor. Smaller. Dwarf Gorse.
W Europe.

Ulmus ul-mus *Ulmaceae*. The L. name.
Deciduous trees. Elm.
angustifolia ang-gus-ti-*fo*-lee-a. Narrow-
leaved. Goodyer's Elm. S England,
N France.
cornubiensis kor-new-bee-*en*-sis. Of
Cornwall. Cornish Elm. SW England.
'Belgica' *bel*-gi-ka. Belgian. Belgian Elm.
'Camperdownii' kăm-per-*down*-ee-ee. Of
Camperdown House, Dundee, where it
was raised. Camperdown Elm.
carpinifolia kar-peen-i-*fo*-lee-a. *Carpinus*-
leaved. Smooth Elm. Europe, N Africa,
SW Asia.
glabra glăb-ra. Smooth (the bark). Wych
Elm. Europe, W Asia.
'Exoniensis' eks-ō-nee-*en*-sis. Of Exeter
where it was raised. Exeter Elm.
'Hollandica' ho-*lănd*-i-ka. Of Holland.
Dutch Elm.
procera prō-*kay*-ra. Tall. English Elm.
S and C England.
sarniensis sar-nee-*en*-sis. Of Guernsey.
Jersey Elm, Wheatley Elm.

Umbellularia um-bel-ew-*lah*-ree-a
Lauraceae. From L. *umbella* (an umbel)
referring to the inflorescence. Evergreen tree.
californica kă-li-*forn*-i-ka. Of California.
California Laurel. W N America.

Umbrella Pine see *Sciadopitys verticillata*
Umbrella Plant see *Cyperus involucratus*,
Peltiphyllum peltatum
Umbrella Tree see *Magnolia tripetala*
Ear-leaved see *M. fraseri*
Unicorn Plant see *Proboscidea louisianica*
Unicorn Root see *Aletris farinosa*

Urceolina ur-kee-*o*-li-na *Amaryllidaceae*.
From L. *urceolus* (a small pitcher) referring
to the shape of the flowers. Bulbous herb.
peruviana pe-roo-vee-*ah*-na. Of Peru. Peru,
Bolivia.

Urginea ur-*gin*-ee-a *Liliaceae*. From *Beni Urgin*, the name of an Arab tribe in Algeria. Bulbous herb.
   maritima ma-*ri*-ti-ma. Growing near the sea. Sea Squill. Mediterranean region. W Asia.

Urn Gum see *Eucalyptus urnigera*
Urn Plant see *Aechmea fasciata*

Ursinia ur-*si*-nee-a *Compositae*. After Johannes Heinrich Ursinus (1608–67), a botanical author. Annual herbs. S Africa.
   anethoides a-nay-*thoi*-deez. Like *Anethum*.
   anthemoides ăn-them-*oi*-deez. (= *U. pulchra*). Like *Anthemis*.
   cakilifolia ka-ki-li-*fo*-lee-a. With leaves like *Cakile*.

Utricularia ew-trik-ew-*lah*-ree-a *Lentibulariaceae*. From L. *utriculus* (a small bottle) referring to the bladders which trap small, aquatic animals. Aquatic, carnivorous herb. Bladderwort.
   vulgaris vul-*gah*-ris. Common. Europe, Asia, N America.

Uvularia oo-vew-*lah*-ree-a *Liliaceae*. From *uvula* of anatomy, a hanging structure in the throat, referring to the hanging flowers. Perennial herbs. Bellwort, Merrybells. N America.
   grandiflora grănd-i-*flō*-ra. Large-flowered.
   perfoliata per-fo-lee-*ah*-ta. With the base of the leaves pierced by the stem.
   sessilifolia se-si-li-*fo*-lee-a. With sessile leaves.

V

Vaccaria va-*kah*-ree-a *Caryophyllaceae*. From L. *vacca* (a cow), it is said to have been used for forage. Annual herb.
   pyramidata pi-ra-mi-*dah*-ta. (= *Saponaria vaccaria*). Pyramidal. Cow Herb. S and C Europe.

Vaccinium va-*keen*-ee-um *Ericaceae*. A L. name, variously stated to apply to either *V. myrtillus* or a hyacinth. Deciduous and evergreen shrubs.
   arctostaphylos ark-tō-*stă*-fi-los. Gk. name for a plant from *arktos* (a bear) and *staphyle* (a bunch of grapes). Caucasian Whortleberry. Caucasus, SE Europe.
   corymbosum ko-rim-*bō*-sum. With flowers

Vaccinium (continued)
   in corymbs. Highbush Blueberry. E N America.
   delavayi de-la-*vay*-ee. After Delavay who discovered it, see *Abies delavayi*. SW China, Burma.
   deliciosum day-li-kee-*ō*-sum. Delicious (the fruits). NW North America.
   glaucoalbum glow-kō-*ăl*-bum. Glaucous-white (the undersides of the leaves). Himalaya.
   macrocarpon see *Oxycoccus macrocarpus*
   mytillus *mur*-ti-lus. A small myrtle. Bilberry, Whortleberry. Europe.
   nummularia num-ew-*lah*-ree-a. Coin-shaped (the leaves). Himalaya.
   ovatum ō-*vah*-tum. Ovate (the leaves). Box Blueberry. W N America.
   oxycoccus see *Oxycoccus palustris*
   vitis-idaea vee-tis-ee-*die*-a. Grape of Mt Ida. Cowberry. Europe, Asia, N America.

Valerian, Red see *Centranthus ruber*

Valeriana va-le-ree-*ah*-na *Valerianaceae*. The medieval L. name, probably from L. *valere* (to be healthy), referring to medicinal properties. Perennial herbs.
   arizonica ă-ri-*zo*-ni-ka. Of Arizona. SW United States, Mexico.
   montana mon-*tah*-na. Of mountains. Europe.
   officinalis o-fi-ki-*nah*-lis. Sold as a herb. Common Valerian. Europe, W Asia.
      sambucifolia săm-bew-ki-*fo*-lee-a. (= *V. sambucifolia*). *Sambucus*-leaved.
   phu foo. From Gk. *phou* (a kind of valerian). N Anatolia.
   saxatilis săks-*ah*-ti-lis. Growing among rocks. Alps, SE Europe.
   supina su-*peen*-a. Prostrate. Alps.

Valerianella va-le-ree-a-*nel*-la *Valerianaceae*. Diminutive of *Valeriana* q.v. a related genus. Annual herb.
   locusta lo-*kus*-ta. L. name for locust. Corn Salad, Lamb's Lettuce. Europe, N Africa, W Asia.

Vallisneria vă-lis-*ne*-ree-a *Hydrocharitaceae*. After Antonio Vallisnieri de Vallisnera (1661–1730). Aquarium plants.
   asiatica ah-see-*ah*-ti-ka. Asian. E and SE Asia.
   gigantea gi-*găn*-tee-a. Very large. SE Asia to Australia.
   spiralis spee-*rah*-lis. Spiralled (the female scape in fruit). S Europe, W Asia.

Vallota va-*lō*-ta *Amaryllidaceae*. After Pierre Vallot (1594–1671), French botanical author.
  speciosa spe-kee-*ō*-sa. (= *V. purpurea*). Scarborough Lily. S Africa.

Vancouveria vang-koo-*ve*-ree-a *Berberidaceae*. After Captain George Vancouver (1758–98), British explorer. Perennial herbs. W United States.
  hexandra heks-*ăn*-dra. With six stamens.
  planipetala plah-ni-*pe*-ta-la. With flat petals.

Vanda *văn*-da *Orchidaceae*. From the Hindi name. Greenhouse orchids.
  amesiana aymz-ee-*ah*-na. After Frederick Lothrop Ames (1835–93) of Boston, an amateur orchid grower. India.
  coerulea kie-*ru*-lee-a. Deep blue. Himalaya, SE Asia.
  coerulescens kie-ru-*les*-enz. Bluish. Burma.
  cristata kris-*tah*-ta. Crested, the mid-lobe of the lip. Himalaya, Tibet, Assam.
  sanderiana sahn-da-ree-*ah*-na. After the Sander orchid nursery. Philippines.
  teres te-reez. Cylindrical. Himalaya, N India, Burma.
  tricolor tri-ko-lor. Three-coloured. Java.

Veltheimia vel-*tie*-mee-a *Liliaceae*. After August Ferdinand von Veltheim of Brunswick (1741–1801). Tender, bulbous herbs. S Africa.
  capensis ka-*pen*-sis. Of the Cape of Good Hope.
  viridiflora vi-ri-di-*flō*-ra. Green-flowered. Forest Lily.

Velvet Leaf see *Kalanchoe beharensis*
Velvet Plant see *Gynura aurantiaca*
  Trailing see *Ruellia makoyana*

× **Venidio-Arctotis** vay-*ni*-dee-ō-ark-*tō*-tis *Compositae*. Intergeneric hybrids, from the names of the parents. *Arctotis* × *Venidium*. Annual herbs.

Venidium vay-*ni*-dee-um *Compositae*. From L. *vena* (a vein) referring to the ribbed fruits. Annual herbs. S Africa.
  decurrens day-*ku*-renz. (= *V. calendulaceum*). With the petioles running into the stem.
  fastuosum făs-tew-*ō*-sum. Proud. Monarch of the Veldt, Namaqualand Daisy.

Venus's Fly Trap see *Dionaea muscipula*
Venus's Looking Glass see *Legousia speculum-veneris*

Veratrum vay-*rah*-trum *Liliaceae*. The L. name. Perennial herbs. False Hellebore.
  album ăl-bum. White. Europe, N Asia.
  nigrum nig-rum. Black. Europe, Asia.
  viride vi-ri-dee. Green. N America.

Verbascum ver-*băs*-kum *Scrophulariaceae*. The L. name. Biennial and perennial herbs. Mullein.
  arcturus ark-*tew*-rus. (= *Celsia arcturus*). From Gk. *arktos* (a bear) and *oura* (a tail). Cretan Bear's Tail. Crete.
  blattaria bla-*tah*-ree-a. The L. name from *blatta* (a moth). Moth Mullein. Europe, Asia.
  bombyciferum bom-bi-*ki*-fe-rum. Silky. W Asia.
  chaixii shay-zee-ee. After Dominique Chaix (1730–99), a French botanist. Nettle-leaved Mullein. S Europe.
  creticum kray-ti-kum. (= *Celsia cretica*). Of Crete. Cretan Mullein. Mediterranean region.
  densiflorum den-si-*flō*-rum. (= *V. thapsiforme*). Densely flowered. Europe.
  dumulosum dew-mew-*lō*-sum. Like a small shrub. SW Turkey.
  longifolium long-gi-*fo*-lee-um. Long-leaved. Italy, SE Europe.
  nigrum nig-rum. Black. Dark Mullein. Europe.
  olympicum o-*lim*-pi-kum. Of Mt Olympus. Turkey.
  phlomoides flo-*moi*-deez. Like *Phlomis*. C and S Europe.
  phoeniceum foy-*nee*-kee-um. Purple-red. Purple Mullein. SE Europe, W Asia.
  spinosum spee-*nō*-sum. Spiny. Crete.
  thapsiforme see *V. densiflorum*
  thapsus thăp-sus. A town in Sicily, now Magnisi. Aaron's Rod. Europe, Asia.
  virgatum vir-*gah*-tum. Wand-like. W Europe.

Verbena ver-*bay*-na *Verbenaceae*. L. name for the foliage of ceremonial and medicinal plants. Annual and perennial herbs. Vervain.
  bonariensis bo-nah-ree-*en*-sis. Of Buenos Aires. S America.
  × *hybrida* hi-bri-da. Hybrid. Rose Vervain.
  officinalis o-fi-ki-*nah*-lis. Sold as a herb. Common Vervain. Europe.
  peruviana pe-roo-vee-*ah*-na. Of Peru.

Verbena (continued)
Argentina, Brazil.
rigida ri-gi-da. Rigid. Lilac Vervain. Brazil,
Argentina.

Vernonia ver-*non*-ee-a *Compositae*. After
William Vernon (died c 1711), English
botanist who collected in Maryland. Perennial
herbs.
altissima ăl-*ti*-si-ma. Tallest. E United
States.
crinita kree-*nee*-ta. Long-haired. United
States.
noveboracensis no-vee-bo-ra-*ken*-sis. Of
New York. United States.

Veronica ve-ro-ni-ka *Scrophulariaceae*. After
St Veronica. Perennial herbs. Speedwell. See
also *Hebe*.
allionii ă-lee-ō-nee-ee. After Carlo Allioni
(1705–1804), Italian botanist. SW Alps.
armena ar-*me*-na. Of Armenia.
austriaca ow-stree-*ah*-ka. Austrian. C and
E Europe.
    teucrium toyk-ree-um. (= *V. teucrium*).
    From the resemblance to *Teucrium*.
    Europe.
bombycina bom-bi-*keen*-a. Silky. Lebanon.
cinerea ki-*ne*-ree-a. Grey. W Asia.
exaltata see *V. longifolia*
filiformis fee-li-*form*-is. Thread-like (the
pedicels). W Asia.
fruticans froo-ti-kănz. Shrubby. Europe.
gentianoides gen-tee-a-*noi*-deez. *Gentiana*-
like. W Asia, Caucasus.
incana see *V. spicata incana*
longifolia long-gi-*fo*-lee-a. (= *V. exaltata*).
Long-leaved. Europe, Asia.
pectinata pek-ti-*nah*-ta. Comb-like (the
leaves). W Asia, SE Europe.
prostrata pros-*trah*-ta. Prostrate. Europe.
saturejoides săt-ew-ray-*oi*-deez. Like
Satureja. Balkan peninsula.
spicata spee-*kah*-ta. With flowers in spikes.
Europe, Asia.
    incana in-kah-na. (= *V. incana*). Grey-
    hairy.
teucrium see *V. austriaca teucrium*
virginica see *Veronicastrum virginicum*

Veronicastrum ve-ro-ni-*kăs*-trum
Scrophulariaceae. From *Veronica* q.v. and
L. -*aster* (somewhat like or an inferior sort
of). Perennial herb.
virginicum vir-*jin*-i-kum. (= *Veronica
virginica*). Of Virginia. E N America.

Vervain see *Verbena*
  Common see *V. officinalis*
  Lilac see *V. rigida*
  Rose see *V.* × *hybrida*

Viburnum vee-*bur*-num *Caprifoliaceae*. The
L. name for *V. lantana*. Deciduous and
evergreen shrubs.
× *bodnantense* bod-nant-*en*-see. *V. farreri* ×
V. grandiflorum. Of Bodnant where the
most popular forms were raised.
× *burkwoodii* burk-*wud*-ee-ee. *V. carlesii* ×
V. utile. After Albert Burkwood of
Burkwood and Skipwith, the raisers.
carlesii karlz-ee-ee. After W. R. Carles who
collected the type specimen. Korea,
Japan.
cassinoides kă-si-*noi*-deez. Like *Ilex cassine*.
Withe-rod. E N America.
cinnamomifolium kin-a-mō-mi-*fo*-lee-um.
Cinnamomum-leaved. W Sichuan.
davidii dă-*vid*-ee-ee. After David, see
Davidia. China.
farreri fă-ra-ree. (= *V. fragrans*). After
Reginald John Farrer (1880–1920),
horticultural writer who collected in China,
Burma and the Alps. Japan, China.
fragrans see *V. farreri*
× *globosum* glo-bō-sum. *V. calvum* × *V.
davidii*. Spherical (the habit).
grandiflorum grănd-i-*flō*-rum. Large-
flowered.
henryi hen-ree-ee. After Henry, who
discovered it in 1887, see *Illicium henryi*.
C China.
× *hillieri* hi-lee-a-ree. *V. erubescens* × *V.
henryi*. After Hillier's who raised it.
× *juddii* jŭd-ee-ee. *V. bitchiuense* × *V.
carlesii*. After the raiser William H. Judd.
lantana lan-*tah*-na. An old name for
Viburnum. Wayfaring Tree. Europe,
N Africa, W Asia, Caucasus.
macrocephalum măk-rō-*kef*-a-lum. Large-
headed. China.
opulus op-ew-lus. L. name for a kind of
maple. Guelder Rose. Europe, N Africa,
W Asia, Caucasus.
    'Xanthocarpum' zănth-ō-*kar*-pum.
    Yellow-fruited.
plicatum pli-*kah*-tum. Pleated (the leaves).
Japan, China.
    'Mariesii' ma-*reez*-ee-ee. After Maries
    who introduced it, see *Davallia mariesii*.
'Pragense' prah-*gen*-see. *V. rhytidophyllum*
× *V. utile*. Of Prague where it was raised.
× *rhytidophylloides* ri-ti-do-fi-*loi*-deez. *V.
lantana* × *V. rhytidophyllum*. Like

Viburnum (continued)
V. rhytidophyllum
rhytidophyllum ri-ti-*do*-fi-lum. With
wrinkled leaves. China.
tinus teen-us. The L. name. Laurustinus.
Mediterranean region.

Vicia *vi*-kee-a *Leguminosae*. L. name for
vetch. Annual herb.
faba fǎ-ba. The L. name. Broad Bean.
Cult.

Victoria vik-*tor*-ree-a *Nymphaeaceae*. After
Queen Victoria. Tender, aquatic perennial
herbs.
amazonica ǎ-ma-*zon*-i-ka. (= *V. regia*). Of
the Amazon. Royal Water Lily.
cruziana krooz-ee-*ah*-na. Of Santa Cruz.
Santa Cruz Water Lily. S America.

Vinca *ving*-ka *Apocynaceae*. From the L.
name *vinca pervinca* from *vincio* (to bind)
referring to the shoots. Evergreen sub-
shrubs. Periwinkle.
major mah-yor. Larger. Greater Periwinkle.
Mediterranean region.
minor mi-nor. Smaller. Lesser Periwinkle.
Europe, W Asia, Caucasus.
rosea see *Catharanthus roseus*

Viola *vee*-o-la *Violaceae*. L. name for several
scented flowers. Annual and perennial herbs.
Pansy, Violet.
aetolica ie-*to*-lika. Of Aitolia, Greece.
Balkan peninsula.
bertolonii ber-to-*lōn*-ee-ee. After Bertoloni,
see *Bertolonia*. S Europe.
biflora bi-*flō*-ra. Two-flowered. Europe to
the Himalaya, China, Japan, W N
America.
cornuta kor-*new*-ta. Horned, referring to
the long spur. Horned Violet. Pyrenees.
cucullata kuk-ew-*lah*-ta. (= *V. obliqua*).
Hood-like. Marsh Violet. E N America.
gracilis grǎ-ki-lis. Graceful. Olympian
Violet. Balkan peninsula, W Asia.
hederacea he-de-*rah*-kee-a. Like *Hedera* (the
leaves). Australian Violet. Australia.
labradorica lǎb-ra-*do*-ri-ka. Of Labrador.
N America.
obliqua see *V. cucullata*
odorata o-dō-*rah*-ta. Scented. Sweet Violet.
Europe, N Africa, Asia.
pedata pe-*dah*-ta. Like a bird's foot (the
leaves). Bird's-foot Violet. E United
States.
rupestris roo-*pes*-tris. Growing among rocks.
Europe.

Viola (continued)
saxatilis see *V. tricolor subalpina*
septentrionalis sep-ten-tree-ō-*nah*-lis.
Northern. E N America.
sororia so-*rō*-ree-a. Sisterly (i.e. resembling
another species). E N America.
tricolor tri-ko-lor. Three-coloured.
Heartsease. Europe.
subalpina sub-ǎl-*peen*-a. (= *V. saxatilis*).
Sub-alpine. N Spain to the Crimea.
× *wittrockiana* vit-rok-ee-*ah*-na. After
Professor Veit Brecher Wittrock
(1839–1914), author of a history of the
cultivated pansy. Garden Pansy.

Violet see *Viola*
  Australian see *V. hederacea*
  Bird's-foot see *V. pedata*
  Horned see *V. cornuta*
  Marsh see *V. cucullata*
  Olympian see *V. gracilis*
  Sweet see *V. odorata*
Violet Cress see *Ionopsidium acaule*
Viper's Bugloss see *Echium vulgare*
Virginia Creeper see *Parthenocissus
quinquefolia*
Virginia Cowslip see *Mertensia maritima*
Viscaria alpina see *Lychnis alpina*

Viscum *vis*-kum *Loranthaceae*. The L. name.
Evergreen semi-parasitic shrub.
album ǎl-bum. White (the fruits).
Mistletoe. Europe, N Africa, temperate
Asia.

Vitaliana vi-tǎ-lee-*ah*-na *Primulaceae*. After
Vitaliano Donati (1717–62). Perennial herb.
primuliflora preem-ew-li-*flō*-ra. (=
Douglasia vitaliana). *Primula*-flowered. SE
Alps.

Vitex *vi*-teks *Verbenaceae*. The L. name for
V. agnus-castus. Deciduous shrubs.
agnus-castus ǎn-yus-*kǎs*-tus. An old name
for this plant, from *agnos* the Gk. name
and L. *castus*, both meaning chaste. Chaste
Tree. Mediterranean region to C Asia.
negundo ne-*gun*-do. The native name.
S Asia.

Vitis *vee*-tis *Vitaceae*. L. name for the grape
vine. Deciduous, woody climbers.
amurensis am-ew-*ren*-sis. Of the Amur
region. NE Asia.
coignetiae kwŭn-*yay*-tee-ie. After Mme
Coignet, who collected seeds of it in its

Vitis (continued)
native Japan.
labrusca lah-*brus*-ka. L. name for the wild
grape vine. Northern Fox Grape. E N
America.
vinifera veen-*i*-fe-ra. Wine-bearing.
Common Grape Vine. W Asia.
   'Apiifolia' ă-pee-i-*fo*-lee-a. *Apium*-leaved,
   apium being the L. for celery and parsley.
   Parsley Vine.

Vriesia *vreez*-ee-a *Bromeliaceae*. After
Willem Hendrick de Vriese (1806–62), a
Dutch professor of botany. Tender,
evergreen herbs.
carinata kă-ri-*nah*-ta. Keeled (the bracts).
Painted Feather. Brazil.
fenestralis fe-ne-*strah*-lis. Window-like (the
net-veined leaves). Netted Vriesia. Brazil.
hieroglyphica hee-e-rō-*gli*-fi-ka. Marked
with hieroglyphs (the banded leaves).
King of Bromeliads. Brazil.
psittacina si-ta-*keen*-a. Parrot-like, the
colour of the bracts and flowers. Dwarf
Painted Feather. Brazil.
regina ray-*geen*-a. Queen. Brazil.
saundersii sawn-*derz*-ee-ee. After William
Wilson Saunders (1809–79), a collector of
rare plants. Brazil.
splendens *splen*-denz. Splendid. Flaming
Sword. French Guiana.

W

Wahlenbergia vah-lan-*berg*-ee-a
Campanulaceae. After Georg Wahlenberg
(1780–1851), Swedish professor of botany.
Perennial herbs.
albomarginata ăl-bō-mar-gi-*nah*-ta. White-
margined. New Zealand.
hederacea he-de-*rah*-kee-a. *Hedera*-like (the
leaves). W Europe.
matthewsii măth-*ewz*-ee-ee. After H. J.
Matthews who discovered it. New
Zealand.
saxicola săks-*i*-ko-la. Growing on rocks.
Tasmania.

Wake Robin see *Trillium grandiflorum*

Waldsteinia văld-*stien*-ee-a *Rosaceae*. After
Count Franz Adam Waldstein-Wartenburg
(1759–1823), an Austrian botanist. Perennial
herbs.
fragarioides fra-gah-ree-*oi*-deez. Like
Fragaria. E United States.

Waldsteinia (continued)
ternata ter-*nah*-ta. In threes (the leaflets).
Europe to Japan.

Wall Rue see *Asplenium ruta-muraria*
Wallflower see *Cheiranthus cheiri*
   Siberian see *C. allionii*
Walnut see *Juglans*
   Black see *J. nigra*
   Common see *J. regia*
   Japanese see *J. ailantifolia*
   Texan see *J. microcarpa*
Wandflower see *Dierama pulcherrimum,
Sparaxis*
Wandering Jew see *Tradescantia albiflora,
Zebrina pendula*
Waratah see *Telopea speciosissima*
   Tasmanian see *T. truncata*
Wart Plant see *Haworthia tesselata*

Washingtonia wo-shing-*ton*-ee-a *Palmae*.
After George Washington (1732–99), first
President of the United States. Tender Palms.
Washington Palm.
filifera fee-*li*-fe-ra. Thread-bearing,
referring to the fibres that separate from
the leaf margins. Desert Fan Palm. SW
United States.
robusta rō-*bus*-ta. Robust. Thread Palm.
Mexico.

Water Avens see *Geum rivale*
Water Caltrop see *Trapa natans*
Water Chestnut see *Trapa natans*
   Chinese see *Eleocharis dulcis*
Water Crowfoot see *Ranunculus aquatilis*
Water Figwort see *Scrophularia auriculata*
Water Hyacinth see *Eichhornia crassipes*
Water Lettuce see *Pistia stratiotes*
Water Lily, Cape Blue see *Nymphaea
capensis*
   Royal see *Victoria amazonica*
   Santa Cruz see *Victoria cruziana*
   White see *Nymphaea alba*
   Yellow see *Nuphar lutea*
Water Melon see *Citrullus lanatus*
Water Milfoil see *Myriophyllum*
Water Plantain see *Alisma plantago-aquatica*
Water Poppy see *Hydrocleys nymphoides*
Water Soldier see *Stratiotes aloides*
Water Sprite see *Ceratopteris thalictroides*
Water Starwort see *Callitriche*
Water Trumpet see *Cryptocoryne*
Water Violet sec *Hottonia palustris*
Watercress see *Nasturtium officinale*

Watsonia wot-*son*-ee-a *Iridaceae*. After Sir
William Watson (1715–87), London

Watsonia (continued)
physician and botanist. Tender and semi-hardy perennial herbs. S Africa.
beatricis bee-*ah*-tri-kis. After Beatrix Hops who discovered it.
fourcadei four-*kahd*-ee-ee. After H. G. Fourcade.
meriana me-ree-*ah*-na. After Sybilla Merian, a Dutch naturalist.
tabularis tăb-ew-*lah*-ris. Of Table Mountain.

Wattle see *Acacia*
  Cootamundra see *A. baileyana*
  Ovens see *A. pravissima*
  Queensland see *A. podalyriifolia*
  Rice's see *A. riceana*
  Silver see *A. dealbata*
  Sydney Golden see *A. longifolia*
Wax Flower see *Stephanotis floribunda*
Wax Myrtle see *Myrica cerifera*
Wax Plant see *Hoya carnosa*
  Miniature see *H. bella*
Wax Privet see *Peperomia glabella*
Wayfaring Tree see *Viburnum lantana*

Weigela vie-ge-la *Caprifoliaceae*. After Christian Ehrenfried von Weigel (1748–1831), a German botanist. Deciduous shrubs.
  florida flō-ri-da. Flowering. N China, Korea.
    'Foliis Purpureis' *fo*-lee-is pur-*pew*-ree-is. With purple leaves.
  praecox prie-koks. Early (flowering). NE Asia.

Wellingtonia see *Sequoiadendron giganteum*
Western Red Cedar see *Thuja plicata*
Western White Cedar see *Thuja occidentalis*
Whin see *Ulex*
White Paint-brush see *Haemanthus albiflos*
White Raintree see *Brunfelsia undulata*
White Snake-root see *Eupatorium rugosum*
White Velvet see *Tradescantia sillamontana*
Whitebeam see *Sorbus aria*
  Swedish see *S. intermedia*
Whorl Flower see *Morina longifolia*
Whortleberry see *Vaccinium myrtillus*
  Caucasian see *V. arctostaphylos*

Wilcoxia wil-*koks*-ee-a *Cactaceae*. After General Timothy E. Wilcox, American soldier and amateur botanist. Slender-stemmed cacti.
  poselgeri po-zal-*ge*-ree. After Heinrich Poselger (died 1883), a German cactus grower. Texas, N Mexico.

Wilcoxia (continued)
  schmollii shmol-ee-ee. After F. Schmoll. Lamb's-tail Cactus. Mexico.

Willow see *Salix*
  Bay see *S. pentandra*
  Crack see *S. fragilis*
  Creeping see *S. repens*
  Cricket Bat see *S. alba* 'Caerulea'
  Dragon's Claw see *S. matsudana* 'Tortuosa'
  Goat see *S. caprea*
  Golden see *S. alba* Vitellina
  Musk see *S. aegyptiaca*
  Peking see *S. matsudana*
  Silver see *S. alba* Argentea
  Violet see *S. daphnoides*
  Weeping see *S.* 'Chrysocoma'
  White see *S. alba*
  Woolly see *S. lanata*
Willow Herb see *Epilobium*

× *Wilsonara* wil-son-*ah*-ra *Orchidaceae*. Intergeneric hybrids, after Mr Alfred Gurney Wilson (c 1878–1957), Chairman of the RHS orchid committee. *Cochlioda* × *Odontoglossum* × *Oncidium*. Greenhouse Orchids.

Wineberry see *Rubus phoenicolasius*
Wingnut see *Pterocarya*
  Caucasian see *P. fraxinifolia*
Winter Aconite see *Eranthis hyemalis*
Winter Cherry see *Solanum capsicastrum*
Winter Heliotrope see *Petasites fragrans*
Winter Sweet see *Chimonanthus praecox*
Wintergreen see *Pyrola*
Winter's Bark see *Drimys winteri*
Wishbone Flower see *Tovenia fournieri*

Wisteria wis-*te*-ree-a *Leguminosae*. After Caspar Wistar (1761–1818), an American professor of anatomy. Deciduous climbers.
  floribunda flō-ri-*bun*-da. Profusely flowering. Japan.
  sinensis si-*nen*-sis. Of China.

Witch Grass see *Panicum capillare*
Witch Hazel see *Hamamelis*
  Chinese see *H. mollis*
  Japanese see *H. japonica*
Withe-rod see *Viburnum cassinoides*
Woad see *Isatis tinctoria*
Wolf's Bane see *Aconitum vulparia*
Wonga-wonga Vine see *Pandorea jasminoide*:
Wood Sorrel see *Oxalis acetosella*
Woodbine see *Lonicera periclymenum*
Wormwood see *Artemisia absinthium*

Wulfenia wul-*fen*-ee-a *Scrophulariaceae*.
After Franz Xavier Freiherr von Wulfen
(1728–1805). Perennial herbs.
  baldaccii băl-*dăk*-ee-ee. After Antonio
  Baldacci of the Bologna Botanic Garden.
  Albania.
  carinthiaca ka-rinth-ee-*ah*-ka. Of Carinthia
  Austria. E Alps, Balkan peninsula.
  orientalis o-ree-en-*tah*-lis. Eastern. W Asia.

X

Xantheranthemum zănth-e-*rănth*-e-mum
Acanthaceae. From Gk. *xanthos* (yellow) and
Eranthemum, a related genus. Tender
perennial herb.
  igneum ig-nee-um. Fiery red. Peru.

Xanthoceras zănth-*o*-ke-ras *Rosaceae*. From
Gk. *xanthos* (yellow) and *keras* (a horn)
referring to the yellow, horn-like appendages
between the petals. Deciduous shrub or tree.
  sorbifolium sor-bi-*fo*-lee-um. *Sorbus*-leaved.
  N China.

Xanthorhiza zănth-o-*reez*-a *Ranunculaceae*.
From Gk. *xanthos* (yellow) and *rhiza* (a root),
the roots are yellow. Deciduous shrub.
  simplicissima sim-pli-*ki*-si-ma. Most simple
  i.e. unbranched. Yellowroot. E United
  States.

Xeranthemum ze-*rănth*-e-mum *Compositae*.
From Gk. *xeros* (dry) and *anthos* (a flower)
referring to the dry, everlasting flowers.
Annual herb.
  annuum ăn-ew-um. Annual. Immortelle.
  S Europe.

Y

Yam see *Dioscorea*
  Ornamental see *D. discolor*
Yarrow see *Achillea millefolium*
Yellow Adder's Tongue see *Erythronium
americanum*
Yellow Archangel see *Galeobdolon luteum*
Yellow Elder see *Tecoma stans*
Yellow Flag see *Iris pseudacorus*
Yellow Jessamine see *Gelsemium sempervirens*
Yellow Sage see *Lantana camara*
Yellow Wood see *Cladrastis sinensis*
Yellowroot see *Xanthorhiza simplicissima*

Yesterday Today and Tomorrow see
Brunfelsia calycina
Yew see *Taxus*
  Common see *T. baccata*
  Irish see *T. baccata* 'Hibernica'
  West Felton see *T. baccata* 'Dovastoniana'
Youth and Old-age see *Aichryson* ×
domesticum

Yucca *yoo*-ka *Agavaceae*. The Caribbean
name for Cassava (*Manihot esculenta*)
originally thought to apply to *Y. gloriosa*.
Evergreen, hardy and tender perennial herbs
and shrubs.
  aloifolia ă-lō-i-*fo*-lee-a. *Aloe*-leaved.
  Spanish Bayonet. S United States,
  Mexico, W Indies.
  elephantipes e-le-*făn*-ti-pays. Like an
  elephant's foot. Spineless Yucca. SW
  United States, Mexico.
  filamentosa fee-lah-men-*tō*-sa. With
  filaments (on the leaf margins). Adam's
  Needle. SE United States.
  flaccida *flăk*-ki-da. Flaccid (the leaves). SE
  United States.
  glauca *glow*-ka. Glaucous (the leaves). SE
  United States.
  gloriosa glō-ree-*ō*-sa. Glorious. SE United
  States.
  smalliana smawl-ee-*ah*-na. After John
  Kunkel Small (1869–1938). SE United
  States.
  whipplei *wi*-pal-ee. After Lieut. Amiel
  Weeks Whipple (1818–63), who explored
  N America. Our Lord's Candle. California,
  N Mexico.

Yulan see *Magnolia denudata*

Z

Zantedischia zăn-te-*dis*-kee-a *Araceae*. After
Francesco Zantedischi (born 1797), Italian
botanist. Tender and semi-hardy perennial
herbs.
  aethiopica ie-thee-ō-pi-ka. African. Arum
  Lily. Transvaal.
  albomaculata ăl-bō-măk-ew-*lah*-ta. (= *Z.
  melanoleuca*). White-spotted (the leaves).
  Black-throated Arum, Spotted Arum.
  S Africa.
  elliottiana e-lee-o-tee-*ah*-na. After Elliott.
  Golden Arum. S Africa.
  pentlandii pent-*lănd*-ee-ee. Of Pentland
  House, home of Mr White who introduced

Zantedischia (continued)
it. Yellow Arum. S Africa.
rehmannii ray-*mahn*-ee-ee. After Rehmann.
Pink Arum. S Africa.

Zanthoxylum zănth-*oks*-i-lum *Rutaceae*.
From Gk. *xanthos* (yellow) and *xylon* (wood)
referring to the yellow wood of some species.
Deciduous trees.
ailanthoides ie-lănth-*oi*-deez. Like
Ailanthus. Japan, China.
americanum a-me-ri-*kah*-num. American.
Toothache Tree. E N America.
piperitum pi-pe-*ree*-tum. Pepper-like (the
taste of the seeds). Japan Pepper. Japan,
China.
planispinum plah-ni-*speen*-um. With flat
spines. Japan, Korea, China.
schinifolium skeen-i-*fo*-lee-um. With leaves
like *Schinus*. China, Korea, Japan.

Zauschneria zowsh-*ne*-ree-a *Onagraceae*.
After Johann Baptist Zauschner (1737–99),
professor of Natural History at Prague. Sub-
shrubs.
californica kă-li-*forn*-i-ka. Of California.
W United States.
cana kah-na. Grey. California.

Zea *zee*-a *Gramineae*. From the Gk. name for
a related plant. Annual grass.
mays mayz. From the Mexican name. Corn,
Maize. Tropical America.

Zebra Basket Vine see *Aeschynanthus
marmoratus*
Zebra Plant see *Aphelandra squarrosa*,
Calathea zebrina, *Cryptanthus zonalis*

Zebrina ze-*breen*-a *Commelinaceae*. From
zebra referring to the striped leaves. Tender,
perennial herb.
pendula pen-dew-la. Pendulous. Wandering
Jew. Mexico.

Zelkova zel-*kō*-va *Ulmaceae*. From the
Caucasian name. Deciduous trees.
carpinifolia kar-peen-i-*fo*-lee-a. *Carpinus*-
leaved. Transcaucasus, N Iran.
serrata se-*rah*-ta. Saw-toothed (the leaves).
E Asia, Japan.

Zenobia zay-*no*-bee-a *Ericaceae*. After
Zenobia, a Queen of Palmyra. Deciduous or
semi-evergreen shrub.
pulverulenta pul-ve-ru-*len*-ta. Powdered,
referring to the glaucous-bloomed leaves.
SE United States.

Zephyranthes ze-fi-*rănth*-eez *Amaryllidaceae*.
From Gk. *zephyros* (the west wind) and
anthos (a flower), they are natives of the
W hemisphere. Tender and semi-hardy,
bulbous herbs.
atamasco ă-ta-*măs*-kō. A native name. SE
United States.
candida kăn-di-da. White (the flowers).
Flower of the Western Wind. Argentina,
Uruguay.
citrina ki-*tree*-na. Lemon-yellow.
S America.
grandiflora grănd-i-*flō*-ra. Large-flowered.
Mexico, Guatemala.
robusta see *Habranthus robustus*
rosea ro-see-a. Rose-coloured. W Indies,
Guatemala.

Zigadenus zi-ga-*day*-nus *Liliaceae*. From
Gk. *zygos* (a yoke) and *aden* (a gland)
referring to the paired glands. Perennial herb.
elegans ay-le-gahnz. Elegant. N America.

Zingiber zing-gi-ber *Zingiberaceae*. Of
ancient origin probably from pre-roman
snga (a horn) and *ver* (a root) giving L.
zingiber and eventually English *ginger* which
is obtained from the root. Tender herb.
officinale o-fi-ki-*nah*-lee. Sold as a herb.
Ginger. SE Asia.

Zinnia *zin*-ee-a *Compositae*. After Johann
Gottfried Zinn (1727–59). Annual herbs.
Mexico.
angustifolia ang-gus-ti-*fo*-lee-a. Narrow-
leaved.
elegans ay-le-gahnz. Elegant.

Zygocactus truncatus see *Schlumbergera
truncata*

Zygopetalum zi-gō-*pe*-ta-lum *Orchidaceae*.
From Gk. *zygos* (a yoke) and *petalon* (a
petal), the callous at the base of the lip
appears to be pulling the perianth segments
together. Greenhouse orchids. Brazil.
crinitum kree-*nee*-tum. Long-haired.
intermedium in-ter-*me*-dee-um.
Intermediate.
mackayi ma-*kay*-ee. After Mr Mackay of
Dublin College Botanic Garden who
imported it.

Acknowledgements

I am grateful to the Director, Royal Botanic Gardens, Kew for use of library facilities and to Nigel Taylor and Susyn Andrews of that institution for help on several, mainly nomenclatural, points. I would also like to thank Dr M. J. Harvey of the Department of Biology, Dalhousie University, Halifax, Canada for interesting and useful discussions on the subject of pronunciation.

The following publications (excluding taxonomic works) were freely consulted during the preparation of this work and would be interesting to anyone wishing to pursue the subject.

Desmond, R. 1977. *Dictionary of British and Irish Botanists and Horticulturists*. Taylor and Francis.

Glare, P. G. W. (ed.) 1982. *Oxford Latin Dictionary*. Oxford University Press.

Johnson, A. T. & Smith, H. A. 1972. *Plant Names Simplified*. Landsmans Bookshop Ltd.

Kidd, D. A. 1957. Collins Gem *Latin–English English–Latin Dictionary*. Collins.

Lewis, C. T. & Short, C. 1879. *A Latin Dictionary*. Oxford University Press.

Liddell, H. G. & Scott, R. 1843. *A Greek–English Lexicon*. Oxford University Press. New (9th) Edition revised by H. Stuart Jones & R. Mackenzie 1940.

Smith, A. W. & Stearn, W. T. 1972. *A Gardener's Dictionary of Plant Names*. Cassell.

Stearn, W. T. 1966. *Botanical Latin*. 2nd Edition 1973. David & Charles.